Augmented and Virtual Reality in Industry 5.0

Augmented and Virtual Reality

Edited by
Vishal Jain

Volume 2

Augmented and Virtual Reality in Industry 5.0

Edited by
Richa Goel, Sukanta Kumar Baral, Tapas Mishra
and Vishal Jain

DE GRUYTER

Editors

Dr. Richa Goel
SCMS Noida, Symbiosis International University
Sector 62
Noida
India
richasgoel@gmail.com

Prof. Sukanta Kumar Baral
Department of Commerce
Faculty of Commerce and Management
Indira Gandhi National Tribal University
Amarkantak, Madhya Pradesh
India
sukanta.baral@igntu.ac.in

Prof. Tapas Mishra
Southampton Business School
University of Southampton
Building 2
12 University Rd
Highfield, Southampton SO17 1BJ
Great Britain
t.k.mishra@soton.ac.uk

Dr. Vishal Jain
Department of Computer Science & Engineering (CSE)
Sharda University
Greater Noida, Uttar Pradesh
India
vishal.jain@sharda.ac.in

ISBN 978-3-11-078999-7
e-ISBN (PDF) 978-3-11-079014-6
e-ISBN (EPUB) 978-3-11-079048-1

Library of Congress Control Number: 2023901379

Bibliographic information published by the Deutsche Nationalbibliothek
The Deutsche Nationalbibliothek lists this publication in the Deutsche Nationalbibliografie; detailed bibliographic data are available on the internet at http://dnb.dnb.de.

Cover image: Thinkhubstudio/iStock/Getty Images Plus
Typesetting: Integra Software Services Pvt. Ltd.
Printing and binding: CPI books GmbH, Leck

www.degruyter.com

Foreword

Application of virtual reality (VR) and augmented reality (AR) in Industry 5.0 is a comprehensive reference source that will provide personalized, accessible, and well-designed experiences. AR and VR are finding accelerated adoption in this new normal. It will be a key part of how people access the metaverse – the next iteration of the internet. A few breakthrough use cases have entered early consumer adoption, such as gaming, fitness, and social, as hardware devices have become more affordable.

Covering topics such as redefining travel and tourism with AR and VR, use of VR and AR to grow across the smart cities, education, e-commerce, e-gaming, healthcare, manufacturing, automotive, and design will have the potential to contribute to sustainability in two ways: directly by making certain processes more sustainable, and indirectly by encouraging people to live more environmentally friendly lives.

I assume that this book is anticipated to be highly valuable for a wide range of readers with a variety of interests, and it is not only limited to academics, postgraduate students, and research associates but also to corporate executives, entrepreneurs, and other professionals and masses in all fields who can improve and expand their knowledge with the learning of the basic trends and activities in this book. As AR/VR is a new field which has the potential to transform entire industries and can revolutionize any industry as and when required, in the next years, it will have an influence on every single person on the planet.

This book offers a valuable guide to the intellectual and practical work and calls for a need to rethink and examines the consequences for future management of innovation. I am pleased to write this foreword as the editors of this book have given full-hearted effort for a great solution and innovation. All chapters in this book have been selected based on peer review where reviewers were very much experts in the sector.

Prof. (Dr.) Mohammad Saeed
Professor Emeritus, Department of Business Administration,
Minot State University, Minot, ND 58707, USA

https://doi.org/10.1515/9783110790146-202

Preface

There isn't a single industry that does not employ augmented reality (AR) or virtual reality (VR) in some way. This technology's applicability has already spanned a wide range of industries. It will soon infiltrate regions that no one could have predicted. Technology behemoths are already putting money, effort, and time into AR/VR integration. As this technology advances and expands, it is vital that everyone understands the numerous uses of AR and VR, as well as their entire potential. This book of research on "Application of Virtual Reality (VR) and Augmented (AR) in Industry 5.0" reveals new and inventive aspects of technology, as well as how they might aid in increasing economic efficiency and efficacy for improved production. It is an excellent resource for researchers, academics, politicians, business executives, corporations, and students, as it covers a wide range of issues. Numerous applications of AR and VR that improve industrial skills and decision-making are gaining traction. The exponential growth of AR-enabling technology boosts support for Industry 5.0 service enhancement. Meanwhile, in many industry-wide advances and real-time applications, practical difficulties of AR and VR play a key role in building cognitive tools and analytics.

The influence of AR/VR on artificial intelligence (AI) appears to be set as significant. The next generation of AR/VR gadgets will deliver customized, accessible, and well-designed experiences. However, more effort is needed before these technologies become broadly used, which is addressed in this book via multiple chapters under Industry 5.0, as future AR/VR gadgets will give customized, accessible, and well-designed experiences. In this new normal, AR and VR are gaining traction. People's access to the metaverse – the next generation of the internet – will be aided by AR/VR. As hardware devices are more inexpensive, a few groundbreaking use cases, including gaming, fitness, and social, have seen early market acceptance. This book is written in a reader-friendly manner, with a significant material that has been thoroughly examined, allowing for simple comprehension of the subject. The book gives readers with resources, allowing them to conduct more in-depth research. The case studies will present a tried-and-true method for resolving common challenges in the field of study. The reader will be able to assimilate the material briefly, thanks to the essential concepts and simplified substances of the chapters.

This book reveals new and innovative features of AR/VR and how it can help in promoting sustainability among various sectors for raising economic efficiency at both microlevel and macrolevel and provides a deeper understanding of the relevant aspects of AI impacting efficacy for better output. It is an ideal resource for researchers, academicians, policymakers, business professionals, companies, and students. Numerous practical aspects of AR/VR that enhance industry skills as well as decision-making are gaining momentum. This book is a solid step forward and will be of great use for the people in corporate, business professionals, sociology, political science, public administration, mass media and communication, information system, development studies, as well to the business studies. The models discussed in the book will have a huge replication and

https://doi.org/10.1515/9783110790146-203

practice potential across the world, and the field is one of the most important growing fields across the globe. On the other hand, this book will serve as an excellent reference source to the practitioners working in the field of stakeholders and their strategies.

Secondly, this book is laid out in a reader-friendly format where important information duly analyzed is highlighted, thus facilitating easy understanding of the content. The book provides resources to the readers, thus providing an opportunity for further detailed studies. The case studies will provide a tried and tested approach to the resolution of typical problems in the area of study. The key concepts and summarized content of the chapters will enable the reader to absorb contents at glance.

It talks about the following enlisted chapters:

Chapter 1 talks about the application of AR and VR for supply chain: a macro-perspective approach. Post-COVID-19, global manufacturers' attention has suddenly shifted toward supply chain networks to enhance their efficiency and effectiveness. This necessitated redesigning and incorporating novel viewpoints into the service and production processes. Manufacturing has undergone a new industrial shift with the introduction of Industry 4.0, which uses cutting-edge technologies to enhance all aspects of operations and the supply chain. VR knowledge permits customers to interrelate with and immerse themselves in a computer-generated situation, recreate the actual world, or create an unreal world. The goal of this chapter is to investigate the existing literature, available data, and concepts to provide future research directions to help academicians and managers systematically understand the various applications and growth of AR and VR in managing the supply chain.

Chapter 2 talks about an intelligent traffic control system using machine learning techniques. The ever-increasing traffic jams in urban areas makes it necessary to make use of cutting-edge technology and equipment in order to advance the state of the art in terms of traffic control. The currently available solutions, like time visitors or human control, are not adequate to alleviate the severity of this crisis. The findings of this investigation have led to the proposal of a system for the control of traffic that makes use of canny edge detection and digital image processing to determine, in real time, the number of vehicles present. The above imposing traffic control advanced technologies offer significant advantages over the existing systems in real time, transportation management, robotization, reliability, and efficiency. In addition, the complete process of digital image acquisition, edge recognition, as well as green signal assignment is demonstrated with accurate blueprints, and the final outcome are stated by hardware. All of this is done with four separate photographs of various traffic scenarios.

Chapter 3 talks about proficient implementation toward detection of thyroid nodules for AR and VR environment through deep learning methodologies. A disease that commonly exists on a global scale is thyroid nodule. It is identified by unusual thyroid tissue development. Radiologists sometimes may not notice minor elements of an ultrasound image, resulting in an incorrect diagnosis. To help physicians and radiologists to better diagnose, several deep learning (DL)-based models which can accurately classify the nodules as benign and malignant have been implemented. After performing a comparative study of several DL-based models implemented with different classification algorithms on an open-source data set, it has been found that Inception Net V3 gave the best accuracy (~96%), F_1 score (0.957), sensitivity (0.917), and so on. A simple and easy-to-use graphical user interface has been implemented.

Chapter 4 talks about the convergence of AR and VR with IoT in manufacturing and their novel usage in IoT. Augmented reality is an interactive experience of a real-world environment where the objects that reside in the real world are enhanced and superimposed by computer-generated information. When AR and VR technology is combined with IOT, there are endless use cases and industries which benefit from this convergence. AR/VR + IOT can be used to display a machine's performance in real time, using sensors we can get the real-time data of an object, that data can be superimposed and shown to the user in AR or VR. This benefits the manufacturing or service industries to know about the status of the machine or a section visually, which helps in finishing the tasks sooner than conventional methods of working. AR + IOT is more suited for fixing day-to-day operations. VR + IOT is suited for training the employees on a given situation or virtual training of a machinery. This chapter explains how combining AR and IOT can increase the productivity, and how VR and IOT can help in reducing the training time of employees. Finally, touching upon the data we extract out of this convergence of AR/VR and IOT which helps the companies to get a meaningful insight of their operations and further optimize their production.

Chapter 5 talks about the proficiency of metaverse using VR for industry and user's perspective. Amid the widespread scenario, the world of web has seen a gigantic climb within the advancing and digitizing world. The paradigm of the digital world is just beginning to take shape and reality as a metaverse. It is still emerging, and many key components have started to take shape and are a major part of almost every field. This chapter presents an effort to showcase the ecosystem and elements which metaverse consists of. It examines the capabilities and application of the existing virtual worlds. Every tech-giant and field have begun to implement VR to come up with positive innovative results. Therefore, we discuss economical and industrial impact and negative aspects of implementing metaverse.

Chapter 6 talks about the retailing and e-commerce riding on technology. Today's world and economy are driven by technology and technology-savvy people across all sectors. This chapter gives the reader idea about the use of AR and VR in the e-commerce and retail sectors. The retail industry has grown considerably over the last two decades with the advent of e-commerce. It is no longer just a "purchasing" activity-based industry. Instead, it has become an "interactive" industry that provides customers with a "shopping experience." Shortly, AR and VR could be critical factors in establishing customer engagement and product promotion. The Indian retail business is close to $1 trillion in size. Some were secular developments, whereas the Covid-19 tailwind helped others. As a result, the old-world order of organized retail, the value of location and scale, and the power of shelf space are all up for grabs.

Some upcoming advances (supported by tech innovations) are VR, AR, robots, facial recognition, and 3D printing. They are showing their presence. As a result, retail has become more transformative.

Chapter 7 talks about the inclusive education through AR and VR in India. With the advancement of technology, the education imparting scenario also changed drastically. This study lays the framework to understand how far AR and VR have contributed to inclusive edification in India. The sample for this research was collected through nonprobabilities sampling, precisely through the convenient sampling technique. The findings of the study conclude that with the help of AR and VR there were transformations in the education system, leading to inclusive education by improved learning, improved motivation for study, enhanced students engagement, authentic learning opportunity, better communication, and learning opportunity.

Chapter 8 talks about exploring practical use-cases of AR using photogrammetry and other 3D reconstruction tools in the metaverse. In today's world, people rely more and more on mobile apps to do their day-to-day activities, like checking their Instagram feed and online shopping from websites like Amazon and Flipkart. People depend on WhatsApp and Instagram stories to communicate with local businesses and leverage the said platforms for online advertising. Using Google Maps to find their way when they travel and finding out the immediate road and traffic conditions with digital banners around the road has obviously led to a boom in advertising and marketing. Thus, the proposed framework in this chapter would be a new kind of system that may develop a socio-meta platform powered by AR and other technologies like photogrammetry and LiDAR.

Chapter 9 talks about an empirical analysis of conspicuous consumption for luxury cosmetic products via AR technology concerning India's workforce. Status and conspicuous purchasing pattern help in developing links between customers with certain products and brands toward establishing status. AR allows consumers to virtually try goods and provide a "try before you buy" experience while shopping online. Within the national boundaries of a culturally diverse India, this study sought to examine the homogeneity of the luxury industry. The study aimed to decipher the behavioral pattern of the workforce in India in terms of conspicuous purchase intentions for luxury cosmetic products with the assistance of AR-driven technology, emphasizing brand antecedents and psychology.

Chapter 10 talks about the application of VR and AR technologies to boost technical capabilities for constructive effective learning in education for a performing virtual ecosystem: a design-thinking approach. This chapter introspects into the use of the technical capabilities of VR and AR to boost constructive learning in the domain of effective education. It examines a model approach for instructional and developmental design for creating a virtual educational ecosystem. It creates a team for participation employing VR/AR technology to combat the obstacles in ease of learning and assessing its feasibility of performance. The study has employed an exploratory study with critical literature review in a mixed method approach. It engaged the method for a deeper understanding of the constructivism learning theory applied to the VR technologies for better knowledge.

Chapter 11 talks about the role of AR and VR in sports. Computer-generated reality and expanded truth are part of comparative studies of HMDs. The two advances use high-goal shows and following sensors to show stereoscopic pictures that change sufficiently to the client's head. In VR, the client sees the virtual climate. It preferably shuts out impacts from this present truth to amplify inundation. The increased fact utilizes cloudy presentations. It overlays virtual articles inside reality impeccably. The client cannot recognize genuine and virtual items while employing an expanded reality headset. The mix of AR and principal component vision is applications. It includes game diversion applications. Enlisting the AR and PC vision innovation into sports diversion applications gives novel open doors and new difficulties. The work is a survey article discussing various contributions.

Chapter 12 talks about space traffic management: a simulation-based educational app for learners. Due to the swift creation of human-centric applications and technologies, educational and training programs are progressively transforming toward newer ecosystems of modernization through the adoption of AR and VR technologies. Simulations and gamifications in engineering education and training could play a huge role in imparting knowledge with entertainment and amusement, thereby bringing in positive and effortless understanding of concepts among students. The game helps in stimulating strategic understanding among players by providing them an invigorating environment consisting of animations, graphics, sound effects, and music. This chapter reports the inception of this game prototype with the mathematical theory related to its creation.

Chapter 13 talks about the integration of VR in the e-learning environment. In our ever-changing world, learning is an essential part of everyone's daily routine. The traditional approach to education relies on students applying what they have learned from textbooks and professors in the classroom to real-world circumstances. The use of cutting-edge technology in teaching and learning techniques is critical in today's digital world. Information and communication technology) is a major focus for universities. They are important scientific instruments with the potential to have an impact on how people learn and teach science. Three-dimensional VR interfaces give an experience via e-learning activities, software games, and simulated labs. Using VR, users may see, modify, and interact with computer systems and massive amounts of data. The term "visualization" refers to a computer's ability to provide the user with sensory inputs such as visual, aural, or any other combination of these. E-learning may benefit from the realistic virtual environments that VR and web technologies can create.

Chapter 14 talks about the resurgence of AR and VR in construction: past, present, and future directions. For a long time, AR and VR have been used in gaming and entertainment. However, construction is turning to AR and VR in an increasing number of applications, so it is not the only place where these technologies are gaining traction. The most current developments allow complete teams to plan a project meticulously, from enhancing safety to creating intricate designs and choosing the best materials for the job. In addition, project managers can accurately convey their vision to stakeholders since VR and AR make construction projects come to life. The study examines how AR and VR can be used more frequently in the construction industry. The objectives are to determine the amount of knowledge about AR hardware and software, look into AR application areas, and spot construction companies lagging. The study explores how AR and VR might be used in the building process.

Thus, this book intends to give a quality publication with unique insights and methods of application for current scholars and users. This book offers a great overview of how AR and VR transforms organizations and organizes innovation management in Industry 4.0.

Acknowledgments

First and foremost, we want to praise our Almighty God. We learned how true is this talent of writing in the process of putting this book together. You have given us the ability to believe and follow our ambitions in our passion. Without trust in you, we could never have done this. Our most sincere gratitude to our family, who have provided us support during the difficult process of creating this book.

All people involved in this project and in particular the writers and reviewers who participated in the review process would be grateful to the editors. This book would not have been a reality without their assistance. We owe an enormous debt of gratitude to those who gave us the detailed and constructive comments on chapters. They gave freely of their time to discuss and clarify concepts, explore facets of insight work, and explain the rationales for specific recommendations. First, each of the authors wishes to thank them for their efforts. Our heartfelt appreciation goes to the authors of these chapters who have given this book their time and expertise. Second, the editors recognize the evaluators' important contributions in improving the quality, consistency, and substance of chapters. Most of the authors were also reviewers and we respect their enormous work.

We owe a tremendous debt of appreciation to our experts who have commented on several chapters in length and have constructively urged us to clarify concepts, to examine certain features of insight, and to explain reasons for specific recommendations.

We also like to thank many people who have helped us learn and practice both the art and science of networking throughout the years.

Richa Goel
Sukanta Kumar Baral
Tapas Misha
Vishal Jain

https://doi.org/10.1515/9783110790146-204

Contents

List of contributors

Saurabh Tiwari
School of Business
University of Petroleum and Energy Studies
Dehradun, India
E-mail: tiwarisaurabht@gmail.com

Sukanta Kumar Baral
Department of Commerce
Indira Gandhi National Tribal University
Amarkantak, India
E-mail: sukanta.baral@igntu.ac.in

Richa Goel
SCMS Noida, Symbiosis International
University
Sector 62
Noida, India
E-mail: richasgoel@gmail.com

P. Santosh Kumar Patra
Principal, St. Martin's Engineering College,
Secunderabad, India

B. Rajalingam
Computer Science and Engineering, St. Martins
Engineering College, Secunderabad, India
E-mail: rajalingam35@gmail.com

R. Santhoshkumar
Computer Science and Engineering, St. Martins
Engineering College, Secunderabad, India
E-mail: santhoshkumar.aucse@gmail.com

P. Deepan
Computer Science and Engineering, St. Peter's
Engineering College, Secunderabad, India
E-mail: deepanp87@gmail.com

Avani K. V. H
Department of Electronics and
Telecommunication Engineering, BMS College of
Engineering, Bengaluru, India

Deeksha Manjunath
Georgia Institute of Technology, Atlanta, USA

C. Gururaj
Senior Member IEEE
Department of Electronics and
Telecommunication Engineering, BMS College of
Engineering, Bengaluru, India

Vijayarajan Ramanathan
Hogarth Worldwide, Chennai, Tamil Nadu, India
E-mail: Vijayvrar1@gmail.com

Gnanasankaran Natarajan
Department of Computer Science, Thiagarajar
College, Madurai, Tamil Nadu, India
E-mail: sankarn.iisc@gmail.com
Orcid: 0000-0001-9486-6515

Sundaravadivazhagan Balasubramanian
Faculty of IT
Department of Information Technology,
University of Technology and Applied Sciences
Al Mussanah, Oman
E-mail: bsundaravadivazhagan@gmail.com
Orcid: 0000-0002-5515-5769

Deepa B. G
School of Computer Science and Applications
REVA University, Bangalore, India
E-mail: deepabg03@gmail.com

Senthil S
School of Computer Science and Applications
REVA University
Bangalore, India
E-mail: senthil_udt@rediffmail.com

Zaiba S
School of Computer Science and Applications
REVA University, Bangalore, India
E-mail: zaibaunique786@gmail.com

Danish D. S
School of Computer Science and Applications
REVA University, Bangalore, India
E-mail: danishdeepasoman02@gmail.com

https://doi.org/10.1515/9783110790146-206

Kshitiz Tripathi
School of Computer Science and Applications
REVA University, Bangalore, India
E-mail: kshitiz.vns94@gmail.com

Alessandro Bruno
Department of Biomedical Sciences, Humanitas University, Via Rita Levi Montalcini
PieveEmanuele (Milan), Italy
E-mail: alessandro.bruno@hunimed.eu

Vijay Joshi
Dr. Ambedkar Institute of Management Studies and Research (DAIMSR), Deekshabhoomi
Nagpur, Maharashtra, India
E-mail: vijayjoshi62@gmail.com
Cell: +91 9049975365

Sukanta Kumar Baral
Department of Commerce, Faculty of Commerce and Management, Indira Gandhi National Tribal University – IGNTU (A Central University)
Amarkantak, Madhya Pradesh, India
E-mail: drskbinfo@gmail.com
Cell: +91 9437163942 / 9124393942

Manish Pitke
Freelance Faculty
Prin. L. N. Welingkar Institute of Management Development and Research (WeSchool)
Matunga, Mumbai, Maharashtra, India
E-mail: manishpitke@gmail.com
Cell: +91 9960305485

Rocky J. Dwyer
D. B. A. Contributing Faculty Member
College of Management and Human Potential (CMHP), Walden University, Canada
E-mail: rocky.dwyer@mail.waldenu.edu

Ipseeta Satpathy
D.Litt, KIIT School of Management, KIIT University, Odisha, India
E-mail: ipseeta@ksom.ac.in

B.C.M. Patnaik
KIIT School of Management, KIIT University
Odisha, India
E-mail: bcmpatnaik@gmail.com

S.K. Baral
Department of Commerce, Indira Gandhi National Tribal University, MP, India
E-mail:drskbinfo@gmail.com

Majidul Islam
John Molson School of Business, Concordia University, Montreal, Canada
E-mail: majidul.islam@concordia.ca

Aditya Singh
Sharda University, School of Engineering and Technology, Greater Noida, Uttar Pradesh, India 201310

Siddharth Mishra
Sharda University, School of Engineering and Technology, Greater Noida, Uttar Pradesh, India 201310

Shubham Jain
Sharda University, School of Engineering and Technology, Greater Noida, Uttar Pradesh, India 201310

Sandeep Dogra
Sharda University, School of Engineering and Technology, Greater Noida, Uttar Pradesh, India 201310

Anubhav Awasthi
Sharda University, School of Engineering and Technology, Greater Noida, Uttar Pradesh, India 201310

Nihar Ranjan Roy
Sharda University, School of Engineering and Technology, Greater Noida, Uttar Pradesh, IN 201310

Kunwar Sodhi
Dominus Labs LLC 5900 Balcones Dr. STE 100
Austin, Texas, US, 73301

Shreya Soman
Amity School of Economics, Amity University
Noida, India
E-mail: shreya.soman07@gmail.com

Neeru Sidana
Amity School of Economics, Amity University
Noida, India
E-mail: nsidana@amity.edu

Abhiraj Malia
School of Management, KIIT University
Bhubaneswar, India
E-mail: abhirajmalia75@gmail.com

Prajnya Paramita Pradhan
School of Management, KIIT University
Bhubaneswar, India
E-mail: prajnyapradhan11@gmail.com

Biswajit Das
Professor
School of Management, KIIT University
Bhubaneswar, India
E-mail: biswajit@ksom.ac.in

Ipseeta Satpathy
School of Management, KIIT University
Bhubaneswar, India
E-mail: ipseeta@ksom.ac.in

Sambit Lenka
International Business School, Jonkoping
University, Sweden
E-mail: sambit.lenka@ju.se

Ambika N
Department of Computer Science and
Applications, St. Francis college, Bangalore, India
E-mail: Ambika.nagaraj76@gmail.com

Mansaf Alam
Department of Computer Science, Jamia Millia
Islamia, New Delhi, India
E-mail: malam2@jmi.ac.in

Bibhorr
IUBH University, Mülheimer Str. 38, Bad Honnef
53604, Germany
E-mail: bibhorr@zoho.com
https://orcid.org/0000-0003-0404-4601

Raj Gaurang Tiwari
Chitkara University Institute of Engineering and
Technology, Chitkara University, Punjab, India
E-mail: rajgaurang@chitkara.edu.in
Mobile: 9415561502

Ambuj Kumar Agarwal
Department of Computer Science and
Engineering, Sharda University, Greater Noida
India
E-mail: ambuj4u@gmail.com

Mohammad Husain
Department of Computer Science, Islamic
University of Madinah, Saudi Arabia
E-mail: dr.husain@iu.edu.sa

Shantanu Trivedi
Centre for Continuing Education, University of
Petroleum and Energy Studies, Dehradun, India
E-mail: s.trivedi@ddn.upes.ac.in

Saurabh Tiwari
School of Business
University of Petroleum and Energy Studies
Dehradun, India
E-mail: tiwarisaurabht@gmail.com

Saurabh Tiwari, Sukanta Kumar Baral, Richa Goel

1 Application of AR and VR for supply chain: a macro perspective approach

Abstract: Post COVID-19, the attention of global manufacturers has suddenly shifted towards supply chain networks to enhance their efficiency and effectiveness. This has necessitated redesigning and incorporating novel viewpoints into service and production processes. Manufacturing has undergone a new industrial shift with the introduction of Industry 4.0, which uses cutting-edge technologies to enhance all aspects of operations and the supply chain. Virtual reality (VR) permits customers to interrelate with and immerse themselves in a computer-generated situation, recreate the actual world, or create an unreal world. Augmented reality (AR) technology is any method that instantly inserts virtual objects, information, or elements into the real world. There are many ways that AR and VR can be used in the supply chain. AR applications could help companies increase the productivity and performance as regards their operations, such as planning of the facility, transportation, order management, maintenance activity, and last but not least, last-mile delivery. On the contrary, the application of VR could help organizations manage freight operations, layout planning of production units, warehousing, management of workforce, and managing deliveries of the last mile. The goal of this paper is to investigate the existing literature, available data, and concepts to provide future research directions to help academicians and managers systematically understand the various applications and growth of AR and VR, in managing the supply chain.

Keywords: logistics, supply chain, augmented reality, virtual reality, emerging technology

1 Introduction

The world has faced unfortunate encounters in global operations and supply chains since 2019, owing to the COVID-19 pandemic. Due to the quick worldwide shift toward greater productivity and effectiveness across the supply chain network, processes for manufacturing and for providing services have had to be redesigned and approached

Saurabh Tiwari, School of Business, University of Petroleum and Energy Studies, Dehradun, India, e-mail: tiwarisaurabht@gmail.com
Sukanta Kumar Baral, Department of Commerce, Indira Gandhi National Tribal University, Amarkantak, India, e-mail: sukanta.baral@igntu.ac.in
Richa Goel, Symbiosis Centre for Management Studies, Symbiosis International University, Noida, India, e-mail: richasgoel@gmail.com

https://doi.org/10.1515/9783110790146-001

in novel ways [1]. Manufacturing processes and cutting-edge technology, such as cyber-physical systems and widespread internet use, are included in Industry 4.0 [2]. To maintain a constant competitive advantage [3] and attain the maximum heights of achievement in terms of quality, delivery, cost, and flexibility, businesses must develop core capabilities [4]. Implementing I4.0 improves connectivity and interaction between people, machines, and systems [5, 6], strengthening the relationship between production and supply chains. To manage the pandemic, many preventative measures were implemented to deal with the crisis, including travel restrictions, physical seclusion, and social lockdowns [7]. However, the maturity of the supply chain and its integration with I4.0 technologies will determine how effective these preventative measures are. AR and VR are two important I4.0 technologies that can help supply chains develop and mature in the post-COVID-19 era.

> An intermediary in which digital content is overlaid over the actual world and is interactive in real-time, as well as both spatially and temporally within the actual world, is AR [8].

> Any system that instantaneously includes virtual objects, elements, or data into the actual world is mentioned to as AR [9].

> Using virtual reality (VR) technology, users can interact with and immerse themselves in a computer-generated situation, a recreation of the actual, or an unreal world [10].

In order for users to feel the virtual world is realistic, VR aims to simulate them [11].

The way for AR was paved in 1966 with the invention of the first head-mounted display by Sutherland [12]. However, it took decades to investigate the practical applications of AR in business, thoroughly. By combining the actual world and virtual surroundings, AR technologies "improve a customer's experience of an interface with the actual world" [13]. All activities that have the primary objective of enhancing human senses and abilities by overlapping virtual elements over a physical environment are included in AR. AR is speedily gaining traction and is considered a cutting-edge technology within the I4.0 manufacturing archetype [14, 15]; as a result of this, I4.0 is quickly adopting AR technology [14–16]. AR expertise offers comprehensive backing to industries in managing the complete product life cycle and conceptual planning, production, and maintenance activities.

AR systems, for instance, have been applied in the product design phase, in order to provide quicker and increased efficient design operation [17]. R&D groups can quickly influence, examine, and measure active models using sophisticated AR visualization. Using this competency, industries can streamline complex engineering project jobs and reduce the time and cost of constructing physical models. Furthermore, combining virtual and actual items in the same surroundings allows developers to give consumers a mechanism for creating, visualizing, and contextualizing objects [17]. Several studies have been done on applying AR technologies in manufacturing and service delivery. Participants range from manufacturers of machine tools and power plants to suppliers to the aerospace and automotive industries [18]. In this case, AR technologies, more specifically AR glasses, enable workers to envision the placement of individual components on the production line and exhibit production data in their area of vision, making the line prone to fewer disasters and offering enhanced excellence mechanism [19].

The information and capabilities of operators can also be greatly enhanced by using AR devices, because they can see the jobs being completed in their occupied environment. There are many fields in industry where AR and VR are being functional and implemented; prominent among them are pharmaceuticals and hospitals, R&D, military operations, gaming industry, and science & technology. These days, due to the advancements in AR and VR technologies, they have been prominently used in managing the operations in logistics, production, and supply chains [20]. VR applications, for example, could assist companies with various tasks, including planning factory layouts, last-mile deliveries, warehousing, human resource management [21], etc. Unfortunately, the COVID-19 supply chain experienced numerous disruptions due to travel restrictions. This article seeks to promote the existing awareness of AR and VR in the supply chain, while presenting a novel angle for future research.

RQ1. What is the present awareness and understanding of AR and VR in logistics and supply chains?
RQ2. What are the primary research directions for AR/VR in the supply chain?

This paper has been broadly structured and planned, with six sections. The first section delivers a summary of the introduction; the literature analysis and evaluation on AR and VR are introduced in the second section. The research method is captured in the third section. The fourth section contains the results and discussions that were reported. Finally, section five presents the discussions and conclusions based on the findings, and section six addresses and presents potential guidelines for upcoming research.

2 Literature review

2.1 Augmented reality

In recent decades, the development of cutting-edge tools, like AR, has frequently been viewed as an active step for boosting the effectiveness of many supply chain functions [22]. Numerous industries, including manufacturing, military operations, education, and health care, have used AR [23]. Although AR has a history dating back to 1960, it has only recently come to be regarded as a research field unto itself [24]. AR has made providing an interactive and immersive surrounding for the users possible, by producing computer-generated content in the real world [25]. Digital or computer-generated content is being superimposed over the customer outlook of the actual world to augment reality to work [24, 26]. A real-time operating system, blending virtual and real elements, and integration into a 3D environment are, thus, necessary components of an AR [27]. The application of AR is frequently included in ideas for more comprehensive technological or digital transformation, though just adding new technical features will not work. As a substitute, it provides a paradigm shift by providing new methods of producing and delivering value by streamlining processes, enhancing process coordination, and raising customer satisfaction [28]. For companies looking to execute I4.0, AR technology is essential due to its capacity to produce immersive user experiences and increase process efficiency [16]. Technologies that combine the actual and virtual worlds are quoted to be "augmented reality" (Rohacz and Strassburger 2019), [29, 30]. It speaks of a surrounding that is assisted by exclusive hardware and software and is visualized [31]. It offers an imagined physical environment that is supported by specialized hardware and software [32]. Using simulations, customers can produce a world that is distinct from reality using virtual environments. The sensations in an augmented reality environment are "augmented," or made to come to life by combining virtual and 3D computer-generated images [33]. The users are provided feedback, instructions, and direction while interacting in a dynamic environment.

The complication in the supply chain has increased due to its complex structure consisting of multiple stakeholders, planning layers, complex modalities, and comprehensive information sharing at every level; and layers can benefit from AR [6, 34]. Additionally, new business models with a focus on mass customization are being created. As a result, in order to support specialization and customization, manufacturing and operational processes must evolve. To support such an ecosystem and provide workplace flexibility requires an entire reconfiguration, and managing the operations includes changing the ergonomics of plant layout, stock replenishment level, placement of machines and the configuration of the production line, and the positioning of the conveyor belt [35]. According to Lupu and List [36], AR can play a critical part in an organization's technological landscape to ensure effective supply chain and logistics operations as well as to add value and enhance user experiences [37]. AR-based industrial systems should strengthen the supply chain and service functions like warehousing, maintenance, and assembly. Illa and Padhi [38] looked into how AR technology could support development of intelligent factories and speed up supply chain digitization.

AR and VR help organizations develop and improvise product design in the nascent stages of production [39, 40] along with reconfiguring and planning the factory ergonomics and the design process [41]. AR/VR could support the business by providing support and helping manufacturing process optimize operations [42] through rescheduling the production [43], waste reduction and optimizing energy consumption [44], and managing and maintaining clarity in supply chain [45]. AR and VR help organizations improvise and provide technical aid to enhance the efficiency of distribution and warehousing operations and procedures [46] and order management such as order picking, editing, and processing [47]. Even though AR and VR would not entirely resolve all of the supply chain problems, they would provide helpful tools for boosting sustainability and resilience.

2.2 Virtual reality

Ivan Sutherland first proposed the concept of virtual reality (VR) in 1965, using the phrase "a (virtual) world in the window looks real, feels real, sounds real, and acknowledges realistically to the users' actions" as its definition [10]. Applications for VR can be found in several industries, including production, logistics, supply chain, gaming, science and technology, and the military [20, 48]. VR is "the combination of hardware and software systems that strives to exemplify an all-encompassing, sensory illusion of being present in another surroundings" [49]. Cipresso et al. [11] used the term "enveloping, associated, multisensory, viewer-centered, 3D computer-generated surroundings," in contrast. Even though these statements and words differ, it is important to note that virtual reality (VR) relies on three key elements: visualization, involvement, and collaboration [50]. When contrasting VR and physical reality, the first differs in the categories of content and learning actions. Both are then operationalized into several factors and consequences for using VR and physical reality using learning derived from each element. The category content includes the four factors of complexity, dynamics, networked news, and transparency, whereas the category learning actions consist of reversibility, cost-dependence, and time-dependence.

3 Methods

Using the systematic literature review technique, the objective of research in the areas of AR/VR in the supply chain was met [51]. The issue, as mentioned above, was addressed by using a database to examine the existing literature using the Systematic Literature Review (SLR) technique, which follows the five steps outlined by Denyer and Tranfield [51]. The papers were gathered from the Scopus database, the most significant abstract and citation repository for journals, books, conference proceedings, and book chapters (www.scopus.com). The SLR methodology recommendations made by Denyer and Tranfield [51] were followed in this study. The literature was then analyzed and summarized to determine current findings, potential future research topics, and gaps. The SLR technique is a tried-and-true method for examining bibliographic sources for a specific field to reach an organized conclusion based on the existing body of understanding about the field. The SLR method can be used to categorize and examine the literature offerings to a particular field of study. The most important benefit of this method is that it comprises of several widely used steps that are simple to verify or replicate by other researchers. The main benefits of this method over a simple literature review are its transparency, widely accepted structure, widely accepted steps, and repeatability. Seuring and Muller [52] developed a protocol for locating, choosing, appraising, and incorporating pertinent literature to conduct an SLR.

Defining the research area is the first step. Figure 1.1 portrays the five steps used in this article for an SLR following the methodology put forward by Denyer and Tranfield [51]:

(i) Developing the research questions;
(ii) Study identification;
(iii) Choosing and assessing studies;
(iv) Analysis and synthesis; and
(v) Results, followed by discussion.

Figure 1.1: Research Methodology.
Source: Adapted from Tiwari (2020)

(i) Step 1. Developing the research questions
Defining the research question or identifying the question or questions the research must address is the first step in any SLR [52].

Primary Research Question

RQ1. What is the present awareness and understanding of AR and VR in logistics and supply chains?
RQ2. What are the primary research directions for AR/VR in the supply chain?

(ii) Step 2. Study identification
The significance of Step 2 lies in the fact that it was here that we first began seeking out and identifying appropriate lessons to provide solutions to the research questions, as mentioned above. In this case, a Scopus database search was done. The Scopus database was identified and selected to conduct the search, because it is a well-known and reputable database for researchers to conduct this kind of research. Most articles are from the disciplines of life, health, physical, and social sciences. This makes it possible to conduct interdisciplinary research, which is one of the main justifications for using the Scopus database in this investigation. A preliminary search was done using the terms "augment reality" or "virtual reality" and "supply chain." The Scopus database was searched, and we came up with a list of 579 documents from the past ten years that were written in English (2013–2022). Table 1.1 lists the precise search syntax that was employed in the study.

Table 1.1: Search syntax.

Data collection source	Search syntax
Search performed on Scopus Website: www.scopus.com Year of search (2013–2022)	"augment reality" OR "Virtual reality" AND "Supply chain" AND (LIMIT-TO (PUBYEAR, 2022) OR LIMIT-TO (PUBYEAR, 2021) OR LIMIT-TO (PUBYEAR, 2020) OR LIMIT-TO (PUBYEAR, 2019) OR LIMIT-TO (PUBYEAR, 2018) OR LIMIT-TO (PUBYEAR, 2017) OR LIMIT-TO (PUBYEAR, 2016) OR LIMIT-TO (PUBYEAR, 2015) OR LIMIT-TO (PUBYEAR, 2014) OR LIMIT-TO (PUBYEAR, 2013)) AND (LIMIT-TO (DOCTYPE, "ar")) AND (LIMIT-TO (SUBJAREA, "BUSI")) AND (LIMIT-TO (LANGUAGE, "English")) AND (LIMIT-TO (SRCTYPE, "j"))

Source: Authors' own compilation

(iii) Step 3. Choosing and assessing studies
Following the first search phase, 579 papers were located using methodologies built on the papers' titles, keywords, and abstracts. The 579 identified papers' main findings, conclusions, and abstracts were carefully scrutinized to see if they were relevant to the study's questions. For this purpose, 579 articles were identified and critically examined, based on abstracts, methodologies, and conclusions. Only those papers that show a link between AR/VR and the supply chain were selected. As a result, the research yielded 54 articles relevant to the topic.

(iv) Step 4. Analysis and synthesis
The focus of step 4, also known as the fourth stage, was on carefully analyzing the studies that had emerged from the literature and had been selected and assessed in the third stage. Again, the analysis was evaluated using various criteria, including the most cited paper, author, number of articles published, and so on.

(v) Step 5. Present the results
According to the classification of the literature, the results are discussed in this section. In addition, the results give a direction to future researchers to understand the area in which work has been done and the various gaps that need to be searched.

4 Results and discussion

The analysis of the 54 papers provides information on the gaps and future directions for research, as well as the current state of knowledge and understanding about AR/VR in the supply chain (answer RQ1). Below are the specifics of these results.

4.1 Analysis

As shown in Figure 1.2, studies were initially ranked chronologically by year of publication. According to the early studies published, the literature on AR/VR in supply chain research is still fairly young. Figure 1.2 depicts the increase in research interest over the past six years and the direction that has persisted. For this study, the year 2022 is not excluded for analysis. In May 2022, 106 papers pertinent to the study's field of attentiveness had likewise been published or were "in the press," making the year 2021 the year with the most papers published, with 183 total. The top six articles with the most citations, as determined by the author, are listed in Table 1.2.

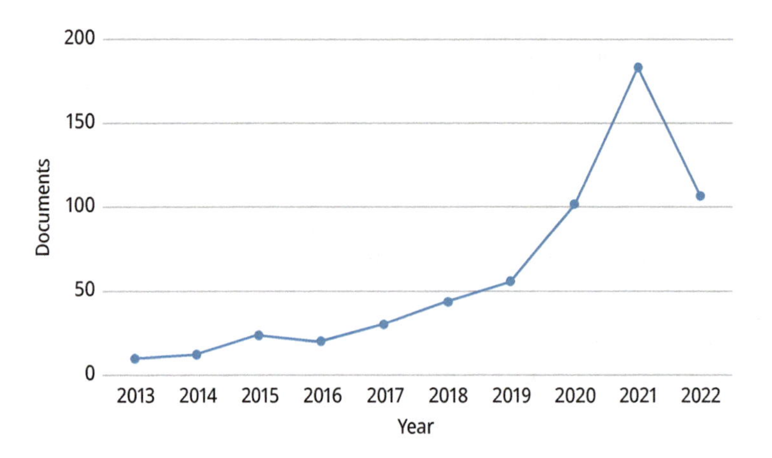

Figure 1.2: Documents by year.
Source: Author's own compilation

According to journal analysis, papers about AR and VR in the supply chain were primarily published in journals like the International Journal of Advanced Manufacturing Technology, the International Journal of Production Research, the International Journal of Computer Integrated Manufacturing, Computers in Industry, the Journal of Business Research, the Journal of Retailing and Consumer Services, and the International Journal on Interactive Design and Manufacturing, among others. Figure 1.3 displays the papers written by various authors; it is clear from the figure that each author publishes three articles.

Table 1.2: Top six authors with the highest number of citations, along with the affiliation of each author.

Author(s), year	Corresponding author's affiliation	Citations
Frank et al. [53]	Department of Industrial Engineering, Universidade Federal do Rio Grande do Sul, Brazil	678
Kamble et al. [54]	National Institute of Industrial Engineering [NITIE], Mumbai	454
Dwivedi et al. [55]	Emerging Markets Research Centre, School of Management, Swansea University, UK	336
Buchi et al. [56]	Management Department, University of Turin, Italy	199
Egger and Masood [57]	Institute for Manufacturing, University of Cambridge	96
Abdel-Basset et al. [58]	Faculty of Computers and Informatics, Zagazig University, Egypt	76

Source: Author's own compilation

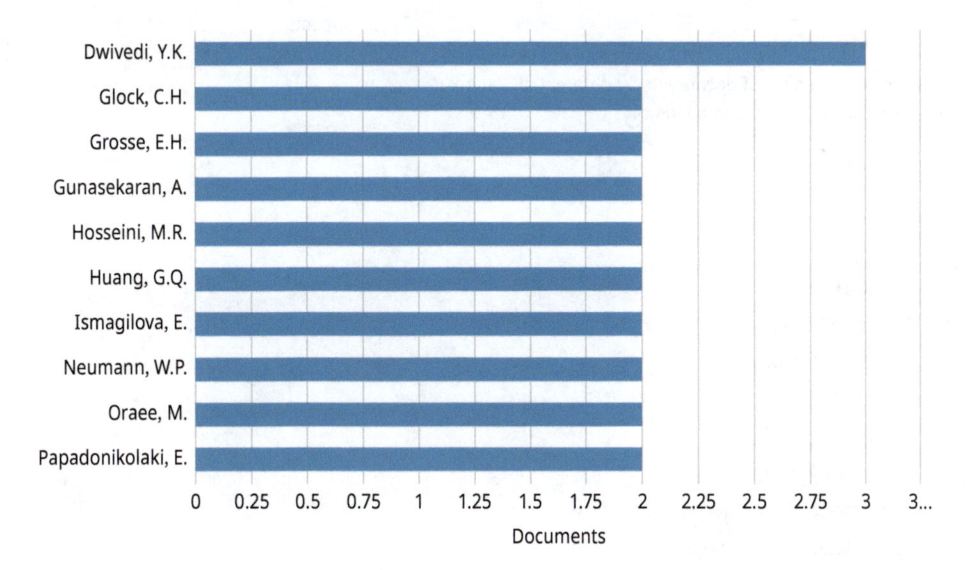

Figure 1.3: Number of documents by different authors.
Source: Author's own compilation

Figure 1.4 shows the papers by different affiliations; it is evident that the highest number of documents is three and, on average, two articles are published by various associations.

Figure 1.5 shows how many papers on AR and VR have been available in the supply chain sector; the manufacturing industry had the most articles published (25), followed by logistics (6), and retail (6). Other contributed industries include e-commerce, marketing, automotive, and so on.

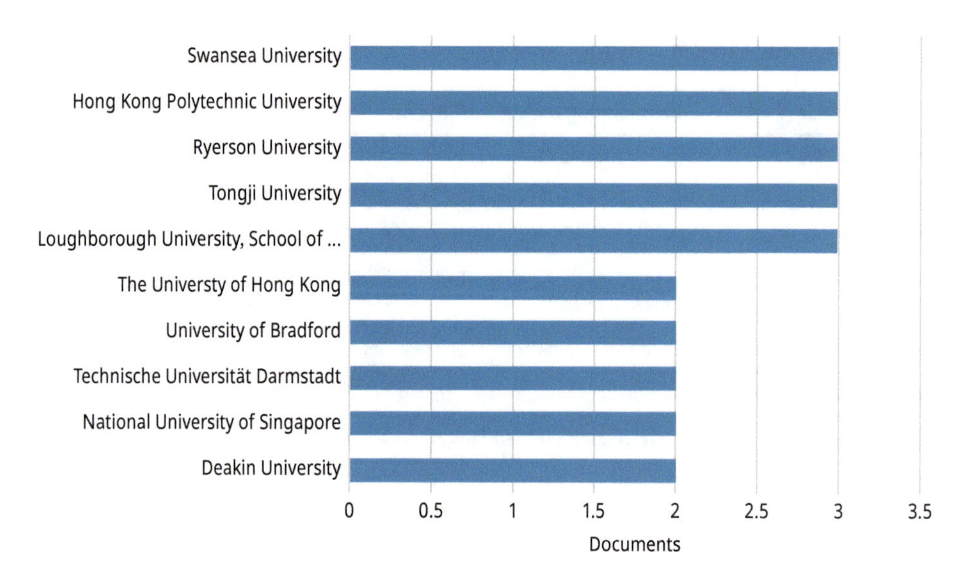

Figure 1.4: Number of documents by different affiliations.
Source: Author's own compilation

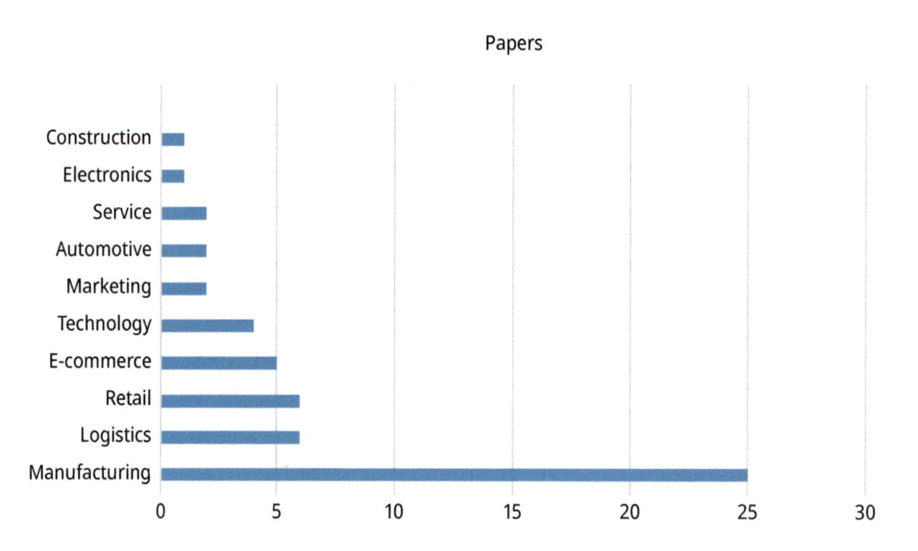

Figure 1.5: Number of documents by Industry.
Source: Author's own compilation

Figure 1.6 displays the number of papers published by each nation. The United Kingdom published the most articles (21), followed by the United States of America (12), China (9), Australia (7), Canada (6), and India (6). France, Germany, Hong Kong, and Finland are some of the other nations that contributed.

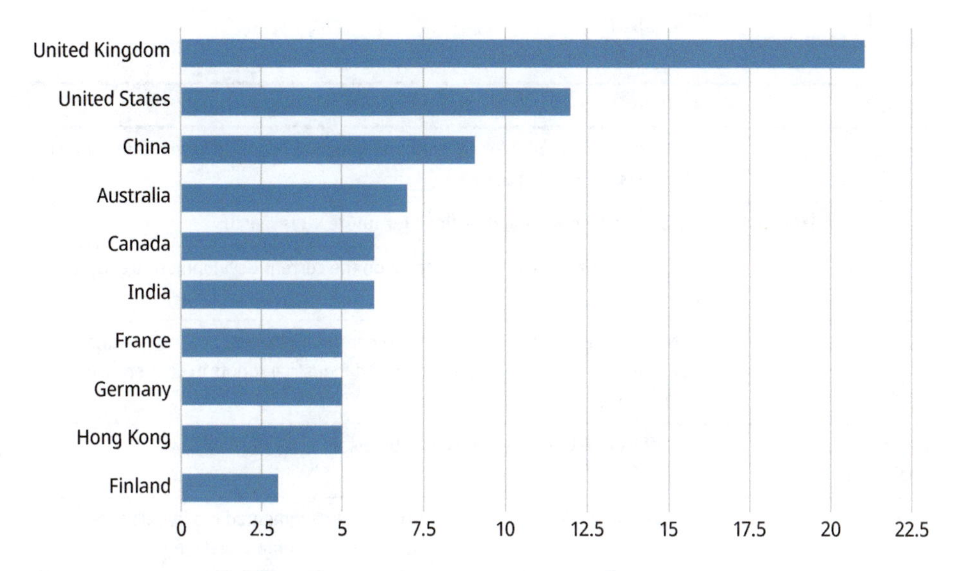

Figure 1.6: Number of documents by countries.
Source: Author's own compilation

Table 1.3 shows the paper summary on AR/VR in the supply chain. Sahu et al. [59] highlight the application of AR in manufacturing [57]. In contrast, Chandra Sekaran et al. [60] focus on implementing VR in the digital factory setting. Vanderroost et al. [61] offer a detailed review on the food package life cycle using digitalization for future research and development. AR impact on retailing was studied [62, 63]. Rejeb et al. [23, 64, 65] examine both the probable and existing development of AR using smart glasses in the supply chain, logistics, and food supply chain in their studies.

Table 1.3: List of recent AR/VR literature review papers.

Author	Methods
Lavoye et al. [62]	Introduces a theoretical framework for understanding consumer actions when using AR in a retail environment.
Rejeb et al. [64]	The probable and existing evolution of AR in the food supply chain is examined using a thorough review of literature.
Rejeb et al. [23]	A thorough analysis of the literature on the most recent AR developments: supply-chain and logistics-focused smart glasses.
Chandra Sekaran et al. [60]	A thorough analysis emphasizes the application of VR in the perspective of the digital factory.

Table 1.3 (continued)

Author	Methods
Sahu et al. [59]	The current application of AR tactics in the manufacturing sector is the subject of a review of the literature.
Xi and Hamari [66]	Introduces 16 potential directions for future VR research.
Bousdekis et al. [67]	An organized review of the literature on the current development trends in I4.0 technologies.
De Pace et al. [60]	A thorough review of the literature to determine the key advantages and disadvantages of augmented reality with industrial robots in cooperative human-robot scenarios.
Perannagari and Chakrabarti [63]	The effect of AR on retailing is the subject of a literature review.
Baroroh et al. [69]	A thorough analysis of the literature on how augmented reality might be used to bridge the gap between computational and human intelligence.
Caboni and Hagberg [70]	A review of the literature on the benefits, uses, and features of augmented reality in the world of retail.
Vanderroost et al. [61]	A thorough examination of the life cycle of food packaging digitization for future research and development.

Source: Author's own compilation

4.2 Prospects for augmented reality/virtual reality for sustainability

According to Vanderroost et al. [61], VR enables businesses to produce, synthesize, and receive a response on designs lacking the need for real models in the food package design stage. Together, VR and other techniques could boost the effectiveness of the food packaging industry's production and pave the way for developing environmentally friendly food packaging. Joerß et al. [71] investigated the efficacy of AR-based recommendation agents in encouraging shoppers to think about sustainability when making purchases. Rocca et al. [44] used robots, IoT, and cyber-physical systems along with AR, VR, and Digital Twin imitation gears to demonstrate how combining AR and VR, with other technologies, can support circular economy practices in the electrical and electronic equipment supply chain. It is interesting to note that Bressanelli et al. [72] developed a model for analyzing the state of the literature and classifying CE enablers in EEESC. The results showed that big data analytics, cloud computing, and IoT were all considered potential enablers in the literary landscape.

On the other hand, innovations like 3D printing, blockchain, and augmented reality were not. According to the results of our review, academics and practitioners have not yet fully understood how valuable and practical AR and VR can be, in promoting sustainability in the field of supply chains. Hence, it would be highly desirable for upcoming research to focus on how AR and VR could support sustainability, as well as the benefits and challenges of doing so. Similar gaps are present in qualitative data collection techniques, where it is still necessary to develop a wealthier and in-depth thought of the low implementation and any potential barriers. This could help the field advance and deepen our understanding of the motivations for and constraints in technological use, providing crucial knowledge for future integration of AR and VR and sustainability. Industry 4.0 not only has the potential, but it also acts as a catalyst for the generation of opportunities, not only in the area of supply chain but also other disciplines as per the new guidelines of United Nations Sustainable Goals, and is getting momentum and focus not only in industry, but also in society. These results suggest that there is still much room for positive momentum, capacity development, and deeper understanding because AR and VR are still in their infancy.

5 Conclusion

The execution of AR and VR in the supply chain over the past ten years is summarized in this paper (2013–2022). The study has been able to give an insight into current developments, their integration, and potential opportunities and challenges for the future, as a result of this thorough systematic review. This in-depth analysis looks at 54 articles that were published in the Scopus databases between 2013 and 2022. The current conversation makes it clear that implementing VR/AR in the supply chain presents challenges. Still, there are also prospects of advancing sustainability [73] and circular economy (CE) in the sustenance of supply chains' post-pandemic retrieval. The research on AR and VR is still nascent and early and requires time to mature, as per the discussions above. The manufacturing sector has done most of the AR and VR work, followed by logistics and retail, according to Figure 1.5. This is a well-recognized consideration in the history of science, where the field is still developing and lacks a robust mainstream to describe underlying societal understanding or awareness. Therefore, it is crucial for the expansion of AR and VR research in the supply chain to have clear guidelines for future work. The Supply Chain 4.0 maturity framework devised by Frederico et al. (2020) [74] can be used as a suggestive study for AR and VR research. They suggested four stages of development: the first, the second, the third, and the cutting-edge. Each maturity level has three sub-levels: requirements for process execution, technology levers, and managerial and capability supporters, all of which support strategic outcomes. When contrasting the early and cutting-edge levels, managerial and capable followers fully use technology levers and integrate them with process performance. In order to achieve sound strategic outcomes, this integration will make it possible to collaborate more effectively and be more flexible and responsive. This study found a void in the discourse that has repercussions and offers chances for creating academic models that consider the various facets from the early stage to development. A deeper understanding and knowledge of the AR and VR technology is required to implement and adapt during the various stages of development and for changes to be suggested during development to use them as strategic and operational tools. Future research directions may also underwrite to the lifespan by contextualizing and assessing the applicability of these stages as well as by identifying additional theoretical and practical implications. Probing the knowledge built on the [16, 74] framework, as opposed to focusing solely on VR and AR in the supply chain in uniqueness, could produce research results that allow for more focused interventions in theoretical and practical applications in AR and VR research to arise. Despite a plethora of papers exploring the uses and advantages of AR and VR in the supply chain, little has been examined in AR and VR from the perspective of undefined environmental surprises. However, it is still unclear and puzzling why this phenomenon occurs. The lack of technology adoption of AR and VR in the supply chain can be easily understood or explained through further research on various aspects and constraints, including obstacles to AR and VR implementation from both aspects, i.e., operational and organizational advantages of AR and VR for suppliers and customers, and from the organizational perspective, the resourcing requirements for AR and VR as related to IT infrastructure and organizational structure.

6 Future research directions

The current condition, industry trends, and prominent research directions in AR and VR have been examined and elaborated upon in this article, from the supply chain perspective. The research conclusions point to the following areas for future research in the area of AR and VR implementation in the supply chain. In terms of sustainability, researchers and academics paid very little attention to AR and VR, how it could be used to promote sustainable practices in the supply chain, and the associated barriers and difficulties in putting such practices into practice, using AR/VR in the supply chain. The competencies that human resources should have in order to implement AR/VR successfully are the other areas that require additional research in the supply chain context. There are obstacles that the organizational structure and IT support systems face during the implementation of AR and VR. Implementing new technology requires careful consideration of the organization's strategic mission and vision.

Similarly, an investigation needs to be done to find the effect of the adoption of AR and VR in an organization and how it is linked with the organization's vision. In the supply chain, coordination plays a significant role since it helps decide effectiveness and efficiency. Therefore, it is essential to thoroughly research the obstacles to supply chain coordination for successfully implementing AR and VR. Publications, databases, and websites must pay more attention to the AR and VR discourse. Studies must be done to support its application to get around these restrictions and offer a route for future research.

References

[1] Caiado, R. G. G., Scavarda, L. F., Gavião, L. O., Ivson, P., de Mattos Nascimento, D. L., and Garza-Reyes, J. A. (2021). A fuzzy rule-based industry 4.0 maturity model for operations and supply chain management. *International Journal of Production Economics*, 231: 107883.

[2] Torbacki, W., and Kijewska, K. (2019). Identifying Key Performance Indicators to be used in Logistics 4.0 and Industry 4.0 for the needs of sustainable municipal logistics by means of the DEMATEL method. *Transportation Research Procedia*, 39: 534–543.

[3] Ward, P. T., Bickford, D. J., and Leong, G. K. (1996). Configurations of manufacturing strategy, business strategy, environment and structure. *Journal of Management*, 22(4): 597–626.

[4] Sarmiento, R., Byrne, M., Contreras, L. R., and Rich, N. (2007). Delivery reliability, manufacturing capabilities and new models of manufacturing efficiency. *Journal of Manufacturing Technology Management*, 18(4): 367–386.

[5] Mourtzis, D., Doukas, M., and Psarommatis, F. (2012). A multi-criteria evaluation of centralized and decentralized production networks in a highly customer-driven environment. *CIRP Annals*, 61(1): 427–430.

[6] Tiwari, S., Bahuguna, P. C., and Walker, J. (2022a). Industry 5.0: A macroperspective approach. In: *Handbook of research on innovative management using AI in Industry 5.0*. IGI Global, 59–73. Pennsylvania, United States: Hershey.

[7] Jamaludin, S., Azmir, N. A., Ayob, A. F. M., and Zainal, N. (2020). COVID-19 exit strategy: Transitioning towards a new normal. *Annals of Medicine and Surgery*, 59: 165–170.

[8] Craig, A. B. (2013). *Understanding augmented reality: Concepts and applications*. San Diego, CA, USA: Newnes.

[9] Stoltz, M. H., Giannikas, V., McFarlane, D., Strachan, J., Um, J., and Srinivasan, R. (2017). Augmented reality in warehouse operations: Opportunities and barriers. *IFAC-PapersOnLine*, 50(1): 12979–12984.

[10] Mandal, S. (2013). Brief introduction of virtual reality & its challenges. *International Journal of Scientific & Engineering Research*, 4(4): 304–309.

[11] Cipresso, P., Giglioli, I. A. C., Raya, M. A., and Riva, G. (2018). The past, present, and future of virtual and augmented reality research: A network and cluster analysis of the literature. *Frontiers in Psychology*, 9: 2086.

[12] Liberati, N. (2016). Augmented reality and ubiquitous computing: The hidden potentialities of augmented reality. *AI & Society*, 31(1): 17–28.

[13] Azuma, R. T. (1997). A survey of augmented reality. *Presence: Teleoperators & Virtual Environments*, 6(4): 355–385.

[14] Hein, D. W., and Rauschnabel, P. A. (2016). Augmented reality smart glasses and knowledge management: A conceptual framework for enterprise social networks. In: A. Roßmann, M. Besch, and G. Stei (Eds.), *Enterprise social networks*. Wiesbaden: Springer Gabler, 83–109.

[15] van Lopik, K., Sinclair, M., Sharpe, R., Conway, P., and West, A. (2020). Developing augmented reality capabilities for industry 4.0 small enterprises: Lessons learnt from a content authoring case study. *Computers in Industry*, 117: 103208.

[16] Tiwari, S. (2020). Supply chain integration and Industry 4.0: A systematic literature review. *Benchmarking: An International Journal*, 28(3): 990–1030.

[17] Elia, V., Gnoni, M. G., and Lanzilotto, A. (2016). Evaluating the application of augmented reality devices in manufacturing from a process point of view: An AHP based model. *Expert Systems with Applications*, 63: 187–197.
[18] Weidenhausen, J., Knoepfle, C., and Stricker, D. (2003). Lessons learned on the way to industrial augmented reality applications, a retrospective on ARVIKA. *Computers & Graphics*, 27(6): 887–891.
[19] Rüßmann, M., Lorenz, M., Gerbert, P., Waldner, M., Justus, J., Engel, P., and Harnisch, M. (2015). Industry 4.0: The future of productivity and growth in manufacturing industries. *Boston Consulting Group*, 9(1): 54–89.
[20] Yao, L. J., Li, J., and Han, J. J. (2020, November). Application of virtual reality technology in intelligent cold chain logistics system. *Journal of Physics: Conference Series*, 1651(1): 012030. IOP Publishing.
[21] Bahuguna, P. C., Srivastava, R., and Tiwari, S. (2022). Two-decade journey of green human resource management research: A bibliometric analysis. *Benchmarking: An International Journal*, 30(2): 585–602.
[22] Davarzani, H., and Norrman, A. (2015). Toward a relevant agenda for warehousing research: Literature review and practitioners' input. *Logistics Research*, 8(1): 1–18.
[23] Rejeb, A., Keogh, J. G., Leong, G. K., and Treiblmaier, H. (2021a). Potentials and challenges of augmented reality smart glasses in logistics and supply chain management: A systematic literature review. *International Journal of Production Research*, 59(12): 3747–3776.
[24] Wang, W., Wang, F., Song, W., and Su, S. (2020a). Application of augmented reality (AR) technologies in inhouse logistics. In *E3S Web of conferences*, Vol. 145, 02018. EDP Sciences.
[25] Le, H., Nguyen, M., Yan, W. Q., and Nguyen, H. (2021). Augmented reality and machine learning incorporation using YOLOv3 and ARKit. *Applied Sciences*, 11(13): 6006.
[26] Tiwari, S., and Bahuguna, P. C. (2022). Digital transformation of supply chains. In: *Artificial intelligence and digital diversity inclusiveness in corporate restructuring*. Nova Science Publishers, Inc., 113–134. doi https://doi.org/10.52305/DPEM1704.
[27] Van Krevelen, D. W. F., and Poelman, R. (2010). A survey of augmented reality technologies, applications and limitations. *International Journal of Virtual Reality*, 9(2): 1–20.
[28] Chase, J. C. W. (2016). The digital revolution is changing the supply chain landscape. *The Journal of Business Forecasting*, 35(2): 20.
[29] Rohacz, A., and Strassburger, S. (2019, April). Augmented reality in intralogistics planning of the automotive industry: state of the art and practical recommendations for applications. In 2019 IEEE 6th International Conference on Industrial Engineering and Applications (ICIEA) (pp. 203–208). IEEE.
[30] Zhang, J., Ong, S. K., and Nee, A. Y. C. (2011). RFID-assisted assembly guidance system in an augmented reality environment. *International Journal of Production Research*, 49(13): 3919–3938.
[31] Cirulis, A., and Ginters, E. (2013). Augmented reality in logistics. *Procedia Computer Science*, 26: 14–20.
[32] Burdea, G. C., and Coiffet, P. (2003). *Virtual reality technology*. Hoboken, NJ, USA: John Wiley & Sons.
[33] Rovaglio, M., Calder, R., and Richmond, P. (2012). Bridging the experience gap-How do we migrate skills and knowledge between the generations?. In *Computer aided chemical engineering*. Elsevier, Vol. 30, 1407–1411.
[34] Sitek, P. (2015, March). A hybrid approach to sustainable supply chain optimization. In *International conference on automation*, 243–254. Springer, Cham.
[35] Duclos, L. K., Vokurka, R. J., and Lummus, R. R. (2003). A conceptual model of supply chain flexibility. *Industrial Management & Data Systems*, 103(6): 446–456.
[36] Lupu, M., and List, J. (2018). Conferences 2017. *World Patent Information*, 52: 68–71.
[37] Tiwari, S. (2022). Supply chain innovation in the era of Industry 4.0. In *Handbook of research on supply chain resiliency, efficiency, and visibility in the Post-pandemic era*. IGI Global, 40–60. Pennsylvania, United States: Hershey.
[38] Illa, P. K., and Padhi, N. (2018). Practical guide to smart factory transition using IoT, big data and edge analytics. *IEEE Access*, 6: 55162–55170.

[39] Guo, Z., Zhou, D., Zhou, Q., Mei, S., Zeng, S., Yu, D., and Chen, J. (2020). A hybrid method for evaluation of maintainability towards a design process using virtual reality. *Computers & Industrial Engineering*, 140: 106227.

[40] Arbeláez, J. C., and Osorio-Gómez, G. (2018). Crowdsourcing Augmented Reality Environment (CARE) for aesthetic evaluation of products in conceptual stage. *Computers in Industry*, 99: 241–252.

[41] Wang, X., Yew, A. W. W., Ong, S. K., and Nee, A. Y. (2020b). Enhancing smart shop floor management with ubiquitous augmented reality. *International Journal of Production Research*, 58(8): 2352–2367.

[42] Deshpande, A., and Kim, I. (2018). The effects of augmented reality on improving spatial problem solving for object assembly. *Advanced Engineering Informatics*, 38: 760–775.

[43] Mourtzis, D., Zogopoulos, V., and Xanthi, F. (2019). Augmented reality application to support the assembly of highly customized products and to adapt to production re-scheduling. *The International Journal of Advanced Manufacturing Technology*, 105: 3899–3910.

[44] Rocca, R., Rosa, P., Sassanelli, C., Fumagalli, L., and Terzi, S. (2020). Integrating virtual reality and digital twin in circular economy practices: A laboratory application case. *Sustainability*, 12(6): 2286.

[45] Barata, J., and da Cunha, P. R. (2021). Augmented product information: Crafting physical-digital transparency strategies in the materials supply chain. *The International Journal of Advanced Manufacturing Technology*, 112(7): 2109–2121.

[46] Plewan, T., Mättig, B., Kretschmer, V., and Rinkenauer, G. (2021). Exploring the benefits and limitations of augmented reality for palletization. *Applied ergonomics*, 90: 103250.

[47] Fang, W., and An, Z. (2020). A scalable wearable AR system for manual order picking based on warehouse floor-related navigation. *The International Journal of Advanced Manufacturing Technology*, 109(7): 2023–2037.

[48] Onyesolu, M. O., Ezeani, I., and Okonkwo, O. R. (2012). A survey of some virtual reality tools and resources. *Virtual Reality and Environments*, 21: 42.

[49] Radianti, J., Majchrzak, T. A., Fromm, J., and Wohlgenannt, I. (2020). A systematic review of immersive virtual reality applications for higher education: Design elements, lessons learned, and research agenda. *Computers & Education*, 147: 103778.

[50] Yung, R., and Khoo-Lattimore, C. (2019). New realities: A systematic literature review on virtual reality and augmented reality in tourism research. *Current Issues in Tourism*, 22(17): 2056–2081.

[51] Denyer, D., and Tranfield, D. (2009). Producing a systematic review. In D. A. Buchanan, and A. Bryman (Eds.), *The SAGE handbook of organizational research methods*. London: SAGE Publications, 671–689.

[52] Seuring, S., and Müller, M. (2008). From a literature review to a conceptual framework for sustainable supply chain management. *Journal of Cleaner Production*, 16(15): 1699–1710.

[53] Frank, A. G., Dalenogare, L. S., and Ayala, N. F. (2019). Industry 4.0 technologies: Implementation patterns in manufacturing companies. *International Journal of Production Economics*, 210: 15–26.

[54] Kamble, S. S., Gunasekaran, A., and Gawankar, S. A. (2018). Sustainable Industry 4.0 framework: A systematic literature review identifying the current trends and future perspectives. *Process Safety and Environmental Protection*, 117: 408–425.

[55] Dwivedi, Y. K., Hughes, L., Ismagilova, E., Aarts, G., Coombs, C., Crick, T., . . . Williams, M. D. (2021). Artificial Intelligence (AI): Multidisciplinary perspectives on emerging challenges, opportunities, and agenda for research, practice and policy. *International Journal of Information Management*, 57: 101994.

[56] Büchi, G., Cugno, M., and Castagnoli, R. (2020). Smart factory performance and Industry 4.0. *Technological Forecasting and Social Change*, 150: 119790.

[57] Egger, J., and Masood, T. (2020). Augmented reality in support of intelligent manufacturing – A systematic literature review. *Computers & Industrial Engineering*, 140: 106195.

[58] Abdel-Basset, M., Chang, V., and Nabeeh, N. A. (2021). An intelligent framework using disruptive technologies for COVID-19 analysis. *Technological Forecasting and Social Change*, 163: 120431.
[59] Sahu, C. K., Young, C., and Rai, R. (2021). Artificial intelligence (AI) in augmented reality (AR)-assisted manufacturing applications: A review. *International Journal of Production Research*, 59(16): 4903–4959.
[60] Chandra Sekaran, S., Yap, H. J., Musa, S. N., Liew, K. E., Tan, C. H., and Aman, A. (2021). The implementation of virtual reality in digital factory – A comprehensive review. *The International Journal of Advanced Manufacturing Technology*, 115(5): 1349–1366.
[61] Vanderroost, M., Ragaert, P., Verwaeren, J., De Meulenaer, B., De Baets, B., and Devlieghere, F. (2017). The digitization of a food package's life cycle: Existing and emerging computer systems in the logistics and post-logistics phase. *Computers in Industry*, 87: 15–30.
[62] Lavoye, V., Mero, J., and Tarkiainen, A. (2021). Consumer behavior with augmented reality in retail: A review and research agenda. *The International Review of Retail, Distribution and Consumer Research*, 31(3): 299–329.
[63] Perannagari, K. T., and Chakrabarti, S. (2020). Factors influencing acceptance of augmented reality in retail: Insights from thematic analysis. *International Journal of Retail & Distribution Management*, 48(1): 18–34.
[64] Rejeb, A., Rejeb, K., and Keogh, J. G. (2021b). Enablers of augmented reality in the food supply chain: A systematic literature review. *Journal of Foodservice Business Research*, 24(4): 415–444.
[65] Tiwari, S., Bahuguna, P. C., and Srivastava, R. (2022b). Smart manufacturing and sustainability: A bibliometric analysis. *Benchmarking: An International Journal*. https://doi.org/10.1108/BIJ-04-2022-0238
[66] Xi, N., and Hamari, J. (2021). Shopping in virtual reality: A literature review and future agenda. *Journal of Business Research*, 134: 37–58.
[67] Bousdekis, A., Lepenioti, K., Apostolou, D., and Mentzas, G. (2021). A review of data-driven decision-making methods for industry 4.0 maintenance applications. *Electronics*, 10(7): 828.
[68] De Pace, F., Manuri, F., Sanna, A., and Fornaro, C. (2020). A systematic review of Augmented Reality interfaces for collaborative industrial robots. *Computers & Industrial Engineering*, 149: 106806.
[69] Baroroh, D. K., Chu, C. H., and Wang, L. (2020). Systematic literature review on augmented reality in smart manufacturing: Collaboration between human and computational intelligence. *Journal of Manufacturing Systems*, 61: 696–711.
[70] Caboni, F., and Hagberg, J. (2019). Augmented reality in retailing: A review of features, applications and value. *International Journal of Retail & Distribution Management*, 47(11): 1125–1140.
[71] Joerss, T., Hoffmann, S., Mai, R., and Akbar, P. (2021). Digitalization as solution to environmental problems? When users rely on augmented reality-recommendation agents. *Journal of Business Research*, 128: 510–523.
[72] Bressanelli, G., Pigosso, D. C., Saccani, N., and Perona, M. (2021). Enablers, levers and benefits of Circular Economy in the Electrical and Electronic Equipment supply chain: A literature review. *Journal of Cleaner Production*, 298: 126819.
[73] Tiwari, S. (2015). Framework for adopting sustainability in the supply chain. *International Journal of Automation and Logistics*, 1(3): 256–272.
[74] Frederico, G. F., Garza-Reyes, J. A., Anosike, A., and Kumar, V. (2020). Supply Chain 4.0: Concepts, maturity and research agenda. Supply Chain Management: An International Journal, 25(2): 262–282.

B. Rajalingam, R. Santhoshkumar, P. Deepan, P. Santosh Kumar Patra

2 An intelligent traffic control system using machine learning techniques

Abstract: The ever-increasing traffic jams in urban areas makes it necessary to make use of cutting-edge technology and equipment in order to advance the state of the art in terms of traffic control. The currently available solutions, such as time visitors or human control, are not adequate to alleviate the severity of this crisis. The findings of this investigation have led to the proposal of a system for the control of traffic that makes use of Canny edge detection and digital image processing to determine, in real time, the number of vehicles present. The above imposing traffic control advanced technologies offer significant advantages over the existing systems in real-time transportation management, robotization, reliability, and efficiency. In addition, the complete process of digital image acquisition, edge recognition, as well as green signal assignment is demonstrated with accurate blueprints, and the final outcomes are stated by hardware. All of this is done with four separate photographs of various traffic scenarios.

Keywords: Intelligent Traffic Control, Density-based Signalization, Edge Enhancement

1 Overview

The ever-increasing number of cars and trucks that are driving around today, combined with the limited resources supplied by the existing infrastructure, is contributing to the worsening of traffic problems. Those making use of a public route for the purpose of travel may include pedestrians, riding or herding animals, automobiles, trolleybuses, or other conveyances. Traffic on roads may move in either direction. The laws that govern traffic and regulate cars are referred to as traffic laws, but the rules of the road include not only the laws but also any unofficial regulations that may have emerged over the course of time, in order to make the flow of traffic more orderly and efficient. Road signs, often known as traffic signs, are signs that are put along the sides of roadways in order to offer information to drivers.

B. Rajalingam, Computer Science and Engineering, St. Martins Engineering College, Secunderabad, India, e-mail: rajalingam35@gmail.com
R. Santhoshkumar, Computer Science and Engineering, St. Martins Engineering College, Secunderabad, India, e-mail: santhoshkumar.aucse@gmail.com
P. Deepan, Computer Science and Engineering, St. Peter's Engineering College, Secunderabad, India, e-mail: deepanp87@gmail.com
P. Santosh Kumar Patra, Principal, St. Martin's Engineering College, Secunderabad, India

https://doi.org/10.1515/9783110790146-002

1.1 Traditional traffic management systems

1.1.1 Human-based control systems

It requires a significant amount of manpower to manually control the instance name, and it also requires manual control of the traffic. The number of traffic police officers that are assigned to a certain city or region is determined by the countries and states in which the location is located. In order to maintain order and control the flow of traffic, police officers will be equipped with tools such as signboards, sign lights, and whistles. In order for them to effectively regulate the traffic, they will be given instructions to dress in particular uniforms.

1.1.2 Mandatory control systems

Timers and electrical sensors work together to regulate the automatic traffic signal. At the start of each new step of the traffic signal, the time limit is programmed with a consistent numerical number. The lamps will switch on and off automatically once the valuation of the clock is adjusted, so that it could be set to the desired duration. It will collect the allocation of the vehicles and also signals on each phase, while using electrical sensors; depending on the sensor, the lights will quickly change between the ON and OFF positions.

1.2 Disadvantages

More man power is required for a system that relies on human controls. Due to a lack of available personnel in our traffic police force, we are unable to manually regulate the traffic flow in any part of a town or city, at this time. Therefore, we require a more effective strategy to manage the traffic. On the other hand, an automatic traffic control system uses a timer for each phase of the traffic signal. Congestion in the traffic also occurs while the electronic sensors are being used to manage the traffic.

1.3 Image processing is essential for effective traffic control

The author developed image-processing-based traffic light control. Instead of implanted electronic sensors, pictures will identify automobiles. Traffic light cameras will be placed. Image sequences are captured. Image processing is better for controlling traffic light states. As it uses traffic photos, it detects vehicles more consistently. It visualizes reality, and thus, works better than devices that detect metal content.

The greatest cities across the world are facing a significant challenge, today, in the form of traffic congestion. As per research conducted by the World Bank over the course of the past ten years, the average speed of vehicles in Dhaka has decreased from 20 km per hour to 7 km per hour. This has caused a decrease in provincial competitiveness and reallocation of economic growth by slowing growth in county, global production, or overall economic growth in metropolitan region employment, because more and more automobiles enter an already congested traffic system. There is an urgent need for a completely new traffic control system that uses sophisticated techniques to improve the most of the already existing infrastructures.

In terms of traffic data collection, these strategies have had good results. However, the estimation of the number of motor vehicles may generate wrong results if the intravehicular distance is very narrow, and it may not consider rickshaws or auto-rickshaws as vehicles, despite the fact that these are the most common forms of transportation in South Asian nations. There are also drawbacks to this approach, such as the fact that it can count insubstantial objects like pavements or pedestrians. Some of the work has suggested that time should be allocated based merely on traffic density. However, this may be a disadvantage to individuals who travel on less-frequently travelled roads. In order to retrieve the necessary traffic data from CCTV camera images, an edge detection approach must be used to separate the desired information from the remaining image. There are a variety of methods for detecting the edges of objects. Their noise reduction, detection sensitivity, and accuracy are all different. Prewitt, Canny, Sobel, Roberts, and LOG are among the most well-known names in the industry. Comparing the Canny edge detector to other methods, it has been found to be more accurate at picking out objects with more entropy, better PSNR, MSE (Mean Square Error), and execution time.

2 Image processing

The term "image processing" applies to any kind of signal analysis that has used image as an input. Some of the most common image-processing methods used to process images are two-dimensional signals in typical signal processing. Although digital image processing is the general standard, optical and analogue image processing are feasible as well. All of them can benefit from the information in this article, which focuses on broad principles. An image's acquisition (the act of creating an input image) is referred to as an imaging process.

2.1 Image acquisition

The procedure of acquiring images is always the very first step in any operation involving image processing. Once the image has been captured, a variety of processing techniques can be applied to it in order to complete the myriad of jobs that are necessary in the modern world. Even with the assistance of various types of picture improvement, the goals may not be attainable if the image has not been acquired in an adequate manner. This is true even if the image has been improved. The generation of digital photographs, often taken from a real-world setting, is referred to as "digital image acquisition." It is common practice to presume that the phrase implies or includes the compression, printing, storage, processing, and showing of images of this kind. Although the use of a digital camera for digital photography is, by far, the most common approach, alternative approaches are also frequently used.

2.2 RGB to gray

Conversion A grayscale or grayscale digital picture is a photograph or digital image in which each pixel's value is a single sample containing, solely, information on the image's intensity. Tones ranging from black to white make up the entirety of a black-and-white image, which is also known as a grayscale image. Photographs in grayscale are unique from those in black-and-white with only two tones: black and white, or one-bit bi-tonal. Many different hues of grey can be shown in grayscale images. It is also known as monochromatic, which denotes the existence of just one color in the image. Enhancement of the image: image-processing techniques allow you to improve the signal-to-noise ratio and draw attention to specific aspects of an image by altering its colors or intensities. In picture enhancement, you perform the following actions:

- Picture denoising
- Gadget hue management
- Reshape an image
- Conversion of an image to another image format

2.3 Persistence of the outer and inner limits

Edge detection, for example, is a key tool in feature detection and feature extraction, which look for areas in digital images where brightness fluctuates sharply or contains singularities. Detecting singularities in a one-dimensional signal is the same as identifying steps. Sharp changes in image intensity are necessary to capture important events and changes in the quality. The general presumptions of a model that are likely to be part of brightness discontinuities in an image are:

- Interferences in depth
- Disparities in the orientation of the surface
- Properties of materials that can change over time
- Differences in the lighting of the scene

Edge detection, with the aid of software, in an image should produce, in an ideal world, a set of connected curves that show object and surface marking borders as well as interferences in surface orientation boundaries. As an edge detection method can exclude information deemed unimportant while keeping the image's structural features, it can drastically reduce the amount of data to be processed. It is possible that the overall response of deciphering the original image's content may be greatly simplified, if the edge detection process is successful. Images of modest richness rarely yield perfect profit margins in real life. It is common for nontrivial images to suffer from separation, which indicates that the side contours are not linked, and there are omitted side segments and also false sides that do not correspond to considerable incidences in the image – making the subsequent task of analyzing the visual data quite difficult.

2.4 Image matching

In edge-based matching, two individuals who are considered to be representatives of the same item are matched up with one another when comparing and analyzing two images, and its illustration on one image is judged in relation to all of the edges on the other image. The Prewitt operator was used in order to perform edge detection on the source image and the current image. After that, these edge-detected photos are compared, and the time span that each traffic light remains on can be adjusted accordingly.

3 Related work

Those who live in non-static cities with heavy traffic know, all too well, the annoyances that come with being stuck in traffic. Accurate vehicle detection is essential for both manual and computerized traffic management. In this paper, a vehicle detection model for traffic control that employs image processing has been proposed. In the beginning, the HSV images are converted from RGB photos. In order to determine whether a picture is day or night, it compares its value readings to a predefined threshold parameter that has been calculated. Vehicles of both day and night can be detected using two separate methods at this point. When it is daylight, a comparison of the front and back views of the same image is made to identify the vehicles. After that, the automobiles are counted using an object counting technique. To distinguish between headlights and ambient light, the intensity of the image is analyzed at night. Finally, the number of automobiles is counted, using another object counting methodology. Tested on a variety of data sets, this model has an average accuracy of 95% for both day and night [1].

The goal of road transportation strategy is to minimize traffic, but it is not clear when this widespread by-product of ancient living actually harms the economy. Based on frame data collected in the United States from 88 census tracts (1993–2008), the impact of overcrowding on employment levels is estimated (2001 to 2007). Job growth appears to be hindered above these levels, if the regional freeways have an ADT of over 11,000 per lane and a commute time of around 4 min, each way. There is no indication that heavy traffic disruptions have such a negative effect on economic growth. A more effective strategy for dealing with congestion's financial drag could be to give priority to economically significant trips or to provide additional travel ability to allow people to get where they need to go, even if there is traffic [2].

Using image processing algorithms, it is possible to learn more about a subject. Traffic congestion on roadways and a model for controlling traffic lights is proposed here, based on images taken by video cameras. The actual sum of pixel volume covered by moving vehicles in a camera shot is calculated, rather than the amount of vehicles on the road. Each route has its own unique traffic cycle and weightings are set as output parameters, and traffic lights are controlled sequentially [3].

The city's population and automobile fleet are both growing at an alarming rate. Controlling roads, highways, and streets has become a big problem, as the population of cities grows. The main cause of today's traffic problem is the usage of traffic management strategies. As a result, current traffic management systems are inefficient, because they do not take into account the current traffic situation. The MATLAB software has been used to construct this project, which attempts to reduce traffic congestion. In addition, image Processing is employed to implement this project. At first, a camera records footage of a roadway. We install a web camera in a traffic lane to record traffic flow on the road we are trying to manage. These photos are then analyzed in an efficient manner, so that traffic density can be determined. The controller will instruct the traffic LEDs to display a specific time on the signal to regulate traffic based on the processed data from MATLAB [4].

In lossless image encoding, the Gradient Adjusted Predictor (GAP) and the Gradient Edge Detection (GED) predictor introduce a new picture edge detection approach with dynamic threshold control that is based on the Multidirectional Gradient Edge Detection Predictor (MGEDP) template. From the center outward, the image is sliced into four equal pieces, and each of these pieces can be run concurrently by the MGEDP template in four different directions to calculate the error values using parallel technology. These feedback values are used by the algorithm to generate a forecast error image, which is then used to determine the threshold values using the Otsu algorithm.

Finally, the algorithm produces a final image representation by classifying and refining the corners of an error image [5]. Tests have shown an algorithm that uses parallel innovation not only diminishes the amount of time it takes to process the image but also generates chamfered edges and many more characteristics. For object extraction using satellite remote sensing photographs from Indian Remotely Sensed sensors LISS 3, LISS 4, as well as Cartosat1 and Google Earth, this paper presents a comparison of numerous corner detections and band-wise analysis of these algorithms [6]. Edges are what define boundaries and are therefore a problem of practical importance. As the challenge of urban congestion worsens, it is imperative that advanced innovation and machinery be deployed to improve traffic control mechanism. Due to the increasing number of vehicles on the road and the limited resources offered by existing infrastructure, today's traffic congestion is getting worse. Traffic lights can be easily programmed to employ a timer for each part of the cycle. Detecting vehicles and generating a signal that repeats in cycles is another option. The traffic light can be controlled using image processing. This will use photographs rather than sensors placed in the pavement to identify automobiles. The traffic light will be equipped with a camera that can record video clips. With a reference image of a deserted road as a starting point, the photographs acquired are compared using image comparison. The edge detection (Prewitt) operator has been used for this purpose, and the proportion of matching traffic light timings can be used to control [7].

When it comes to traffic monitoring and control, there is growing need for sensor networks and accompanying infrastructures. These devices enable authorities to keep tabs on the flow of traffic at detection points, while also providing them with pertinent real-time data (example: traffic loads). Automated tracking units that encapsulate and perceive photos captured by one or more pre-calibrated webcams are illustrated in this research, which uses a network of automated tracking units (ATUs). In addition to monitoring traffic in highway tunnels and airplane parking lots, the suggested system is adaptable, scalable, and suited for a wide range of applications. Other objectives include testing and evaluating a variety of image processing and information fusion methods, which will be used in the actual system. By using the outcome of the image analysis unit, a range of data for each movable object in the scene is transmitted to a distanced traffic control center. This information includes the speed, target ID, location, and classification of the moving vehicle. Through the analysis of these data, real-time output can be obtained.

In image processing and computer vision, the term "edge detection" is critical. There are a number of edge detection operators you need to know in order to be able to identify objects in images. Comparative evaluations of image-processing edge detection operators are offered here. The results of this study show that the Canny edge detection operator outperforms other edge detection algorithms in terms of image presentation as well as the location of object boundaries; MATLAB [9] is the software tool that has been used.

In an attempt to choose a species of shark fish to use as a case study, one must first have a solid understanding of the concepts that lie beneath the surface of the various filters, and then one must use these filters to analyze the shark fish. In this article, the various approaches to edge detection are discussed and analyzed. MATLAB is the tool that is used throughout the development of the program. The two most crucial steps of image processing use the Gradient and Laplacian operators. The Laplacian-based segmentation method, the Canny edge detector, and the gradient-based Roberts, Sobel, and Prewitt edge detection operators are some of the filters that are used in the case study that focuses on the classification of shark fish using image processing. This research [10] examines in-depth, the benefits and drawbacks of using these filters.

The goal of this project is to build a traffic light system that automatically adjusts its timing based on the amount of traffic at a given intersection. In most major cities across the world, traffic congestion is a major concern, and it is time to switch to an automated system that can make decisions. When just one lane of a roadway is active at any given moment in the current traffic signaling system, it can be inefficient. As a result, we have developed a design for an intelligent traffic management system. Occasionally, a higher volume of traffic necessitates a longer green period than the typical allocated time at a junction. As a result, we propose a method in which the amount of time spent with the red light and the green light at any given intersection would be determined by the volume of traffic that is present at any given moment. This can be done using Proximity Infrared (PIR) sensors. After the intensity has been computed, the microcontroller will tell the green light when to start glowing at the appropriate time. The information that is provided to the microcontroller by the sensors on the roadside that detect the presence of vehicles allows the microcontroller to make decisions such as when to swap over the lights or how long a flank will be open. In the following sections, we will explain how this framework works [11].

Congestion in the roads is a pressing issue in any region. A known contributor to traffic jams is the amount of time drivers spend at traffic lights. However, traffic volume has no bearing on when the traffic light will change. Since traffic density is more important than time, traffic control simulations and optimizations are needed. Efforts are made to minimize traffic congestion caused by traffic signals in this system. A density-based traffic control system was created as part of this project to address this issue. Programming an Arduino and using the Arduino enables traffic lights to offer the right of access to the road by picking the lane with the most cars. For example, the sensor can identify an automobile and signal the Arduino to regulate the traffic lights for that vehicle's specific path. As long as none of the four sensors detects a sign, the traffic lights continue to operate under the assumption that nothing is amiss. The sensor's average response time was 0.39 s. A lot more study is needed before the device can be manufactured in large quantities to be used on all of the country's highways [12].

Most cities throughout the world suffer from severe traffic congestion, which has turned into a nightmare for the people who live there. If the signal is delayed or traffic signals are timed incorrectly, this can happen. The traffic light's delay is preprogrammed and does not change based on traffic [13]. As a result, a growing need for systematic, rapid, and automatic traffic control systems has emerged. Dynamic traffic signal control for dense areas is the goal of this work. The timing of the traffic light changes, based on the amount of traffic at the intersection. This project uses an Arduino microcontroller. Whenever a vehicle comes close enough, it activates and receives the signal.

The transportation system relies heavily on traffic light control systems to manage and keep track of its many moving parts. The city's busiest roads are the primary focus of these devices. Multiple traffic signal systems at nearby crossings must be coordinated, but this is a difficult task. We are getting closer to the idea of a traffic volume control system. The traffic system in this area is monitored by counting the number of vehicles on the road. Sensors will be used to count the number of vehicles. Touchline sensors and laser sensors are the two types of sensors available. Heavy vehicles can easily damage the touchline sensors, making it difficult to tally the number of vehicles. Those with high directivity [14] are the ones most often chosen.

The development of a vibrant traffic light system is intended to be the result of this project that is dependent on density. The timing of the traffic light changes, based on the amount of traffic at the intersection. Many major cities across the world suffer from severe traffic congestion, making daily commuting a misery for residents. In the traditional traffic light system, the time allowed to each side of the intersection is set and cannot be adjusted in response to changes in traffic density. The assigned junction times are set in stone. Some junctions require longer green times due to increased traffic density on the same side of the intersection than the typical permitted time. It is necessary to process and convert a color image collected at the signal into a grayscale image before drawing a contour line to determine the image's vehicle count. In order to figure out which side of the road has the most traffic, we need to know how many vehicles are on the road. As a microcontroller, the Raspberry pi provides signal timing dependent on traffic density.

The conventional traffic system assigns a predetermined and fixed green light time to every road lane. When there is a lot of traffic on the road, it does not change. This method is inefficient for dealing with traffic in metropolitan and smart cities. Due to the lack of a priority factor and other considerations, the current method does not take the number of cars in relation to congestion into account. As part of this technique, we submit the current traffic image to the program, which extracts edges from images; if the image has more traffic, there will be more white-colored edges; if the image contains less traffic, there will be fewer white-colored edges.

4 Proposed system

Using this approach, traffic density is estimated by comparing knowledge about the current traffic situation superimposed on an image of an empty road serving as a source image. The percentage of traffic signal duration that matches the authorized time can be modified. Traffic signals will be more effective and practical with this new design. A density-based system means that it will give Priority to the lane that has comparatively greater number of vehicles than the others. Image processing is the method that will be used to calculate the density. **Algorithm**: Canny edge detector,

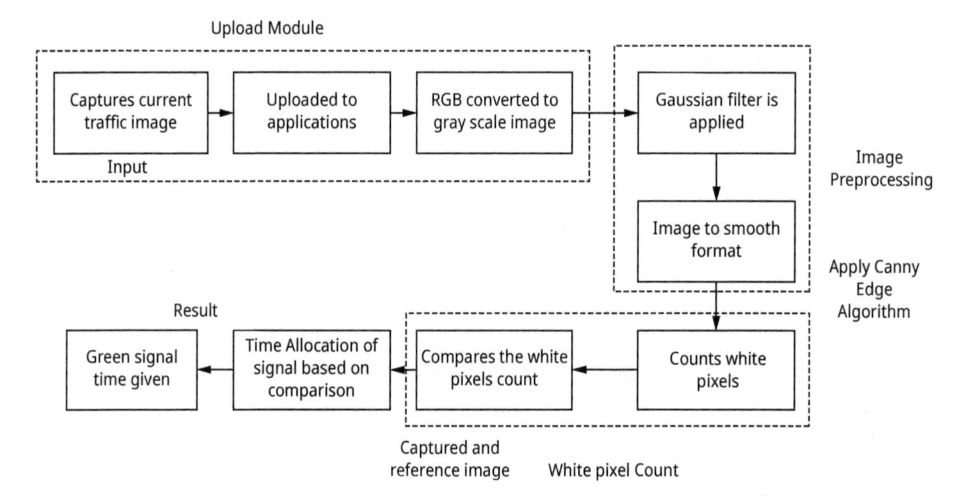

Figure 2.1: Detailed design.

5 Implementations

In this section, we will go through how to formulate a problem as well as the underlying principles and simulation methodology. We used PyCharm to write the code for this project because we needed to work in a Python environment.

5.1 PyCharm IDE

One of the most often used IDEs for computer programming (especially Python) is PyCharm. JetBrains, a Czech firm, developed it. A unit tester that is implemented, in addition to interaction with version control systems (VCSes), are only some of the features it offers, in addition to code analysis and a graphical debugger. There are versions of PyCharm for Windows, Mac OS X, and Linux, making it cross-platform. Both the Community Edition and the Professional Edition come with additional features and are published under a proprietary license. The Apache License governs the Community Edition.

- Many of these operations can be done in Python by just changing the name of an existing function or removing it entirely.
- The web frameworks Django, web2py, and flask are all supported by this module.

- Integrated debugger for the Python programming language line-by-line code coverage for integrated unit testing.
- Python development for Google App Engine.
- User interfaces for Mercurial (Git), Subversion (SVN), Perforce (Perforce), and CVS are all merged into a single user interface.

Aside from the more general Komodo, it competes with different Python IDEs, such as Eclipse's PyDev. It is possible to write your own plugins to extend PyCharm's features, using the PyCharm API. Plugins from other JetBrains IDEs can also be used in PyCharm. Plugins for PyCharm can be found in over a thousand different repositories.
- Several license options are available for PyCharm Professional Edition, each with its own set of features, pricing point, and terms of use.
- The cost of an Academic license is significantly reduced or, perhaps, waived.
- The source code for PyCharm Community Edition may be found on GitHub and is given under the Apache 2 license.

The template is intended for up to six authors, but this is not a requirement. All conference papers must have at least one author. The names of the authors should be listed sequentially from left to right, then down the page. Citations and indexing services will use this sequence of authors in the future. Do not put names in columns or organize them based on affiliation. Please be as concise as possible when describing your affiliations (for example, do not differentiate among departments of the same organization).

5.2 Python

Python is a powerful programming language, which may be handled in a number of different ways and used for a wide variety of applications. The extensive use of underlining, throughout, demonstrates that its design philosophy places a significant emphasis on the accessibility of the code. Its object-oriented methodology and language components are created to aid programmers in generating code that is not only understandable but also sensible for a wide variety of projects. The programming language, Python, is dynamically structured and also collects garbage automatically. It is appropriate for a wide number of development methodologies, such as programming language (in particular, scripting language), object-oriented programming, and functional programming. Due to its comprehensive library that is of very good standard, it is usually referred to as a "battery packs included" language.

In the late 1980s, Guido van Rossum started developing Python as a programming language that would succeed ABC. He first made Python available to the public in 1991, with the version *number 0.9.0. New capabilities* were made available with the release of Python version 2.0, in the year 2000. Python 3.0 was an *important* change of the language that was launched in 2008 and was not fully backward-compatible with earlier versions of the program. Python 2 was officially retired with the release of version 2.7.18, in the year 2020.

A. Upload Image

To produce pixels with black-and-white values, the current traffic image will be uploaded to the program and converted to a grayscale image format.

B. Pre-process

This module uses Gaussian Filtering to smoothen out images that have been uploaded. A Canny edge detection filter will be added to the image after it has been filtered. White-colored pixels will be used by vehicles, whereas black-colored pixels will be used by non-vehicles.

C. White Pixel Count

To achieve a complete traffic count, we will use this method to count white pixels in the Canny image.

D. Canny edge detection algorithm

It is considered by many to be the best edge detector. Clean, narrow edges are well related to the adjacent edges. Image-processing software is likely to include a function that does it all. In this section, we will go into more detail about how they function specifically. It is based on the principle of a multistage edge detection algorithm. Here are the stages:

- Pre-processing
- Gradients calculation
- Non-maximum reduction
- The use of repetition in thresholding

The algorithm's two most important parameters are the higher and lower thresholds. In order to distinguish between real and fake edges, the upper threshold is used. Faint pixels that are actually part of an edge can be found at the lower threshold. The following are some of the most common edge detection criteria: Edge detection with a low error rate indicates that as many of the image's edges as feasible should be reliably caught by the detection process. The operator should be able to properly locate the center of the edge by detecting the edge point. There should be no fake edges in the image if possible, so that a given edge can only be marked once. Canny used the calculus of variations – a method for finding the optimal function for a given functional – to meet these conditions. It is possible to approximate Canny detector's optimal function by taking the first derivative of a Gaussian. The Canny edge detection algorithm is one of the most precisely specified approaches yet developed for edge identification, which delivers good and dependable results. It quickly rose to the top of the list of the most often-used edge detection algorithms due to its high performance in terms of meeting the three edge detection criteria and its ease of implementation.

6 Experimental results

Loss value is also taken into consideration in the evaluation of work prediction accuracy. You may monitor the performance of your classifier as it is being tested on a dataset. Here, True Negative signifies the number of correct negative occurrences, True Positive denotes the number of positive events, and False Positive denotes mistakenly anticipated positive events. The number of accurately anticipated negative events that actually occurred is represented by the term "False Negative." Figure 2.2: General consternation matrices

Predicted_Class \ Actual Class		
	True_Positive	False_Positive
	False_Negative	True_Negative

Figure 2.2: Confusion matrix.

The performance indicators for the proposed categorization model are described on this page. According to eq. (1), the accuracy of classification problems is the total number of valid predictions made by the three machine learning models.

$$Accuracy = \frac{tp + tn}{tn + fp + tp + fn} \tag{1}$$

$$Recall = \frac{tp}{tp + fn} \tag{2}$$

$$FScore = 2\frac{Precision\ X\ Recall}{Precision + Recall} \tag{3}$$

$$Specification = \frac{tn}{tn + fp} \tag{4}$$

$$Precission = \frac{tp}{tp + fp} \tag{5}$$

Video analytics may be used to read license plates and reduce theft with the suggested system, which could restrict traffic, based on density everywhere. A real-time analysis of the movement of a specific vehicle (asset tracking) could be performed, using it for improved vehicle security. It is possible to execute homogeneous emergencies in the long term. We have employed a lean hardware strategy and software capabilities to offer efficiency in order to make the project profitable. The proposed effort has the potential to have a significant impact because of its effectiveness, accuracy, and cost-effectiveness in traffic monitoring. As soon as the system is up and running, we will be able to analyze previous traffic patterns and provide useful information to the police. Using the information, we can locate potholes (with simple tweaks to the cameras) and send that information to the appropriate authorities in a timely manner.

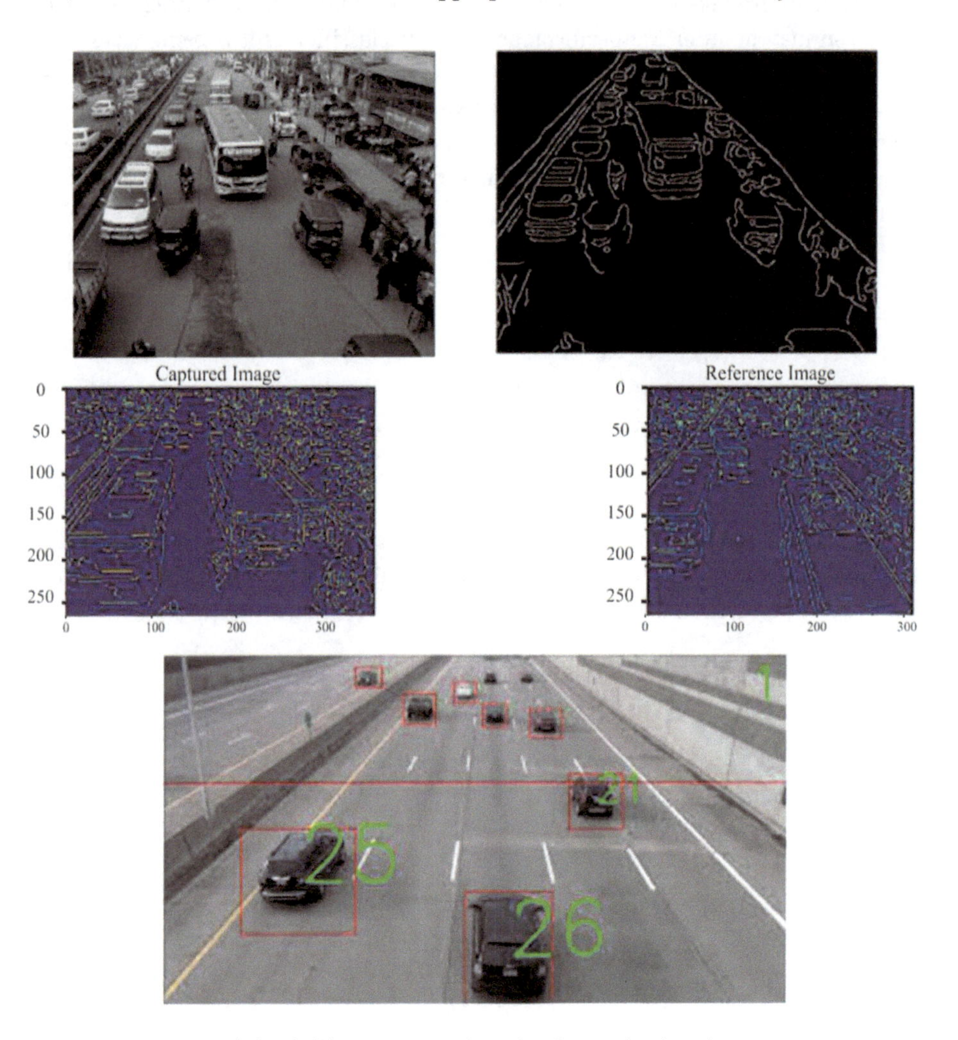

Figure 2.3: Image uploaded and click image processing using Canny edge detection.

7 Conclusion

The home location, the mentioned place, and the tweet location are all taken into account when analyzing tweet data. Geolocation prediction becomes more difficult when Twitter data is taken into account. It is difficult to read and evaluate tweets because of the nature of the language and the limited number of characters. In this study, we used machine learning techniques to identify the location of a person, based on their tweets. For the geolocation prediction challenge, we have implemented three algorithms, and we have chosen the best one. The results of our experimentation show that decision trees may be used to analyze tweets and forecast their location. Other techniques such as user buddy networks and social networks such as Facebook and Tumblr will be used in the future to infer locations. Several fresh avenues for further study have been opened up by the current findings. Allowing people to express themselves in their local language would significantly enrich the system.

References

[1] Hasan, M. M., Saha, G., Hoque, A., and Majumder, M. B. (May 2014). Smart traffic control system with application of image processing techniques. In 3rd International Conference on Informatics Electronics & Vision Dhaka.

[2] Pandit, V., Doshi, J., Mehta, D., Mhatre, A., and Janardhan, A. (January–February 2014). Smart traffic control system using image processing. *International Journal of Emerging Trends & Technology in Computer Science (IJETTCS)*, 3(1): 1–10.

[3] Katiyar, S. K., and Arun, P. V. (2014). Comparative analysis of common edge detection techniques in context of object extraction. *IEEE Transactions on Geoscience and Remote Sensing*, 50(11b): 23–28.

[4] He, J., Hou, L., and Zhang, W. (August 2014). A kind of fast image edge detection algorithm based on dynamic threshold value. *Sensors & Transducers*, 24: 179–183.

[5] Rong, W., Zhanjing, L., and Zhang, W. (August 2013). An improved canny edge detection algorithm. In IEEE International Conference on Mechatronics and Automation.

[6] Sweet, M. (October 2013). Traffic congestion's economic impacts: Evidence from US metropolitan regions. *Urban Studies*, 51(10): 2088–2110.

[7] Shrivakshan, G. T., and Chandrasekar, C. (September 2012). A comparison of various edge detection techniques used in image processing. *International Journal of Computer Science Issues*, 9(5): 121–129.

[8] Acharjya, P. P., Das, R., and Ghoshal, D. (2012). Study and comparison of different edge detectors for image segmentation. *Global Journal of Computer Science and Technology Graphics& Vision*, 12(13): 1–12.

[9] Pallavi, T. C., Sayanti, B., and Muju, M. K. (April 2011). Implementation of image processing in real time traffic light control. In 3rd International Conference on Electronics Computer Technology.

[10] Yang, L., Zhao, D., and Wu, X. (October 2011). An improved Prewitt algorithm for edge detection based on noised image. In 4th International Congress on Image and Signal Processing.

[11] Semertzidis, T., Dimitropoulos, K., Koutsia, A., and Grammalidis, N. (June 2010). Video sensor network for real-time traffic monitoring and surveillance. *The Institution of Engineering and Technology*, 4(2): 103–112.

[12] Gao, W., Zhang, X., and Yang, L. (July 2010). An improved Sobel edge detection. In 3rd International Conference on Computer Science and Information Technology.

[13] Rotake, D., and Kar More, S. (2012). Intelligent traffic signal control system using embedded system. *Innovative Systems Design And Engineering*, 3(5): 23–31.

[14] Santhoshkumar, R., Kalaiselvi Geetha, M., and Arunnehru, J. (2017). Activity based human emotion recognition in video. *International Journal of Pure and Applied Mathematics*, 117(15): 1185–1194. ISSN: 1314-3395.

[15] Santhoshkumar, R., and Kalaiselvi Geetha, M. (2019). Deep learning approach for emotion recognition from human body movements with feed forward deep convolution neural networks. *Procedia Computer Science*, 152: 158–165. Elsevier, ISSN 1877-0509.

[16] Santhoshkumar, R., and Kalaiselvi Geetha, M. (2019). Deep learning approach: emotion recognition from human body movements. *Journal of Mechanics of Continua and Mathematical Sciences (JMCMS)*, 14(3): 182–195. ISSN: 2454-7190.

[17] Abbas, N., Tayyab, M., and Qadri, M. T. (2013). Real Time traffic density count using image processing. *International Journal of Computer Applications*, 83(9): 16–19.

[18] Dharani, S. J., and Anitha, V. (April 2014). Traffic density count by optical flow algorithm using image processing. Second National Conference on Trends in Automotive Parts Systems and Applications (TAPSA-2014). *India International Journal of Innovative Research in Science, Engineering and Technology*, 3(2): 501–507.

[19] Aishwarya, S., and Manikandan, P. Real time traffic light control system using image processing. *IOSR-JECE*, 1(3): 1–5.

[20] Dangi, V., Parab, A., Pawar, K., and Rathod, S. S. (2012). Image processing based intelligent traffic controller. *Undergraduate Academic Research Journal (UARJ)* , 1(1): 1–6.

[21] Kannegulla, A., Reddy, A. S., Sudhir, K. V. R. S., and Singh, S. (November 2013). Thermal imaging system for precise traffic control and surveillance. *International Journal of Scientific & Engineering Research*, 4(11): 464–467.

[22] Rajalingam, B., and Santhoshkumar, R. (2021). Security and privacy preservation using edge intelligence in beyond 5G networks harnessing federated machine learning and artificial immune intrusion detection system. In *Computers and electrical engineering*. Pergamon: Elsevier.

[23] Santhoshkumar, R., Kalaiselvi Geetha, M., and Arunnehru, J. (2017). SVM-KNN based emotion recognition of human in video using HOG feature and KLT tracking algorithm. *International Journal of Pure and Applied Mathematics*, 117: 621–624. ISSN: 1314-3395.

[24] Santhoshkumar, R., and Kalaiselvi Geetha, M. (2020). Vision based human emotion recognition using HOG-KLT feature. *Advances in Intelligent System and Computing, Lecture Notes in Networks and Systems*, 121: 261–272. Springer, ISSN: 2194-5357.

[25] Santhoshkumar, R., and Kalaiselvi Geetha, M. (2020). Emotion recognition on multi view static action videos using Multi Blocks Maximum Intensity Code (MBMIC). In *Lecture notes in computational science and engineering*. Springer, ISSN 1439-7358.

Avani K. V. H., Deeksha Manjunath, C. Gururaj

3 Proficient implementation toward detection of thyroid nodules for AR/VR environment through deep learning methodologies

Abstract: A disease that commonly exists on a global scale is the thyroid nodule. It is identified by unusual thyroid tissue development. The nodules of the thyroid gland are of two kinds namely, benign and malignant. Accurate diagnosis of thyroid nodules is necessary for appropriate clinical therapy.

Ultrasound (US) is one of the most often-employed imaging tools for assessing and evaluating thyroid nodules. It performs well when it comes to distinguishing between the two kinds of thyroid nodules. But diagnosis based on US images is not simple and depends majorly on the radiologists' experience.

Radiologists, sometimes, may not notice minor elements of an US image, leading to an incorrect diagnosis. To help physicians and radiologists to diagnose better, several Deep Learning (DL) based models that can accurately classify the nodules as benign and malignant have been implemented.

After performing a comparative study of several DL-based models implemented with different classification algorithms on an Open Source dataset, it has been found that InceptionNet V3 gave the best accuracy (~96%); others were F1 Score (0.957), Sensitivity (0.917), etc.

A simple and easy-to-use graphical user interface (GUI) has been implemented. A US image of the thyroid gland is to be uploaded by the user following which the classification output is displayed along with a link redirecting the user to a page, which has more information about the reasons for the classification obtained. A downloadable file can also be obtained with the same information as the page, using the link.

Keywords: Malignant, Benign, Ultrasound, Echogenicity, TI-RADS, CAD, VGG-16, ResNet, DenseNet, MobileNet, InceptionNet, KNN, SVM, XGBoost, AdaBoost, DecisionTree, GUI

Avani K. V. H., Department of Electronics and Telecommunication Engineering, BMS College of Engineering, Bengaluru, India
Deeksha Manjunath, Georgia Institute of Technology, Atlanta, USA
C. Gururaj, Senior Member IEEE, Department of Electronics and Telecommunication Engineering, BMS College of Engineering, Bengaluru, India

https://doi.org/10.1515/9783110790146-003

1 Introduction

Conventional diagnostic and treatment procedures of diseases rely heavily on doctors'/clinicians' expert knowledge of the condition at hand. However, this diagnostic procedure has a significant flaw: its effectiveness relies highly on doctors' own experiences and intellect. As a result, diagnostic accuracy varies and is constrained by several factors. Image-based diagnosis procedures have become increasingly popular as digital technology has advanced, allowing clinicians to study abnormalities in organs beneath the skin and/or deep within the human body. The diagnostic performance can be improvised with the use of imaging techniques. However, the ability and expertise of medical practitioners to exploit collected photos is still a factor. To address this issue, computer-assisted diagnostic systems were created to aid doctors in the prognosis and therapeutics process. The CAD systems, as their name implies, is done using a computer to try to improve the performance of human diagnosis and can be used for double verification. This type of technology analyses and interprets (medical) images of various organs, such as US scans. After the analysis is done, the classification output obtained helps doctors diagnose disorders.

The organ in the human body that performs an important function – creates and generates vital hormones that regulate metabolism – is the thyroid. It is located in the human neck. Thyroid disease has become more crucial to diagnose and treat because of its vital role. The formation of nodules that cause thyroid cancer, as previously observed in prior studies, is a prevalent problem in the thyroid region. Anomalous lumps that grow on the thyroid gland are called thyroid nodules. Thyroid nodules are abnormal lumps that form on the thyroid gland. Deficiency in iodine or overgrowth of the tissue (thyroid) or cancer (thyroid) could be among the reasons for the development of nodules. Based on their characteristics, thyroid nodules can be categorized as benign (noncancerous nodules) or malignant (cancerous nodules). Thyroid cancer, constituting 10% of all malignant tumors, is the most common cancer that occurs in the endocrine system. Thyroid nodule detection has increased dramatically as several nodules are found by chance. The CAD classification of thyroid nodules can help the radiologist make an accurate diagnosis.

Clinically, evaluating thyroid nodules is difficult. Thyroid nodules are found out often by chance during neck diagnostic imaging and prevalence increases as a person ages. They are seen in 42–76% of adults. Many of thyroid nodules are considered noncancerous, while 10% of them may be cancerous. Radioactive iodine therapy, thyroidectomy to prevent it from recurring and to prevent death, immunotherapy, or chemoradiotherapy could be recommended. Treatment for this type of cancer varies based on the histological subtype, patient preference, and other factors, as well as comorbidities. US images are not particularly clear, which necessitates the use of a fine needle aspiration biopsy (FNAB (invasive)) procedure, which has unfavorable repercussions. For a cytological study of the nodule, an FNAB of the nodule might be obtained. Patients may have localized pain from FNABs, and while they are mostly safe, there could be a small risk of hematoma. The outcome of cytological analysis is uncertain in 20–30% of FNABs, which implies they do not always produce clinically meaningful data. The assessment of accidental thyroid nodules is usually performed by sonography. Hypo-echogenicity, absence of a halo, micro-calcifications, firmness, intranodular flow, and a taller-than-wide form are some of the sonographic features of thyroid nodules that radiologists have recognized as suggestive indications of cancer (malignant).

Virtual reality (VR) and augmented reality (AR) are types of digital holographic picture technologies that use AI to rebuild clinical data. The concept of VR involves the generation of a virtual image by a computer program. Using this virtual system, surgeons can practice without the risk of major operation failure, allowing them to improve their surgical abilities [1]. VR, however, cannot be used in real surgery owing to the fact that real-world experience is lacking [2]. AR is a combination of intelligent augmented information and the real world. AR differs from VR in terms of its application in the real world.

AR technology can be used to assist surgery preoperatively or intraoperatively. Following the successful translation of patient data and the virtual reestablishment of the vital area, as well as the inclusion of the virtual image to the actual visual world, it works by recognizing anatomical structures that are complex to navigate during the operation [2]. However, there are still limits during surgery due to the bulky equipment for AR navigation systems [2].

1.1 Thyroid imaging reporting and data system (TI-RADS)

TI-RADS is based on a five-point scale that assigns 0–3 points to each of the five US features namely composition, echogenicity, shape, margin, and echogenic foci. A single score is generated for the features of composition, echogenicity, shape, and margin due to mutually exclusive selections. The echogenic foci category might contain many features. The scores generated for all features are added and the sum total obtained determines whether the nodule is harmless (TR1) or lethal (TR5). The TI-RADS score ranges from 1 to 5; nodules that have scores up to 3 can be classified as benign, while nodules that have scores of more than 3 can be classified as malignant. The TR level, along with the nodule's maximal diameter, decides whether an FNAB is needed or no further action is recommended.

Composition

Nodules that are categorized as spongiform in ACR TI-RADS can be directly deemed as benign, requiring no follow-up. If cystic components make up less than half of the nodule, it should not be considered benign. Furthermore, the existence of other characteristics such as outlying calcifications or macro calcifications, which are generally easy to detect, suggests that a nodule is not spongiform and may be malignant. Nodules with obscuring calcifications that impede evaluation of their architecture are assumed to be solid and require additional investigation. Figure 3.1 shows a nodule of size 0.9 cm with

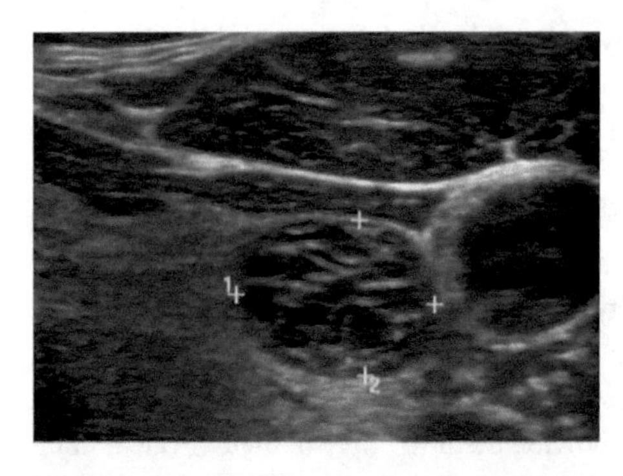

Figure 3.1: Nodule of size 0.9 cm with small cystic spaces and considered benign [3].

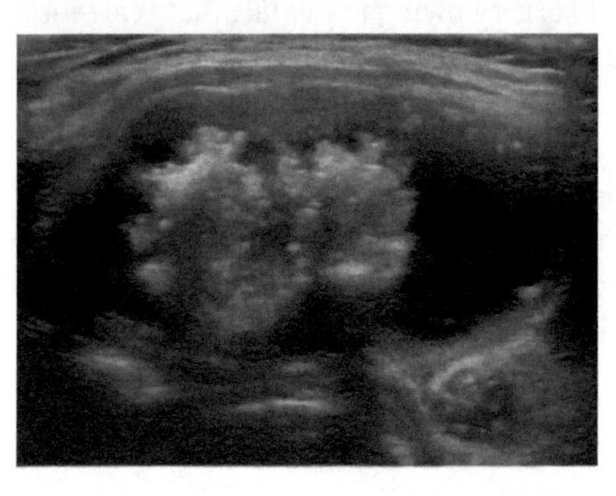

Figure 3.2: Nodule with significant cysts and considered malignant [3].

small cystic spaces which is considered benign. Figure 3.2 depicts a nodule with significant cysts and is considered malignant.

Shape and margin

Whether the nodule has grown more front-to-back than side-to-side can be seen from its shape and margin, which indicates that it has crossed tissue planes and hence is suspicious. Nodules with a fully circular cross-section can also be found. The character of a nodule's interaction with the surrounding intra- or extrathyroidal tissue is identified as its margin. If the nodule's solid component is angled into the surrounding tissue to any extent, the margin should be classed as lobulated or irregular, both of which are deemed safe. A pathognomonic sign of malignancy can be the invasion of the surroundings by extrathyroidal extensions. Figure 3.3 shows a sonogram of a taller-than-wide thyroid nodule. Figure 3.4 depicts a sonogram with smooth margin, which is considered benign.

Echogenicity

The brightness of a thyroid nodule in comparison to the rest of the thyroid tissue is referred to as echogenicity. Hypoechoic nodules that are solid and not fluid-filled are depicted by colors different from the surrounding thyroid tissue. When opposed to cystic or fluid-filled lesions, solid hypoechoic thyroid nodules have a higher risk of turning

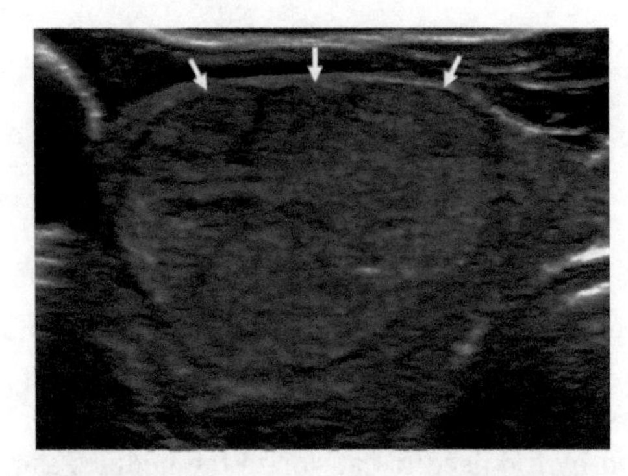

Figure 3.3: Sonogram of a thyroid nodule that is taller-than-wide [3].

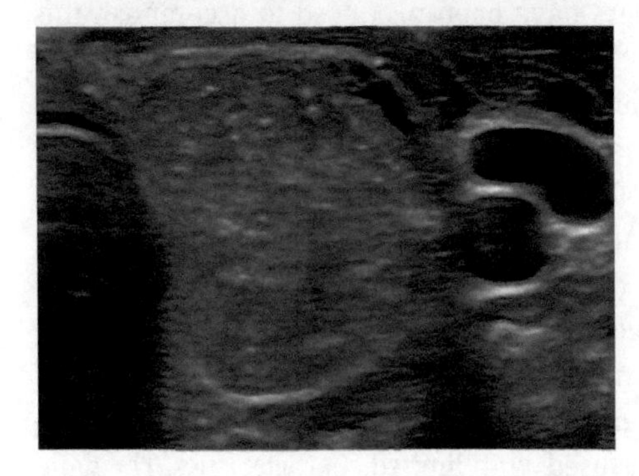

Figure 3.4: Sonogram of a benign nodule with smooth margin (arrows) [3].

malignant. Other factors such as size, may, however, suggest a nodule's likelihood of evolving into thyroid cancer. A fine-needle biopsy may be performed for further investigation, if a doctor feels a nodule is malignant. Figure 3.5 shows a sonogram of a hypoechoic thyroid carcinoma.

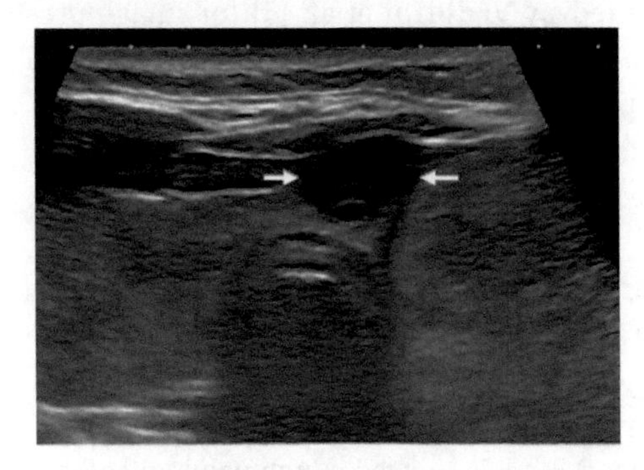

Figure 3.5: Sonogram of a 2.0 cm markedly hypoechoic thyroid carcinoma (arrows) [3].

Techniques based on DL have lately been employed to solve various difficulties in medical image-processing systems, thanks to technological advancements such as the back propagation algorithm, neural networks, and Graphics Processing Units

(GPUs). DL-based technology has had a lot of success in detecting and classifying thyroid nodules.

2 Literature survey

DL methods are used extensively in the realm of healthcare research. This progress has undeniably positive consequences for society. The thyroid nodules identified in US scan images is one such topic that has gotten a lot of attention throughout the world. The endeavor is further complicated by the need to categorize the nodules as benign or cancerous. Many research papers have been produced to accomplish this objective using various methods.

One method involves using standard neural networks to classify thyroid nodules in US pictures. Pretrained GoogleNet and cost-sensitive random forest classifiers have been used [4]. The elements of the course of the nodule itself, as well as the distinctions between consistencies of different tissues, are learned using DCNN (Deep Convolutional Neural Network) architecture. One of the disadvantages was the inclusion of fiducial markers added by radiologists to scan images, which degraded the learning process.

The use of feature vectors in conjunction with Support Vector Machine (SVM) or Fuzzy classifiers by H. Khachnaoui et al. yielded good accuracy [5]. Despite the great accuracy observed, the AdaBoost classifier was used, with limited characteristics. The dataset in this study has a major disadvantage in that it does not sufficiently represent the natural population base in clinical settings. The ROIs (Region of Interests) identifying thyroid lesions are delineated by doctors rather than automatically discovered by the ultrasound CAD system; therefore, the efficacy of the approach is still dependent on physician experience.

In one of the experiments implemented by Vadhiraj et al. [6] for image preprocessing and segmentation, the median filter and picture binarization was used, achieving noise reduction in the raw scan pictures. Seven US picture features were extracted using the gray level co-occurrence matrix (GLCM). The SVM and ANN (Artificial Neural Network) classification methods were used in two different ways. The accuracy of ANN was 75 percent.

Some CAD approaches supplied a clarified segmentation procedure with active outlines operated by local Gaussian distribution, which fit the energy prior image characteristics retrieved from thyroid US pictures, according to experimental data in the work of Q. Yu et al. [7]. For training purposes, the dataset did not include all forms of thyroid nodules. Both the ANN and the SVM models showed a high value in categorizing thyroid nodule samples based on research of previous works. When the two techniques were integrated, the diagnosis of noncancerous nodes was missed. If both procedures indicated that a node was cancerous, the patient was to have a fine needle aspiration cytology (FNAC) test performed at the very least.

Nikhila et al. [8] showed in their work, the Residual U-net approach used to extract attributes from US pictures. Radiologists and clinicians can use CAD to increase diagnostic accuracy and reduce biopsy rates. In comparison to earlier models, which for the purpose of thyroid nodule categorization optimized the feature extraction process using models that were lightweight, the Inception-v3 method, which is the proposed method, had a 90% accuracy rate obtained for validation data. The model had greater accuracy than the Inception-v2 model. The most common way to diagnose thyroid gland problems is via a US scan.

For the categorization of thyroid nodules, Avola et al. [9], in their work, used the transfer learning method. They were able to achieve considerable results using a knowledge-driven learning and feature fusion strategy. They used an ensemble of experts obtained through transfer learning to investigate knowledge-driven techniques. The ensemble provides consultations to a DenseNet during its training, and the two components eventually interact to act as a CAD system. Their knowledge-driven approach which is made up of a data augmentation and feature fusion phase generates detailed nodule images. Transfer learning allows for a reduction in both the time and the quantity of samples required to properly train a network, making it suitable for the proposed ensemble, which is based on difficult-to-find medical images.

The classification of US pictures of thyroid nodules was improved in the DT. Nguyen et al.'s [10] study examined images in both the spatial (using DL) and frequency domains (using Fast Fourier Transform). To learn from data efficiently, a Convolutional Neural Network (CNN) with several weighted layers is necessary. However, when the number of layers grows, the issue of vanishing gradients arises. CNN networks such as AlexNet and VGGNet, for example, have millions of parameters. DT. Nguyen et al. used a ResNet50-based network that has 50 weight layers in total. The size of the nodule had a significant impact on the classification outcome. As the dataset was so small, using Binary Cross Entropy resulted in bias due to the dataset's imbalances. The Weighted Binary Cross Entropy approach, which showed to be effective in reducing overfitting and bias in trained classifiers, was employed.

Abdolali et al. [11] published the first systematic study of CAD systems used for thyroid cancer sonographic diagnosis in this publication. The advancement of Machine Learning (ML) has promoted the construction of complicated and efficient models for thyroid nodule sonographic diagnosis. From a variety of perspectives, the parametrization of thyroid, which is automated, has been approached using various methods, as illustrated in this survey. This paper's categorization of approaches serves as a guide to the present approaches for analyzing thyroid US pictures. To offer the necessary accuracy, more efficient ML models should include the thyroid gland's constitutional anatomical context.

Another research paper by W. Song et al. [12], proposed a new MC-CNN framework for thyroid nodule identification and segregation on ultrasound images. With the intent of identifying benign nodules from malignant nodules and the complex background, the new learning architecture allowed the identification and segregation jobs to share features that are usually needed. In order to achieve this goal, a multi-scale layer was included in the new learning architecture to increase the detection performance of thyroid nodules with scales that vary dramatically. As a result, the nodule candidates discovered were fed back into the spatial pyramid augmented by AlexNet to boost the performance of classification, even further.

The above research papers have obtained accuracies of around 90% and moderate sensitivities and F1 scores. Sensitivity, which is one of the most critical evaluation metrics, indicates whether the model has fewer false negatives obtained on correct prediction of images of malignant nodules as malignant itself. Predicting images of malignant nodules as malignant itself is very crucial, as if predicted as benign, the patient will not be given treatment immediately, hence worsening the condition. F1 scores and accuracies obtained in the previous works can be improved by using an efficient DL model with well-structured architecture and a classifier. This project can also be made accessible to the users by means of a GUI or an application, thus allowing them to analyze their scan images without the intervention of the medical staff.

3 Implementation

3.1 Block diagram

Figure 3.6 is a block diagram that shows end-to-end implementation of the work done.

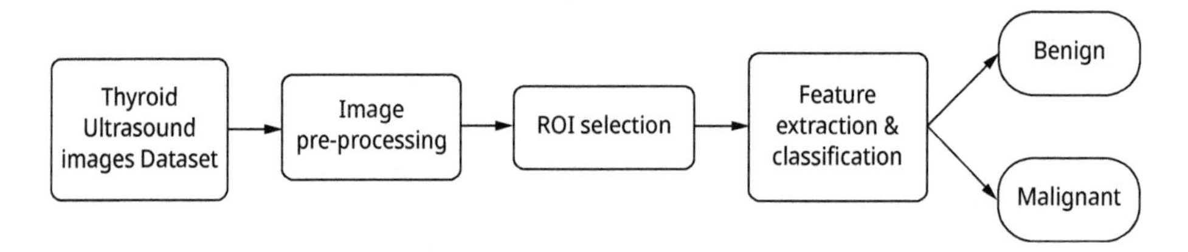

Figure 3.6: Block diagram representing end-to-end implementation.

3.2 Dataset

The dataset used for the implementation of this project is the Digital Database of Thyroid Ultrasound Images (DDTI) which is a free and open-source database from Colombia's Universidad Nacional [13]. A series of B-mode US images are used to create the DDTI, which comprises a detailed explanation along with a description of the diagnosis of suspected thyroid nodules by at least two qualified radiologists. DDTI has 299 cases, of which 270 are of women and 29 are of men. DDTI has a total of 347 photos. The cases included in the DDTI dataset are gathered by a diagnostic center (for imaging) in Columbia – The

IDIME Ultrasound Department. The thyroid US images saved in JPG format (without compressing) were obtained from video sequences taken by US Devices. The devices used were the TOSHIBA Nemio 30 or the TOSHIBA Nemio MX. The nodule polygon and explanation of the diagnosis of the US image of each patient were recorded in a separate file in XML format. TI-RADS was used by experts to classify patients [14]. For all five categories of ultrasonography found in a nodule, points are given according to TI-RADS. More points are given to those features that contribute to a higher risk of malignancy. The sum of the points of all the categories gives the TI-RADS score. The score ranges from TI-RADS Score of 1 (harmless) to a TI-RADS score of 5 (destructive) (high suspicion of malignancy). The XML file includes a detailed explanation of the diagnosis of each patient based on the US image.

Examples of the images in the DDTI dataset:

i. Benign thyroid nodule

a. The XML file provided in the DDTI Dataset for Figure 3.7 states the following: Case number: 600; Gender: female (F); Composition: spongiform appearance; Echogenicity: hyperechogenicity; Margins: well-defined, smooth; Calcifications: microcalcification; TI-RADS score: 2

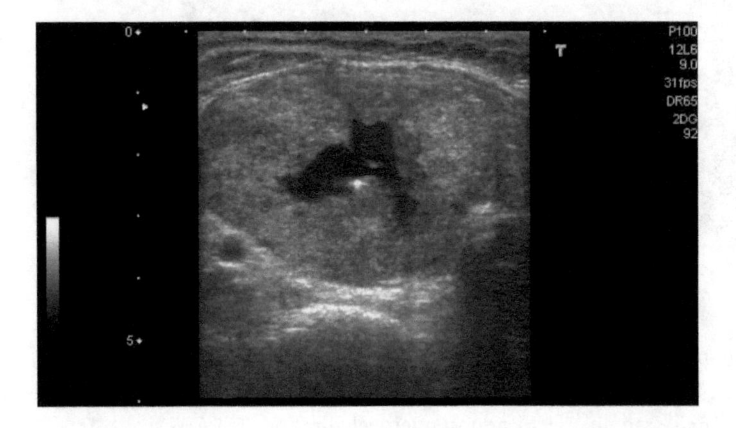

Figure 3.7: US image of a benign thyroid nodule – case number 600 [13].

b. The XML file provided in the DDTI Dataset for Figure 3.8 states the following: Case number: 597; Gender: F; Composition: predominantly solid; Echogenicity: isoechogenicity; Margins: well-defined, smooth; Calcifications: none; TI-RADS score: 2

ii. Malignant thyroid nodule

a. The XML file provided in the DDTI Dataset for Figure 3.9 states the following: Case number: 608; Gender: F; Composition: solid; Echogenicity: hypoechogenicity; Margins: Ill-defined; Calcifications: macrocalcification; TI-RADS score: 5

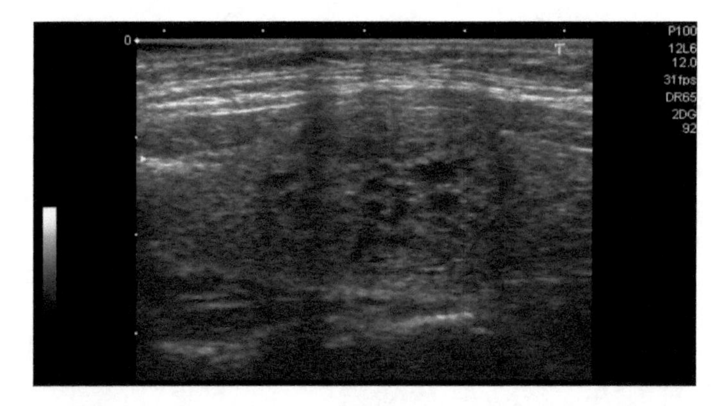

Figure 3.8: US image of a benign thyroid nodule – case number 597 [13].

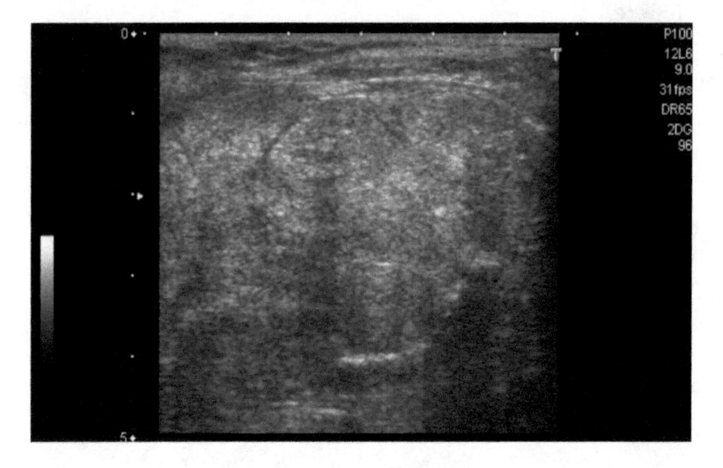

Figure 3.9: US image of a malignant thyroid nodule – case number 608 [13].

b. The XML file provided in the DDTI Dataset for Figure 3.10 states the following: Case number: 612, Gender: F; Composition: solid; Echogenicity: hypoechogenicity; Margins: well-defined, smooth; Calcifications: microcalcification; TI-RADS score: 4c

As depicted in Figures 3.7–3.10, there are regions of high intensity and regions of low intensity. The background depicted by border areas with poor illumination and the thyroid region depicted by the inner brighter part are the two primary parts of the acquired US thyroid images.

Figure 3.7 has a spongiform appearance as stated in the XML file provided. Spongiform appearance is due to the occurrence of very small cysts, similar to spaces that are fluid-filled in a wet sponge. If the cystic components in the image are more than 50%, then the composition is said to be spongiform. An even, progressively curved interface characterizes a smooth margin, which is found in the figure [3]. Tissues that have a

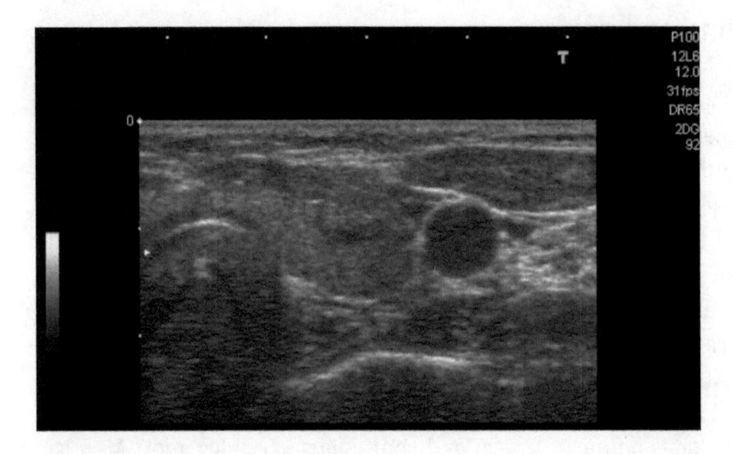

Figure 3.10: US image of malignant thyroid nodule – case number 612 [13].

higher echogenicity are defined as hyperechogenic and are usually represented by higher pixel values (or intensity) in the US images. This image has a TI-RAD score of 2 because it has hyperechogenicity (1 point) and microcalcifications (1 point).

Figure 3.8 has a predominantly solid composition, which signifies that the nodules have small cystic components that take up less than 5% of the total volume. Predominantly solid can be considered as the case that has mixed cystic and solid composition [3]. When the nodule's echogenicity is similar to that of the surrounding thyroid parenchyma, it is defined as isoechogenicity. The TI-RADS score for isoechogenicity is 1 point, and for the composition it is 1 point; hence, the TI-RADS score for this image is 2.

Figure 3.9 has a solid composition, which indicates that the cystic components occupy a volume less than 5% of the total volume [15]. Tissues with lower echogenicity are defined as hypoechogenic and are represented by lower intensity values (or lower pixel values); hence hypoechogenicity is indicated by the darker portion in the image. Huge specks of calcium that can be observed inside a thyroid nodule or in the periphery (so-called eggshell/rim calcifications) as huge bright spots in an US image are called microcalcifications. The TI-RADS score for microcalcification is 1 point, for solid composition it is 2 points, for hypoechogenicity it is 2 points; and hence, the total TI-RADS score is 5 points.

Figure 3.10 is similar to Figure 3.9, but the main difference is in the calcifications that are clearly visible in the image. There are microcalcifications present in this image. The TI-RADS score for this figure is 2 points for hypoechogenicity and 2 points for solid composition making it a total of 4. The XML file for this figure states that the TI-RADS score is 4c, which signifies the presence of highly suspicious nodules (50–85% risk of malignancy) [16].

The dataset has images that vary based on TI-RADS scores calculated by considering factors such as echogenicity, composition, echogenic foci (calcifications), margins, and shape.

3.3 Image loading and pre-processing

The images were first loaded by iterating through the folders of the open-source CAS dataset, using the OpenCV functions. The images were then split into training and testing images in the ratio 4:1. Keras functions were used for pre-processing of the images. The initially loaded image was of the format width × height × channels.

Images obtained from the open-source CAS dataset [13] are of the size or resolution 560 × 360 (width × height). These images were resized to 256 × 256. For ImageNet scale networks, the rule of thumb is to use 256 × 256 sized images [17].

The images were then converted to NumPy format, which stores the images in height × width format using np.array () function.

If only structured data is used for a given data science or ML business problem, and the data obtained includes both categorical and continuous variables, most ML algorithms will not understand categorical variables. ML algorithms can deal better with data in the form of numbers rather than category, thereby providing a better accuracy and better results in terms of other performance measures, as well. Since machine learning algorithms perform better with numeric values and mostly work exclusively with numeric values, the train and test labels of the train and test images are encoded, respectively. The method used is called "label encoding." In Python label encoding, the value of the category is replaced with a number between 0 and the (number of classes − 1). For example, if the categorical variable value comprises 3 unique classes, then the labels are encoded to a range of values between 0 and 2, i.e, (0, 1, and 2). Since the labels used in our project are benign and malignant (number of unique classes = 2), they are encoded to 0 and 1, respectively. Label encoding is done for labels of both the training and testing images. For the implementation of label encoding, the LabelEncoder module from the preprocessing module from sklearn package was imported. Fit and transform functions were then used to assign a numerical value to the categorical values.

The pixel values of the images can have a value ranging from 0 to 256. The computation of high numeric values may get more complex when using the image as is and sending it through a Deep Neural Network. The complexity can be reduced by normalizing the numbers to a range of 0–1. Normalization is done by dividing all pixel values by 255, resulting in pixels in the range 0 to 1. These small values of pixels make computations easier.

The images were in the format (height, width, channels). But the neural network (based on ImageNet) accepts an input which is a 4-dimensional tensor of the format batch size × height × width × channels. Hence, images need to be converted to batch format. As the matrix (input to the network) is required to be of the form (batch size, height, width, channels), an extra dimension is added to the image at a particular axis (axis = 0). The extra dimension is added by using expand_dims () function, where the axis parameter is set to 0. The axis parameter defines at what position the new axis should be inserted. Axis = 0 signifies rows and axis = 1 signifies columns.

3.4 Region of interest

Figures 3.11 and 3.12 are examples of US images provided in the dataset for reference. Figure 3.7 includes the region of interest marked. The area in green represents the nodule. The borders of the nodule have higher intensity (lighter shades – white) values compared to the surrounding background. This is how the nodule region is differentiated from the rest of the image.

Tissues with lower echogenicity are defined as hypoechogenic and are represented by lower intensity values (or lower pixel values); hence, hypoechogenicity is indicated by the darker portion in the image. Tissues that have a higher echogenicity are defined as hyperechogenic and are usually represented by higher pixel values (or intensity) in the US images indicated by the lighter portion of the image. When the nodule's echogenicity is similar to the echogenicity of the surrounding thyroid parenchyma, it is defined as isoechogenicity, that is, there is not much difference in the pixel values or intensity.

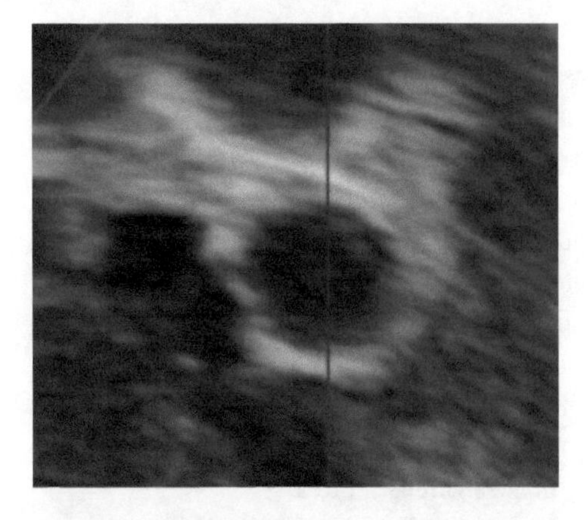

Figure 3.11: US image of a thyroid nodule [13].

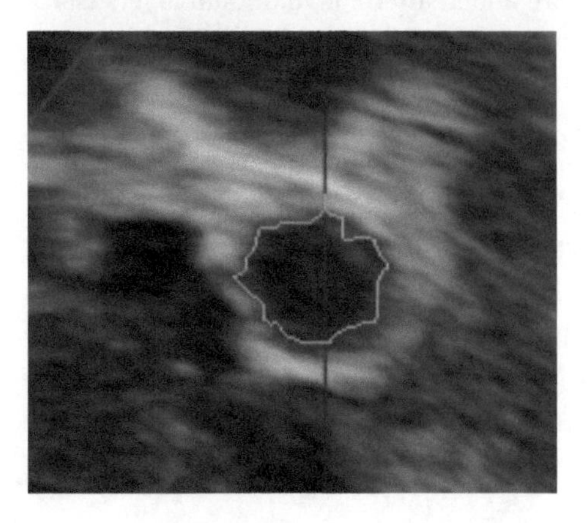

Figure 3.12: US image of thyroid nodule with ROI marked [13].

3.5 Deep learning models

3.5.1 VGG-16

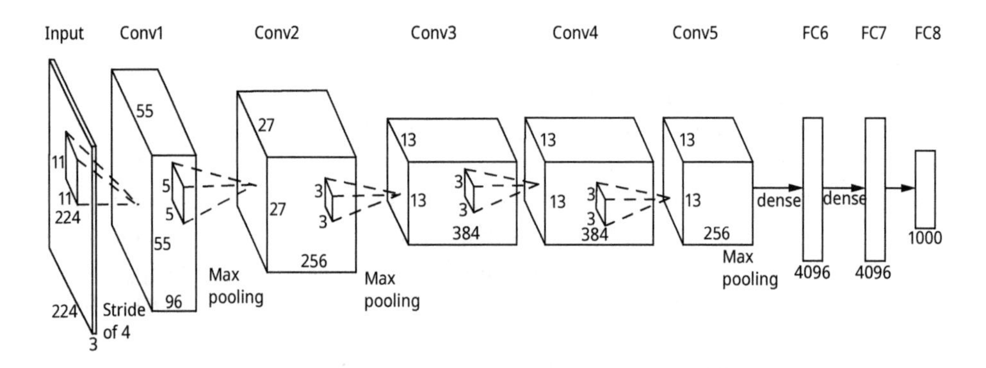

Figure 3.13: VGG-16 architecture layout [18].

Figure 3.13 depicts the VGG-16 architecture, wherein the input to the cov1 layer is an RGB picture with a defined size. The dimensions of the input image are 224 × 224. A set of convolutional layers are used to process the image. The receptive field of the layers is small. VGG-16 includes additional filters (convolutional) in one of the setups. This can be considered as a modification of the input channels wherein the modification is linear. Convolution stride, which is a parameter that decides amount of movement, is set as 1 pixel for 33 convolution layers. Input Spatial Padding (Convolutional layer) is set to the same value. After convolution, this procedure is done for the sake of preservation of spatial resolution. Part of the convolutional layers succeeded by max pooling layers (5 of them) constitutes spatial pooling. For max pooling over a 2 × 2 pixel window, the parameter responsible for movement in an image stride is set to 2.

Three layers that are fully connected are added after convolutional layers (stack of layers). ImageNet Large Scale Visual Recognition Challenge, known as ILSVRC, is used for large-scale evaluation of algorithms. Due to the many deep convolutional layers in the VGG-19, it performs 1000-way ILSVRC classification.

VGG-16 has several advantages, including the fact that it is a particularly suitable architecture for benchmarking on a specific job, and that pretrained networks for VGG are simple to construct and understand. VGGNet, on the other hand, has two shortcomings. For starters, it is quite slow. VGG-19 consumes large memory due to the number of nodes that are connected completely and its depths. Hence, implementation of VGG is a time-consuming process. Second, the network architectural weights are extremely substantial.

3.5.2 Residual network (ResNet)

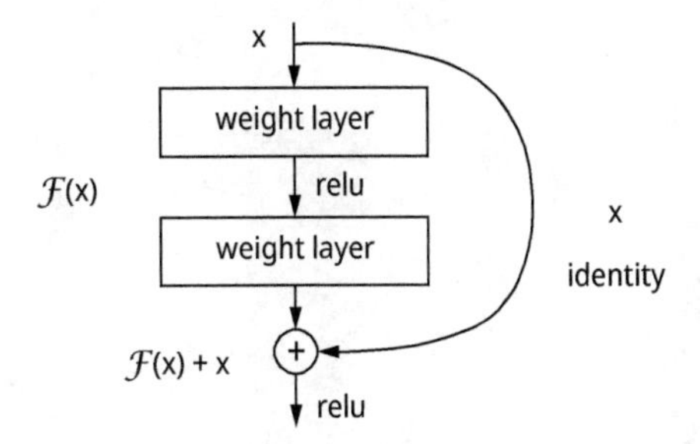

Figure 3.14: A residual block [19].

The idea behind a network with residual blocks is that each layer is fed to the network's next layer as well as directly to the next layers, skipping a few layers in the process, which is shown in Figure 3.14. Considerably deeper neural networks can be trained with residual blocks. As it bypasses one or more levels in between, the connection (curve arrow) is known as a skip connection or shortcut connection. Since an identity function can be learnt from it, it is also known as identity connection. The network fits the residual mapping rather than enabling layers to learn the underlying mapping. Instead of H(x), which is the mapping done initially, the network is made to fit:

$$F(x) = H(x) - x$$

Or,

$$H(x) = F(x) + x$$

The benefit of including skip connection is that any layer that degrades architecture performance will be bypassed by regularization. As a result, very deep neural networks can be trained without the issues caused by vanishing/exploding gradients. The ResNet-50 architecture contains 50 layers of Deep Convolutional network. The ResNet-50 version 2 model has five stages, each with a convolution and identity segment chunk. There are multiple convolution layers in each conv block, and many convolution layers in each identity block. There are around 23 million trainable parameters in the ResNet-150. The aforementioned explanation concludes that ResNet is faster than VGG-16 and gives greater accuracy.

3.5.3 DenseNet

The fact that every layer is connected to every other layer is the reason that this architecture is given the name DenseNet, shown in Figure 3.15. For L layers there are (½) × L × (L + 1) connections which are direct. For each layer, the input includes preceding feature maps, and its feature maps are considered as the input for the succeeding layers. For a layer inside DenseNet, the input is a concatenation of feature maps of the previous layers. Due to this, the network is compact and less bulky,

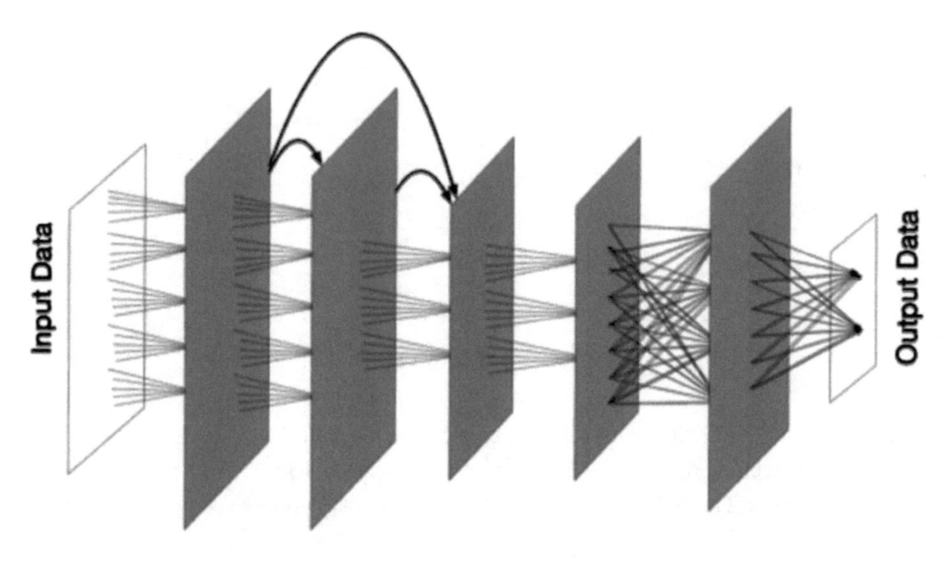

Figure 3.15: DenseNet architecture [20].

thereby resulting in fewer channels and better computational and memory efficiency. Pre-Activation Batch and ReLU constitute the basic architecture of DenseNet. They are used for each layer (composition), after which they are used for 33 convolutional layers with output feature maps of the number of channels.

Some of the advantages of DenseNet over VGG and ResNet are as follows. First, previous levels can get the error signal more directly. This is a type of implicit deep supervision since lower layers can get direct supervision from the top classification layer. Second, whereas DenseNet's parameters are proportional to the product of channel and growth rate for each layer, ResNet's parameters are proportional to the product of channels and growth rate for each layer. Therefore, DenseNet is substantially smaller than ResNet. This aspect is very important in making the computation process easy and subsequently reducing the execution time. DenseNet performs well in the absence of sufficient training data.

3.5.4 MobileNet

MobileNet is a CNN for computer vision applications that are simple and computationally light, the architecture of which is shown in Figure 3.16. Only the first layer is a convolutional layer. All the other convolutional layers are depth wise and separable. Only the last layer is not followed by batch normalization and ReLU nonlinearity as it does not have any nonlinearity and is fully connected. It is followed by a softmax layer that is used for classification. For downsampling, for the first layer, which is a full convolutional layer as mentioned previously, as well as for depth wise convolution, strided

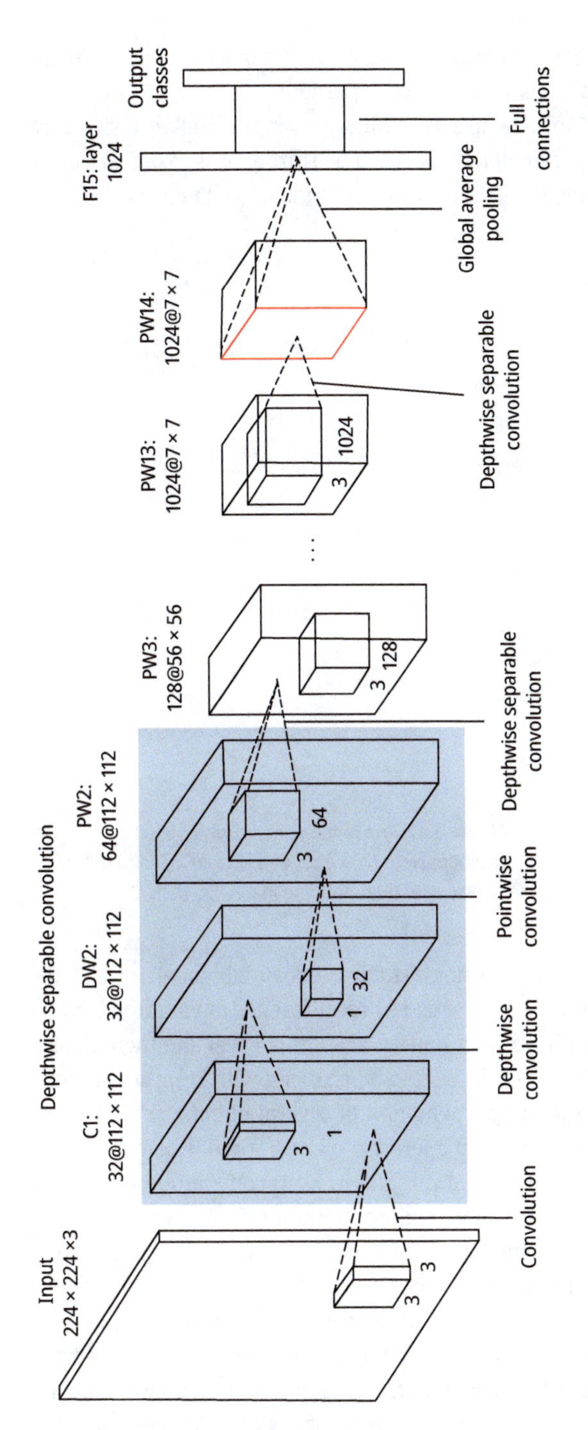

Figure 3.16: MobileNet standard architecture [21].

convolution is used. MobileNet consists of a total of 28 layers, which includes depth wise and point wise convolutional layers considered separately.

Some of the advantages of MobileNet are as follows. Firstly, the network size is reduced. Secondly, the number of parameters is decreased. Thirdly, it is faster in performance and lightweight. Finally, MobileNets are considered small, low-latency CNNs.

3.5.5 InceptionNet

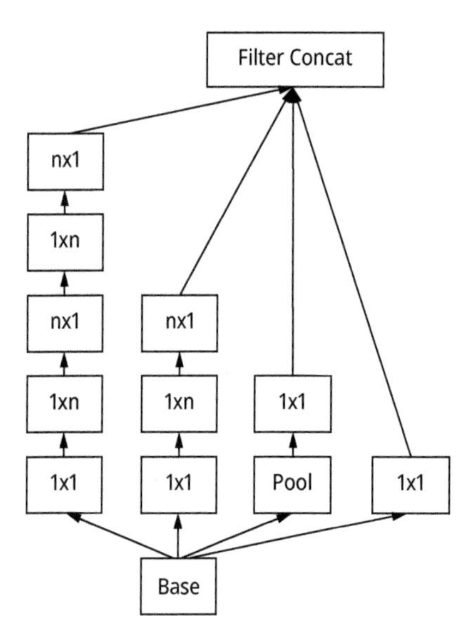

Figure 3.17: The fundamental concept used by InceptionNet: the output of one layer is fed to different layers, all of them training in parallel [22].

InceptionNet‘s architecture, Figure 3.17, does not include a sequential model. Instead, the architecture is such that the output of a layer is fed to different convolutional layers simultaneously. Therefore, the training of the model occurs in parallel and then, finally, all the outputs are concatenated. There could be a loss occurring with respect to certain information or dimensions during the implementation of several options performed simultaneously, which can be ignored, as if one operation (convolution) does not provide a certain feature, then, another operation will provide it. Each convolution or pooling operation pulls a distinct type of data or information; therefore, from each operation, different features are extracted by the module (inception). After the simultaneous completion of individual features, the characteristics obtained will be integrated into a single feature map with all the data that is gathered. The model focuses on several features, simultaneously. This results in improved accuracy. All extracted feature maps will have variable output dimensions, because for each operation, a different kernel size is considered. Using the padding operation, the different feature maps obtained are concatenated to give the same output dimension for all the operations.

The advantages of InceptionNet include greater compatibility and faster training time. However, the drawback of this is that larger InceptionNet models are prone to overfitting, mostly when the number of label samples is limited. The model may be skewed toward particular classes with a higher volume of labels than others.

3.5.6 EfficientNet

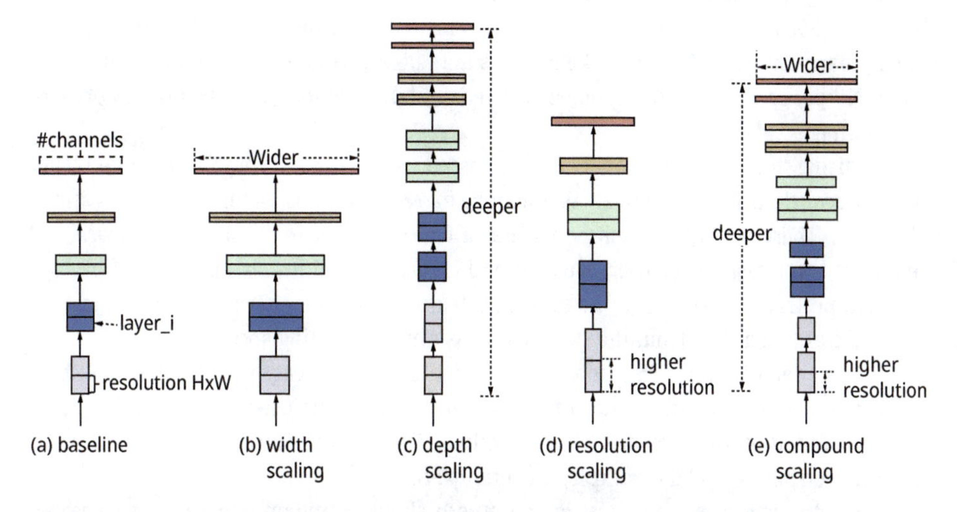

Figure 3.18: Model scaling. (a) baseline network; (b–d) conventional scaling, which can be increased only in one dimension of network width/depth/resolution; (e) EfficientNet in which scaling increases in all three dimensions (combined/compound scaling) [23].

EfficientNet is a unique model scaling strategy that scales up CNNs in a more structured manner, using a simple but extremely effective compound coefficient. This approach scales each dimension uniformly using a predetermined set of scaling factors, as opposed to conventional approaches that scale network dimensions like width, depth, and resolution at random. Figure 3.18 shows a CNN design and scaling technique. This technique employs a coefficient that is compound to uniformly scale parameters such as the depth, breadth and resolution. EfficientNets outperform state-of-the-art accuracy by up to ten times (smaller and faster) [24].

3.6 Feature extraction

Feature extraction for images is done using global feature descriptors such as Local Binary Patterns (LBP), Histogram of Oriented Gradients (HoG), Color Histograms, and so on, as well as local descriptors like SIFT (Scale-invariant Feature Transform), SURF (Speeded Up Robust Features), and ORB (Oriented Fast and Rotated Brief). For these hand-crafted features, knowledge of the respective domain is a must.

But when using Deep Neural Networks like CNN, the Deep Neural Networks automatically learn features in a hierarchical manner from the images inputted. Lower layers learn low-level features like corners and edges, while intermediate layers learn color, shape, and so on, and upper layers learn high-level features like objects present in the image.

By using the activations that can be accessed before the last completely linked layer of the network, CNN is employed as a feature extractor (that is, before the final softmax classifier). These activations will be used as a feature vector in an ML model (classifier) that learns to classify it further. This method is mainly used for classification of images, where a pretrained CNN might be used as a feature extractor – Transfer Learning – instead of training a CNN from the start (which is time-consuming and difficult).

The model, which is pretrained, is loaded from the application module of the Keras library and is built based on the configurations given by the user. Following this, from the layer based on the user given specifications, features are extracted in the ImageNet dataset pretrained model (example, InceptionNet).

ImageNet is a massive image database with about 14 million images. It was created by academics with the objective of doing computer vision research. In terms of scale, it was the first of its kind. ImageNet includes a hierarchy of images that are sorted and labelled. ML and Deep Neural both include training of machines using a large dataset of images. Machines must be able to extract useful features from these training images. Having learned these features, they can use these features to classify images and perform a variety of other computer vision tasks. Researchers can use the set of images provided by ImageNet to compare their models and algorithms.

Example of feature extraction implemented for InceptionV3 Model:

Keras comes with a variety of models. Model architecture and model weights are the two aspects of a trained model. As the weights are big files, they are not included with Keras. On specifying weights = 'imagenet' in the model initialization, the imagenet weights file is imported. If weights is set to none, then the model does not use pretrained weights, and weights are obtained on training the model from scratch, using randomly initialized weights initially.

If an additional output layer and training of the same is required, the parameter, "include top" has to be set to false. This signifies that the model's output layers that are used to make predictions and are fully connected are not loaded. Input_shape = (SIZE, SIZE, 3) is the new output layer, in this case. To exclude the default classifier of the respective model and include a classifier of our choice, the parameter "include top" is set to false.

Fitting the classification model or algorithm –for example, SVM:

From the SVM module of the sklearn package, SVC (Support Vector Classifier) is imported.

The SVM Classifier is implemented using a kernel, which is used to transform low dimensional space (of input data) into a higher dimensional space. It essentially converts a problem which is non-separable into a separable problem by adding a dimension. SVM is known for providing good accuracy for problems that have nonlinear separation. The classifier model is built using SVC with the parameter kernel as linear, wherein a linear kernel is used as a dot product between values. The features extracted (for training and testing images) from the previous stage are used for training the classification algorithm by fitting it on them (features), using the class_model.fit () function.

3.7 Classification algorithms

3.7.1 K nearest neighbors (KNN)

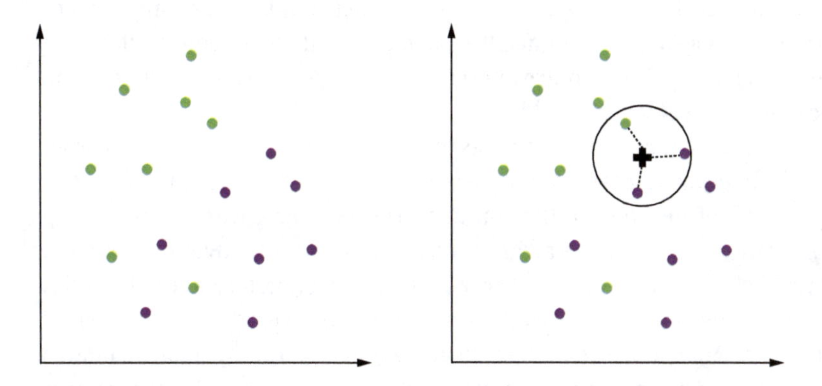

Figure 3.19: (Left) data points of two categories. (Right) application of KNN algorithm on a new data point [25].

For both classification and regression, K-nearest neighbours (KNN), which is mostly a supervised learning approach, is adopted, shown in Figure 3.19. KNN works in the following fashion: the Euclidean distance between the new data point (in this case an image) and the training data is calculated. Following this, if the distance is least between the new data point and one of the classes, then it is categorized into that class. The KNN model compares the new data set to the normal and malignant scans and labels it as one of the two, based on the most similar properties. When there are two categories and a new data point needs to be classified between the two categories, KNN may be required. KNN has a number of advantages, including a short calculation time, ease of implementation, adaptability (it may be used for regression and classification), and high accuracy.

3.7.2 XGBoost

"XGBoost" stands for "Extreme Gradient Boosting." XGBoost is a tool for distributed gradient boosting that has been optimized to be versatile, fast, and portable. To be able to solve a good range of problems related to data science rapidly and precisely, it uses an approach called parallel tree boosting.

The XGBoost approach has shown to have numerous advantages over other classifiers. Due to its simultaneous and distributed method of processing, XGBoost is an algorithm that is faster compared to other algorithms. During the development of XGBoost, the factors that were carefully considered were the optimization of the system and ML methodologies.

3.7.3 AdaBoost

AdaBoost is the abbreviation for Adaptive Boosting. It is one of the efficient boosting designs specifically used for binate classification problems. It is also a fantastic location to learn about boosting algorithms. It is adaptive in that subsequent classifiers are modified in support of occurrences misclassified by previous ones. Outliers and noisy data can make it vulnerable.

Across various cycles, AdaBoost attempts to build a sole composite strong learner. It compiles all the weak learners in one instance to create a strong learner. During every single round of iteration in the training process, a new weak learner is appended to the ensemble, and the weighting vector is changed to pivot on cases that were misclassified in previous rounds. This results in the outperformance of the classifier compared to the weak learner classifiers in terms of accuracy. The most significant advantages of AdaBoost include low generalization error, easy implementation, ability to work with a wide range of classifiers, and nonrequirement of any parameter adjustments. Since this technique is susceptible to outliers, it necessitates close scrutiny of the data.

3.7.4 Decision tree

Decision tree is one of the approaches for supervised learning that is used most commonly to solve classification issues. The nodes of the tree depict the attributes of the dataset, the branches characterize decision rules, and the conclusion is given by the leaf nodes in this tree-structured classifier.

The significant constituents present in a decision tree include leaf nodes and decision nodes. The output of the decisions of the tree characterizes leaf nodes, which further does not contain any branches. On the other hand, the decision nodes have the functionality to make any decision regarding the problem and further contain many branches. Some of the drawbacks of this classifier are: it is complex due to the fact that it has many tiers/layers; overfitting could be one of the issues faced while using this classifier; and the computational complexity is increased because of the class labels that are added additionally. However, these decision trees also have significant benefits which includes: decision trees are designed/developed to reflect and parallel the conventional thinking ability of the human mind when making decisions; this aspect makes them comprehensible.

3.7.5 Support vector machine (SVM)

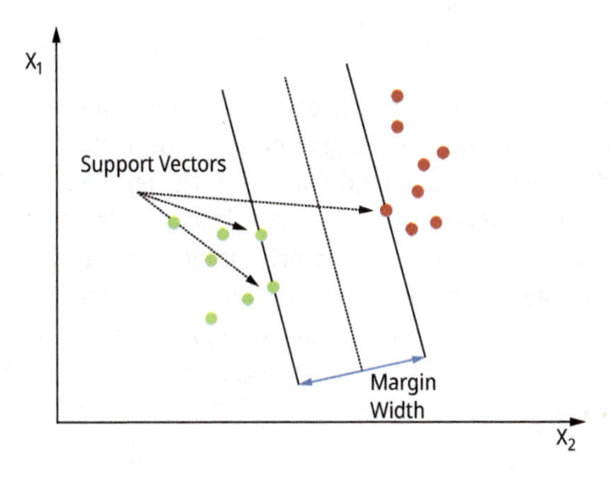

Figure 3.20: SVM classification with two classes and a hyperplane in between [26].

Support Vector Machine, abbreviated as SVM, is one of the most widely adopted methods for supervised learning and, more specifically, for classification issues. The SVM algorithm's sole purpose is to determine the optimum line, also called a decision boundary. This is used to easily group or categorize a new data point in *n*-dimensional space containing different categories. The ideal boundary distinguishing different classes is termed as the hyperplane, depicted in Figure 3.20. SVM helps in choosing the extreme points that assist in creating the hyperplane. These extreme cases are called support vectors.

The concept that a linear SVM classifier is based on involves making a direct line connecting two classes. Data points on different sides of the line represent different categories, which suggests the possibility that there exists an infinite number of lines to pick from. Since it chooses the best line (the line that splits the data and is farthest from the nearest data points), the SVM algorithm (linear) is superior to others like KNN in classifying the data points. This may be fine for data that is linearly distributed.

The reason for use of SVMs is that without having to make a lot of manual modifications, they can uncover intricate associations between the data. When working with smaller datasets that contain thousands of features, SVM is the best solution, compared to any other classifiers.

The pros of SVM are: it is very effective for datasets with multiple features like the medical data; it saves memory by using only a subset of training data points, which are called support vectors, present in the decision function. However, the downsides of SVM include: if the frequency of features is greater than the number of data points, then overfitting may be the consequence. Due to the long training time, it works best with tiny sample sets. On noisy datasets with overlapping classes, it is less effective.

3.8 Prediction

After the network is loaded and initialized, the next step involves using the model.predict () function. The predict function is used on the features obtained from the test images. The output vector obtained from model.predict () consists of the encoded labels that are difficult for humans to comprehend. Hence, output is decoded to convert the encoded labels back to the original categorical values, which can be easily understood by the user. Decoding of the encoded labels is implemented using the inverse_transform method. The encoded labels 0, 1 are, hence, decoded to give benign and malignant. This stage gives the final output in the detection and classification of thyroid nodules into malignant and benign nodules.

4 Evaluation

4.1 Evaluation metrics

The capability of a predictive model is quantified by evaluation metrics. Typically, it requires training of a model on a dataset, following which, predictions are generated by the model on a dataset (not used before). The predictions are then compared to the predicted values in the dataset on which the model generated predictions. Metrics used for the purpose of classification issues include comparing the expected and predicted class labels or the interpretation of the predicted probabilities of the problem for class labels. Model selection and procedures for the preparation of data is a search issue led by the evaluation metric. The outcome of any experiment is quantified with the help of a metric.

4.1.1 Accuracy

The accuracy of a model can be obtained from the ratio of the frequency of correctly predicted classifications to the total count of predictions made by the model. It is a quantitative factor for the evaluation of the performance of a DL model. It is one the most significant metrics that has to be accounted for or considered.

$$\text{Accuracy} = \frac{\textit{True Positive}}{\text{True Positive} + \text{True Negative}} \text{x}\,100 \tag{3.1}$$

4.1.2 Confusion matrix

A confusion matrix is an N-by-N matrix adopted to evaluate the performance of a DL model, where N represents the total count of the target classes present. The matrix attempts to compare the correct target values with those predicted by the classification model. Confusion matrix is used to depict how a model becomes ambiguous with respect to generation of predictions. Large values across the diagonal and small values off the diagonal characterize a good matrix (model). Measuring or finding the confusion matrix gives us a clearer idea of the model performance and the different kinds of errors generated by it.

Components of confusion matrix:

1. True Positive (TP): A result in which the positive class is predicted accurately by the model.
2. False Negative (FN): A result where the negative class is predicted wrongly by the model.
3. False Positive (FP): A result where the positive class is predicted wrongly by the model.
4. True Negative (TN): A result where the negative class is predicted accurately by the model.

4.1.3 Sensitivity or recall or false positive rate (FPR)

Sensitivity is defined as the proportion of TP to actual positives in the data. The ability of a test to correctly identify patients with a condition is referred to as sensitivity.

If a model has high sensitivity, it implies that the model will have fewer FNs, which is crucial when the model is used for the detection of a medical condition.

$$\text{Sensitivity} = \frac{\text{True Positive}}{\text{True Positive} + \text{False Negative}} \tag{3.2}$$

4.1.4 Specificity

Specificity (or True Negative Rate (TNR)) is the percentage of actual negatives that are projected to be negative. Specificity is the proportion of actual negatives that were projected to be negatives (or TN). As a result, a small number of genuine negatives will be forecasted as positives (sometimes called false positives). Specificity and FP rate always add up to one. If a model has high specificity, it implies that it is good at predicting TNs.

$$\text{Specificity} = \frac{\text{True Negative}}{\text{True Negative} + \text{False Positive}} \tag{3.3}$$

4.1.5 Precision

Precision is defined as the proportion of positive data points accurately categorized to total positive data points correctly or wrongly categorized.

Precision tends to be smaller if the DL model makes very few positive classifications or multiple wrong positive classifications. In contrast, the precision tends to be high when the developed classification model makes few incorrect positive classifications or multiple correct positive classifications.

$$\text{Precision} = \frac{\text{True Positive}}{\text{True Positive} + \text{False Positive}} \tag{3.4}$$

4.1.6 F1 score

The mean used between precision and recall is harmonic in the case of F1 score calculation. It is mainly used as a measure for the evaluation of the performance of a DL model DL model. F1 score is employed when FNs and FPs are significant. One of the options to address class imbalance issues is to use more accurate metrics, such as the F1 score, which consider not only the number of prediction mistakes made by the model but also the type of errors made.

$$\text{F1 Score} = \frac{\text{True Positive}}{\text{True Positive} + 0.5\,(\text{False Positive} + \text{False Negative})} \tag{3.5}$$

4.1.7 Negative predictive value (NPV)

The proportion of projected negatives that are true negatives is denoted as the negative predictive value. It expresses the likelihood that a projected negative is indeed a true negative.

$$\text{NPV} = \frac{\text{True Negative}}{\text{True Negative} + \text{False Negative}} \tag{3.6}$$

4.1.8 False positive rate (FPR)

The false positive rate (FPR) is determined by the number of FNs predicted inaccurately by the model. True positive rate or recall complements FPR.

The FPR is best described as the likelihood of incorrectly rejecting the null hypothesis, which is one of the important terms in the medical diagnostic domain. It is one of the metrics used to assess how well ML models perform when used for the purpose of classification. Lesser the FPR, better is the DL model.

$$\text{FPR} = \frac{\text{False Positive}}{\text{Total no. of Negatives}} = \frac{\text{False Positive}}{\text{False Positive} + \text{True Negative}} \tag{3.7}$$

4.1.9 False negative rate (FNR)

FNR is a measure of the percentage of FNs against all negative predictions. True Negative Rate (TNR) complements FNR.

Lesser the false negative rate, better and more accurate is the DL model.

$$\text{FNR} = \frac{\text{False Negative}}{\text{Total no. of Positives}} = \frac{\text{False Negative}}{\text{False Negative} + \text{True Positive}} \tag{3.8}$$

4.1.10 False discovery rate (FDR)

FDR is the (predicted) percentage of FPs among all significant variables.

FDR is the percentage of hypotheses incorrectly believed to be true. A test that passes the acceptability criteria based on threshold is referred to as a “discovery.” The FDR is useful since it calculates how enriched the accepted discoveries are for actual discoveries.

$$\text{FDR} = \frac{\text{False Positive}}{\text{False Positive} + \text{True Negative}} \tag{3.9}$$

4.1.11 Receiver operating characteristic curve (ROC)

The plot that evaluates the performance of a DL (classification) model over all categorization levels is the ROC Curve. The parameters plotted in this curve include TPR and FPR.

The ROC Curve is obtained by plotting TPR vs. FPR at different categorization levels. As the threshold of the classification is reduced, multiple data points are classified as positive, which results in an increase in both false and true positives.

4.1.12 Area under the curve (AUC)

AUC represents the area under an ROC curve. As the name suggests, it evaluates a 2D area of an ROC Curve. AUC gives the outline of the ROC curve and further determines the ability to distinguish or segregate different classes given to the implemented classifier, that is, it tends to give a better view of how the model distinguishes between the present positive and negative classes. The greater the AUC, more accurate is the deep learning model.

5 Graphical user interface

A user interface or computer program that facilitates interaction or communication between people and electronic devices like computers, phones etc. via visual indicators, and symbols is called graphical user interface (GUI).

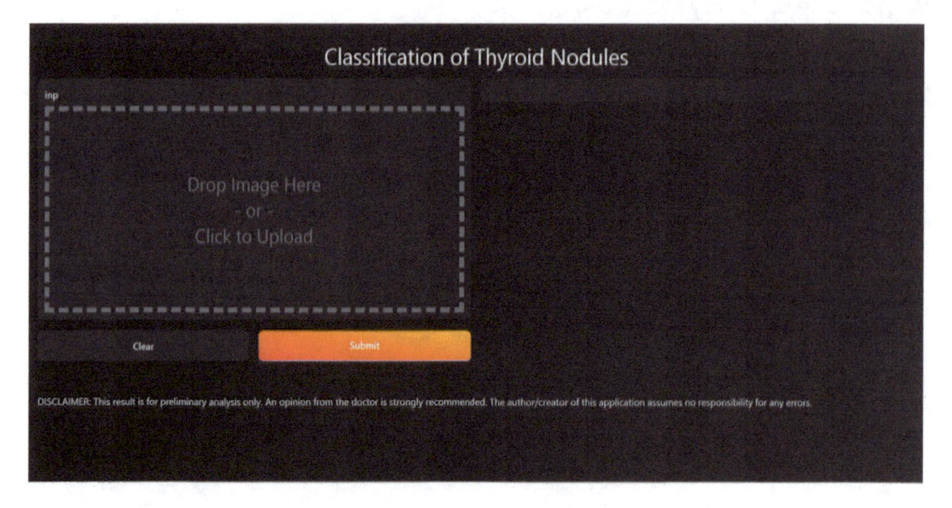

Figure 3.21: Graphical User Interface.

Figure 3.21 is a simple GUI that has been implemented with Gradio (Python package). Using Gradio, it is possible to easily generate customizable User Interface (UI) components for an ML model, any API, or any function using only a few lines of code. The GUI can be either embedded into the Python notebook, or the URL provided by Gradio can be opened by anyone.

Gradio is implemented using the function. Interface (). The user is prompted to upload or drag a US scan image that is to be classified. Internally, the image inputted by the user is first resized to the dimensions 256 × 256, in order to maintain uniformity with the preprocessed source images. The images are in the format (height × width × channels). But the neural network (based on ImageNet) accepts an input, which is a 4-dimensional tensor of the format batch size × height × width × channels. Hence, a dimension is added to the resized image. Following this, features are extracted by using. predict () function and the DL-based model. The features thus obtained are further used by the classifier to predict which kind of nodule the US image (provided by user) consists of. The obtained classification output is then displayed. The output displayed will be red if the output is "MALIGNANT" and green if the output is "BENIGN". A link, which when clicked opens a new tab with more information about the classification, is also displayed. A file with the same contents is also displayed. On clicking the file option, the file gets downloaded. To view more information about TI-RADS classification, either the link can be used or the file can be downloaded.

6 Results

6.1 Confusion matrix and evaluation metrics

Figure 3.22 includes the confusion matrix obtained for the six DL models implemented with the SVM classifier.

The results shown in Table 3.1 are obtained when SVM is used as the classifier.

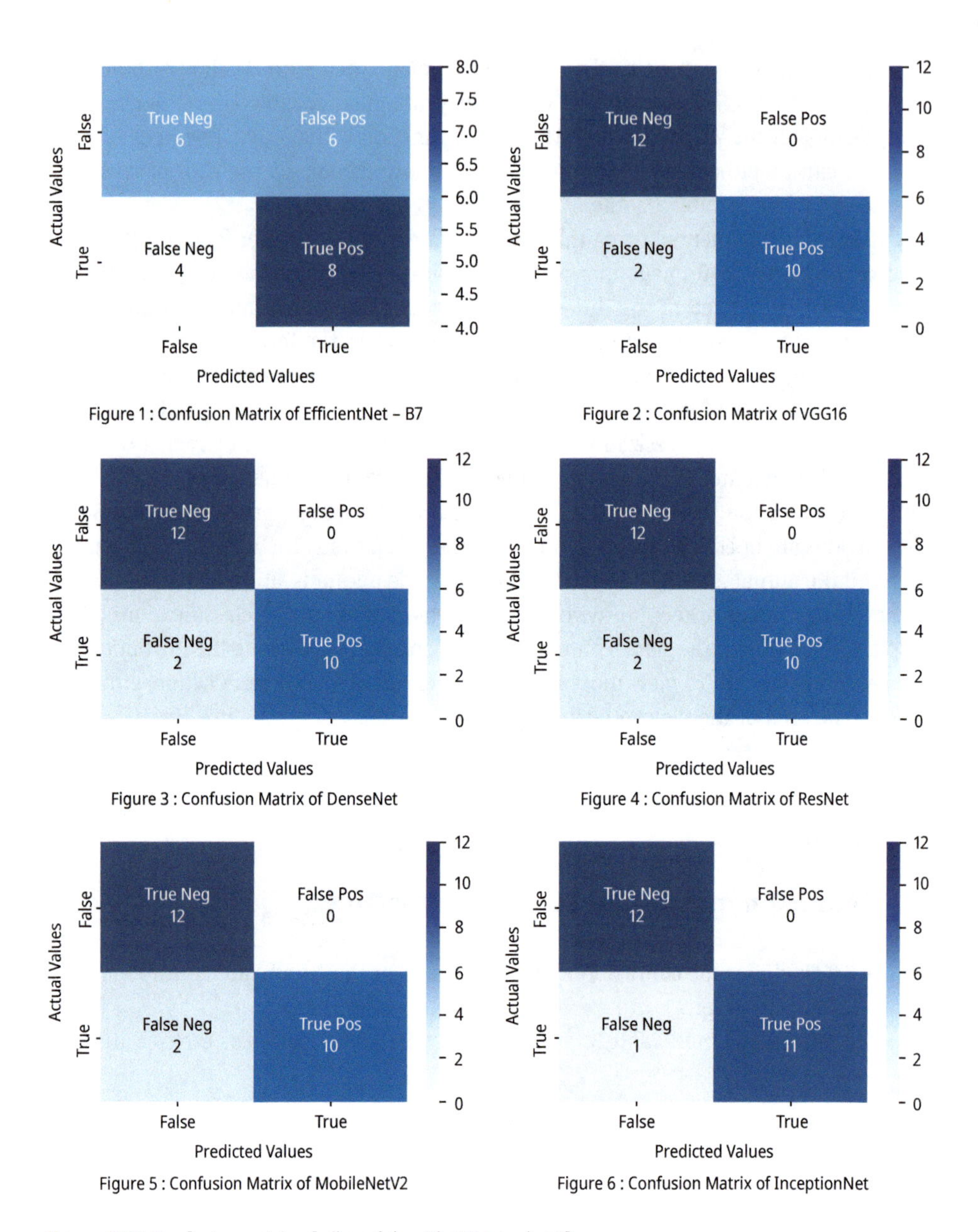

Figure 1 : Confusion Matrix of EfficientNet – B7

Figure 2 : Confusion Matrix of VGG16

Figure 3 : Confusion Matrix of DenseNet

Figure 4 : Confusion Matrix of ResNet

Figure 5 : Confusion Matrix of MobileNetV2

Figure 6 : Confusion Matrix of InceptionNet

Figure 3.22: Confusion matrix of all models with SVM as classifier.

Table 3.1: Evaluation metric values obtained for all the DL models.

	VGG16	DenseNet	Resnet	MobileNetV2	InceptionNet	EfficientNet-B7
Sensitivity	0.834	0.833	0.833	0.833	0.916	0.667
Specificity	1.0	1.0	1.0	1.0	1.0	0.5
Precision	1.0	1.0	1.0	1.0	1.0	0.571
NPV	0.857	0.857	0.857	0.857	0.923	0.6
FPR	0.0	0.0	0.0	0.0	0.0	0.5
FNR	0.167	0.166	0.166	0.166	0.083	0.333
FDR	0.0	0.0	0.0	0.0	0.0	0.428
F1 score	0.909	0.909	0.909	0.909	0.956	0.615
Overall Accuracy	91.667	91.667	91.667	91.666	95.834	58.333

6.2 Comparative study of results obtained from all the models implemented

6.2.1 F1 Score bar graph

The F1 Score is a metric showing how accurate a model is on a given dataset. Therefore, a higher F1 Score implies that the model gives a more accurate classification output. From Figure 3.23, it can be seen that the InceptionNetV3 model has the highest F1 score. Table 3.2 shows the F1 scores obtained when different DL models are used with different classifiers.

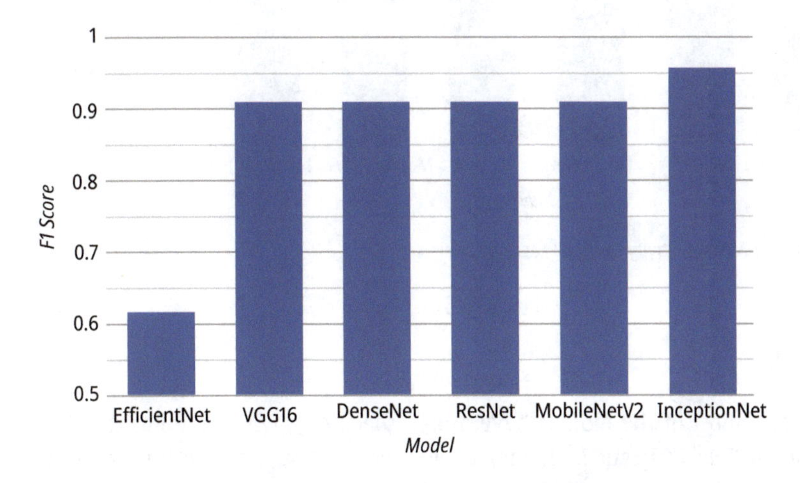

Figure 3.23: Bar graph of F1 scores of all models.

Table 3.2: F1 Scores of the DL models with different classifiers.

	VGG16	DenseNet50	ResNet201	MobileNetV2	InceptionNetV3
KNN	0.451	0.5	0.559	0.4	0.344
Decision Tree	0.857	0.857	0.956	0.857	0.857
SVM	0.909	0.909	0.909	0.909	0.909
XGBoost	0.857	0.909	1.0	0.909	0.909
AdaBoost	0.909	1.0	1.0	0.909	0.956

6.2.2 Sensitivity or recall or true positive rate

If a model has high sensitivity, it implies that the model will have fewer FNs, which is crucial when the model is used for the detection of a medical condition. InceptionNetV3 has the highest sensitivity (from Figure 3.24); hence, it is the best model that can be applied for the detection and classification of thyroid nodules.

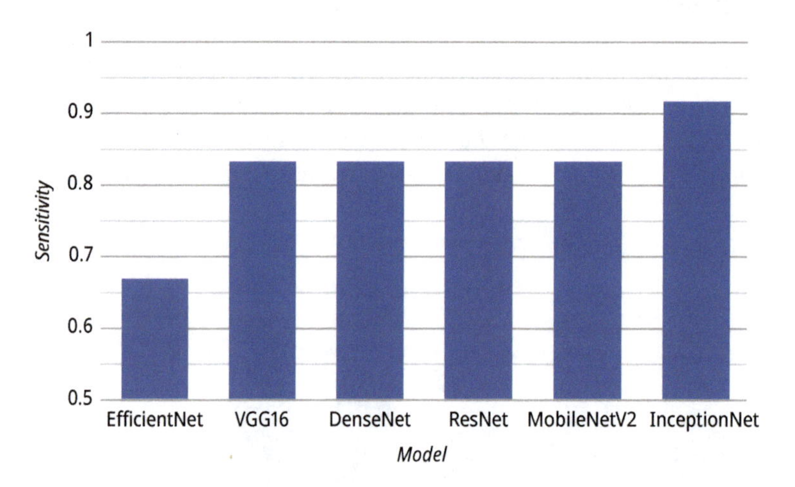

Figure 3.24: Bar graph of sensitivity or recall for all models.

6.2.3 Specificity

From Figure 3.25, most of the models have a high specificity, which implies that it is good at predicting TNs. EfficientNet having a low specificity implies that it is not accurate in predicting TNs.

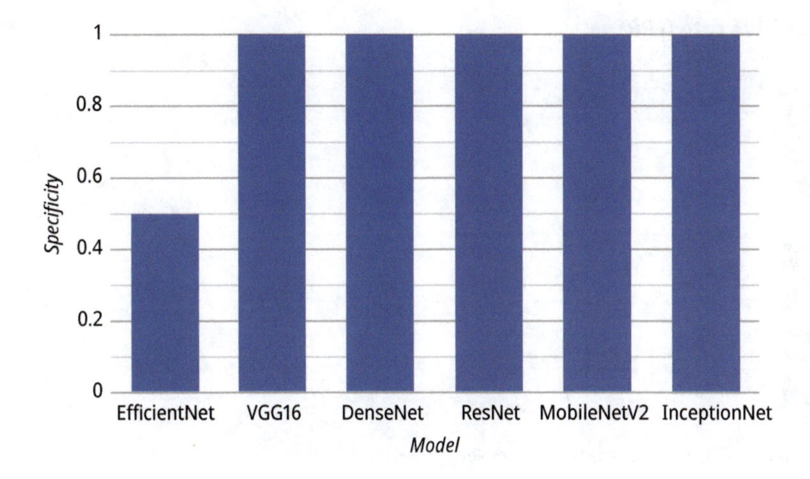

Figure 3.25: Bar graph of specificity for all models.

6.2.4 Negative predictive value (NPV)

If a model has a high value of NPV, when it classifies a US image as negative (or benign, in this case), then the probability of the nodule in the image actually being benign is high. As seen in Figure 3.26, InceptionNetV3 has the highest NPV. Hence, this model can be trusted the most to give an accurate negative (benign) output.

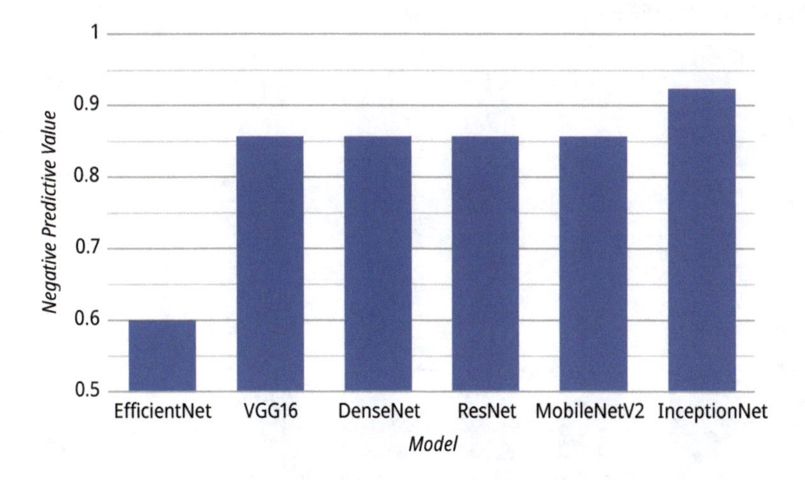

Figure 3.26: Bar graph of negative predictive values (NPVs) of all models.

6.2.5 False positive rate (FPR)

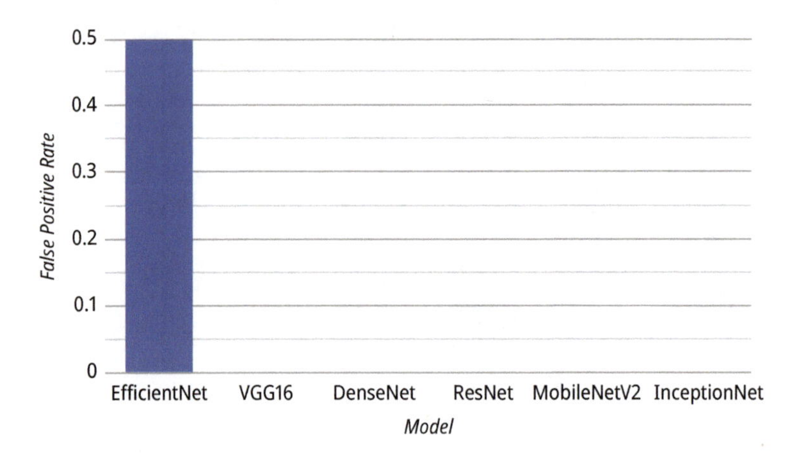

Figure 3.27: Bar graph of false positive rate (FPR) values of all models.

A high value of FPR implies that several images that were predicted positive were actually negative and, hence, were wrongly predicted. From Figure 3.27, EfficientNet has the highest FPR value. Hence, it is not the best choice (of model) for accurate detection and classification of thyroid nodules.

6.2.6 False negative rate (FNR)

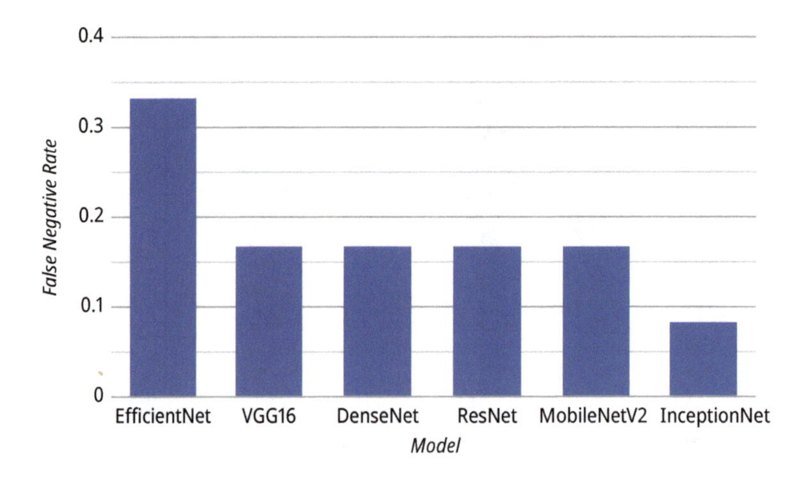

Figure 3.28: Bar graph of FNR values of all models.

A low value of FNR implies that the model classifies an image that is actually positive (or malignant) as negative (benign), rarely. From Figure 3.28, InceptionNet has the lowest value of FNR. Hence, it can be considered as the best model suited for the detection and classification of thyroid nodules.

6.2.7 Accuracy

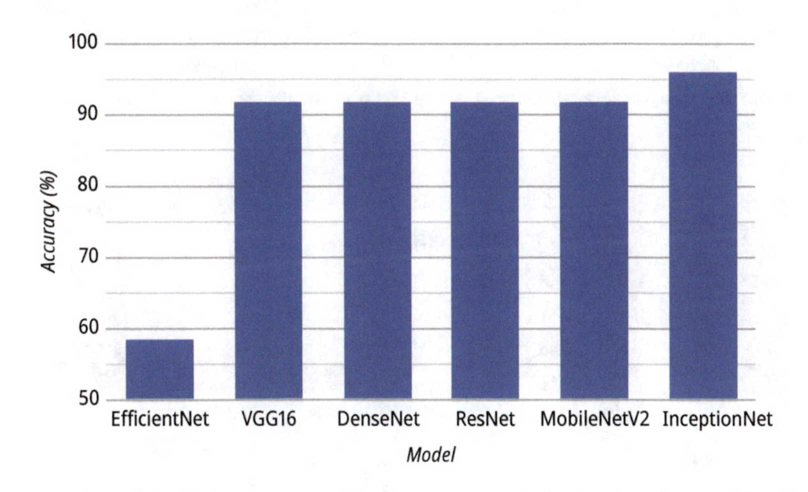

Figure 3.29: Bar graph of accuracies of all models.

From Figure 3.29, it can be seen that InceptionNetV3 has the highest accuracy. InceptionNetV3 expands with respect to the count of layers rather than just increasing the depth. Within the same layer, many kernels of various sizes are implemented. This is the reason for its high Accuracy and F1 score.

6.3 Comparative study of F1 scores of different classifiers for each model implemented

The F1 Score is a metric to evaluate how accurate a model is on a given dataset. Therefore, a higher F1 score implies that the model gives a more accurate classification output. Bar graphs comparing the F1 scores of different classifiers implemented with different DL models are shown in Figure 3.30.

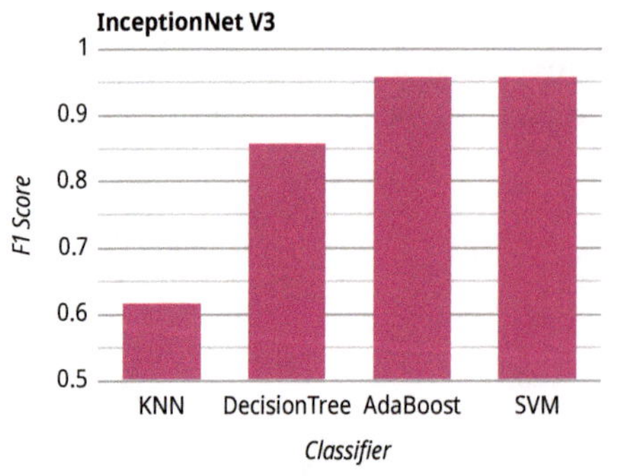

Figure 1 : F1 scores of different classifiers with InceptionNet

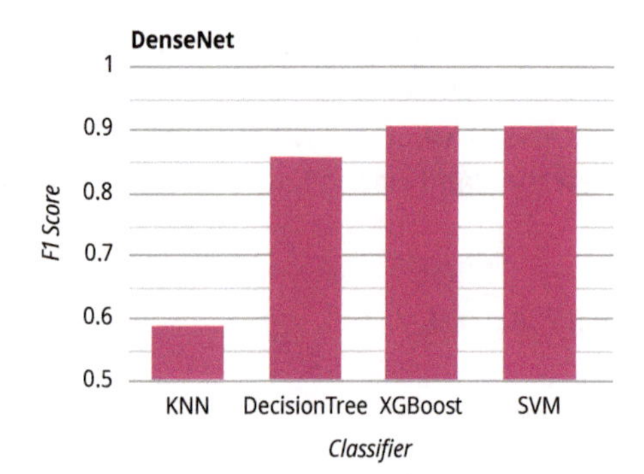

Figure 2 : F1 scores of different classifiers for DenseNet

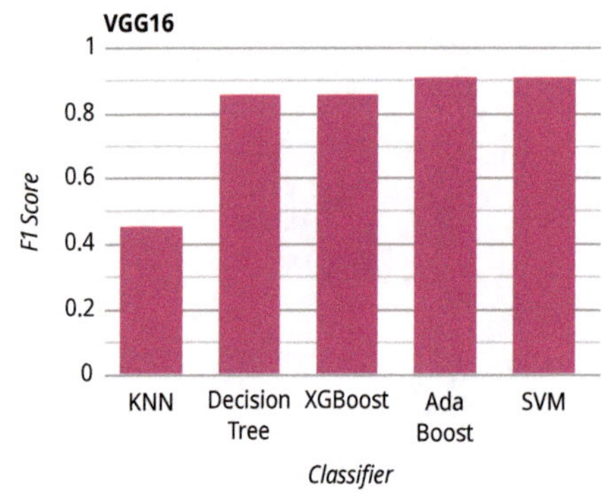

Figure 3 : F1 scores of different classifiers with VGG16

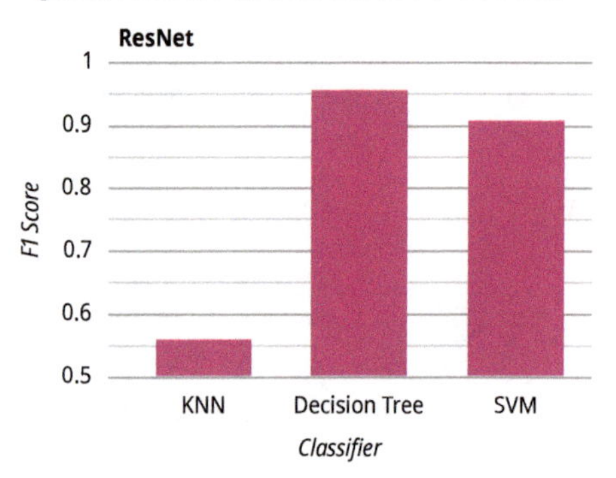

Figure 4 : F1 scores of different classifiers with ResNet

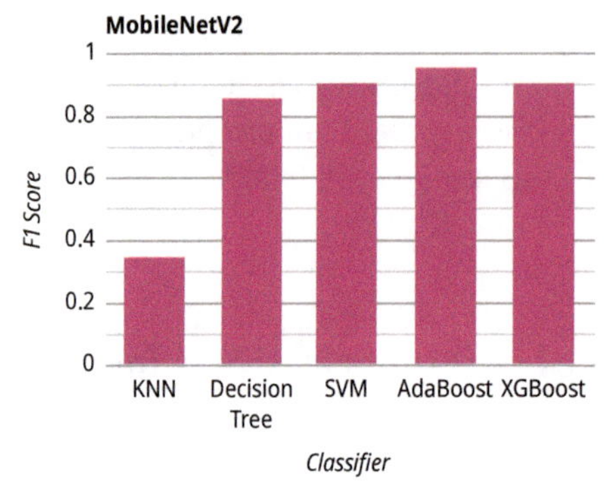

Figure 5 : F1 scores of different classifiers with MobileNetV2

Figure 3.30: Comparison of F1 scores of different classifiers with different DL models.

6.4 Receiver operator characteristic (ROC)curve and Area under the curve (AUC)

The ROC is used mainly for binary classification problems.

The AUC can be considered as a summary of the ROC Curve. AUC determines the ability of a classifier to differentiate across classes. The values of AUC obtained on using different models with the SVM Classifier are shown in Table 3.3.

Table 3.3: AUC values obtained corresponding to various DL models.

DL model	AUC
VGG16	0.916
DenseNet201	0.916
ResNet50	0.916
MobileNetV2	0.916
InceptionNetV3	0.958
EfficientNet-B7	0.583

The ROC Curves are obtained for different models implemented with SVM as the classifier, as shown in Figure 3.31.

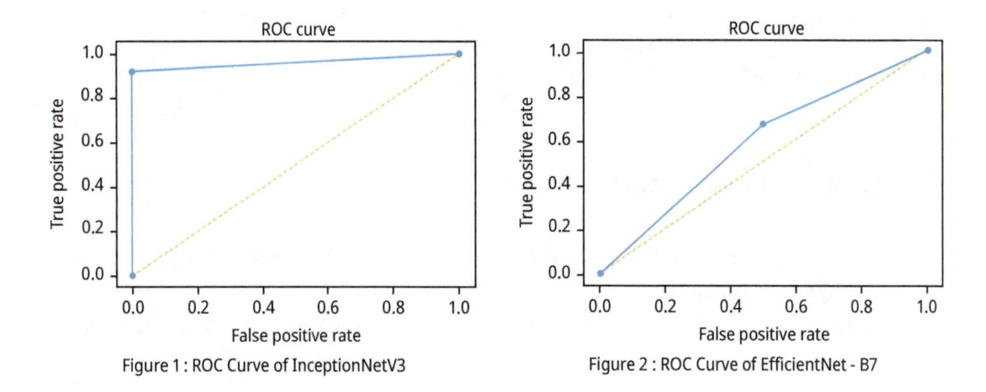

Figure 3.31: ROC curves of all models with SVM as the classifier.

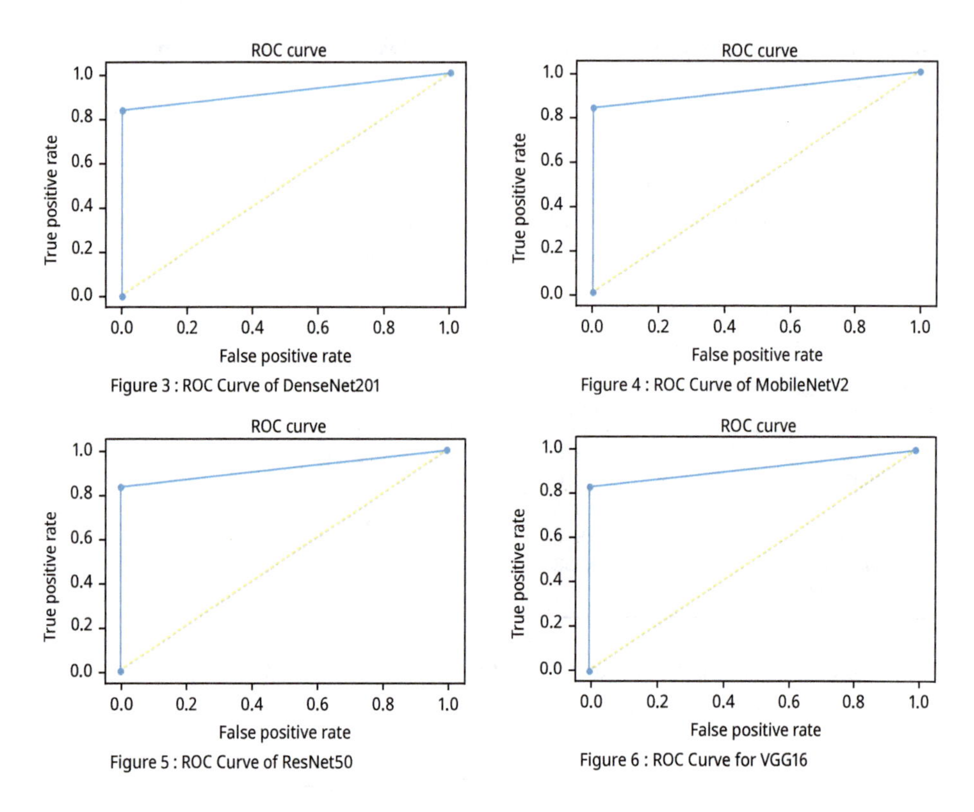

Figure 3.31 (continued)

6.5 Graphical user interface

Figure 3.32 shows the result "BENIGN" obtained when a US image of a thyroid nodule that is benign was uploaded.

Figure 3.33 shows the result "MALIGNANT" obtained when a US image of a malignant thyroid nodule was uploaded.

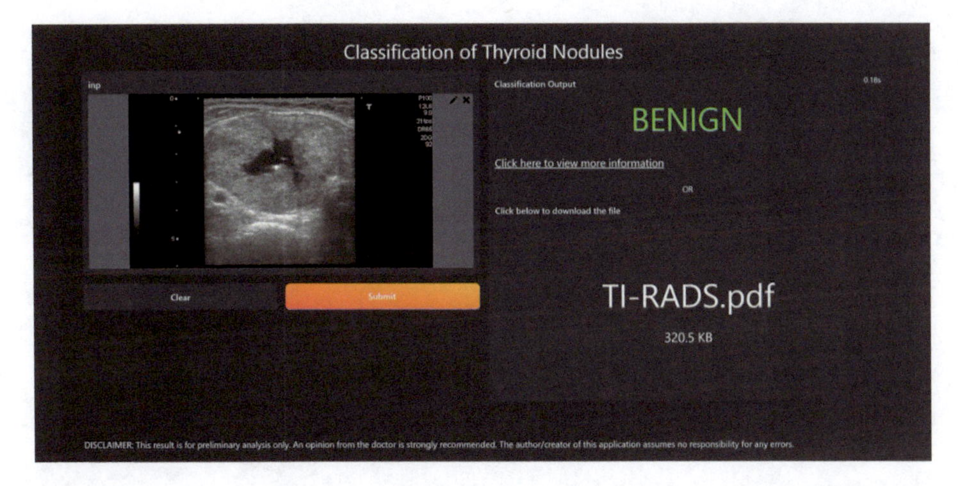

Figure 3.32: When a US image of a benign thyroid nodule was uploaded, the output obtained was benign.

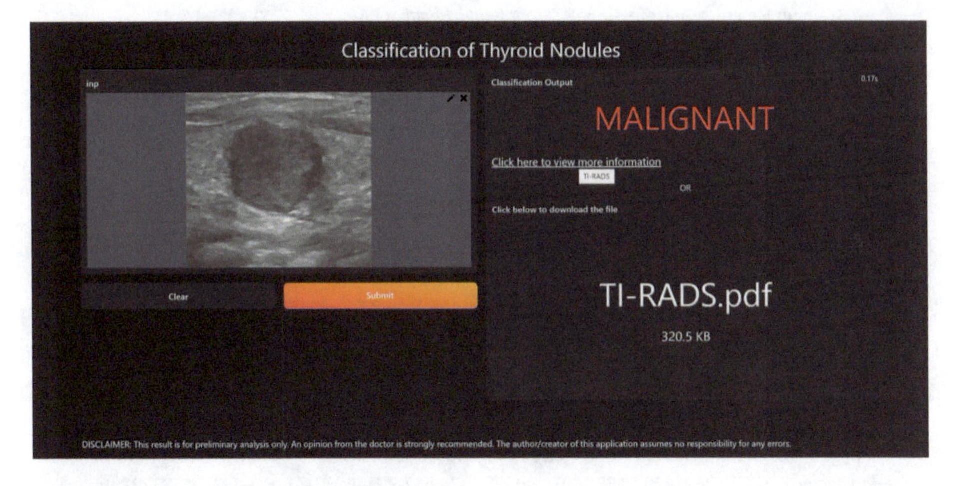

Figure 3.33: When a US image of a malignant thyroid nodule was uploaded, the output obtained was malignant.

7 Conclusion and future scope

A comparative study of different DL-based models and classifiers for the detection and classification of thyroid nodules into benign and malignant nodules has been done.

One of the challenges faced was the lack of a good GPU configuration. This was overcome by using Google Colab, which is an open-source platform specifically used for notebooks written in Python. Colab makes it very easy to use Keras and TensorFlow libraries. After performing a comparative study of different models and classifiers, it has been found that InceptionNet gave the best accuracy – 95.83%, F1 score – 0.957, and Sensitivity – 0.917, when used with the classifier SVM.

To make this project accessible and understandable to everyone, a simple, easy-t-use GUI that displays the classification output after the user uploads a US scan image of the thyroid nodule has been implemented.

Use of real-time datasets acquired from hospitals would help improve the accuracy of the classification output given by our model.

Another enhancement that could be made to this project is implementing it on hardware with GPU.

References

[1] Fertleman, C., Aubugeau, W. P., Sher, C., et al. (2018). A Discussion of Virtual Reality as a New Tool for Training Healthcare Professionals. *Front Public Health*, 6: 44.

[2] Pr., L., Lu, L., Jy, Z., et al. (2021). Application of Artificial Intelligence in Medicine: An Overview. *Current Medical Science*, 41: 1105–1115. https://doi.org/10.1007/s11596-021-2474-3.

[3] FN, T., WD, M., and EG, G. (2018 Apr). Thyroid Imaging Reporting and Data System (TI-RADS): A User's Guide. *Radiology*, 287(1): 29–36. 10.1148/radiol.2017171240.

[4] Chi, J., Walia, E., Babyn, P., et al. (2017). Thyroid Nodule Classification in Ultrasound Images by Fine-Tuning Deep Convolutional Neural Network. *The Journal of Digital Imaging*, 30: 477–486. https://doi.org/10.1007/s10278-017-9997-y.

[5] Khachnaoui, H., Guetari, R., and Khlifa, N. (2018). A review on Deep Learning in thyroid ultrasound Computer-Assisted Diagnosis systems. *2018 IEEE International Conference on Image Processing, Applications and Systems (IPAS)*, 291–297. 10.1109/IPAS.2018.8708866.

[6] Vadhiraj, V. V., Simpkin, A., O'Connell, J., Singh Ospina, N., Maraka, S., and O'Keeffe, D. T. (2021 May 24). Ultrasound Image Classification of Thyroid Nodules Using Machine Learning Techniques. *Medicina (Kaunas)*, 57(6): 527. 10.3390/medicina57060527.

[7] Yu, Q., Jiang, T., Zhou, A., Zhang, L., Zhang, C., and Xu, P. (2017 Jul). Computer-aided Diagnosis of Malignant or Benign Thyroid Nodes based on Ultrasound Images. *European Archives of Oto-rhino-laryngology*, 274(7): 2891–2897. 10.1007/s00405-017-4562-3 Epub 2017 Apr 7. PMID: 28389809.

[8] Nikhila, P., Nathan, S., Gomes Ataide, E. J., Illanes, A., Friebe, M., and Abbineni, S. (2019 Nov). Lightweight Residual Network for The Classification of Thyroid Nodules. *IEEE EMBS*.

[9] Avola, D., Cinque, L., Fagioli, A., Filetti, S., Grani, G., and Rodolà, E. (2021). Knowledge-Driven Learning via Experts Consult for Thyroid Nodule Classification. ArXiv, Abs/2005.14117.

[10] Nguyen, D. T., Kang, J. K., Pham, T. D., Batchuluun, G., and Park, K. R. (2020 Mar 25). Ultrasound Image-Based Diagnosis of Malignant Thyroid Nodule Using Artificial Intelligence. *Sensors (Basel)*, 20 (7): 1822. 10.3390/s20071822 PMID: 32218230; PMCID: PMC7180806.

[11] Abdolali, F., Shahroudnejad, A., Hareendranathan, A. R., Jaremko, J. L., and Noga, M. (2020). Kumaradevan Punithakumar: A systematic review on the role of artificial intelligence in sonographic diagnosis of thyroid cancer: Past, present and future. CoRR Abs/2006.05861.

[12] Song, W., Li, S., Liu, J., Qin, H., Zhang, B., Zhang, S., and Hao, A. (2019 May). Multitask Cascade Convolution Neural Networks for Automatic Thyroid Nodule Detection and Recognition. *IEEE Journal*

of Biomedicine and Health Informatics, 23(3): 1215–1224. 10.1109/JBHI.2018.2852718 Epub 2018 Jul 3. PMID: 29994412.

[13] Pedraza, L., Vargas, C., Narvaez, F., Duran, O., Munoz, E., and Romero, E. (28 January 2015). An Open Access Thyroid Ultrasound-image Database. Proceedings of the 10th International Symposium on Medical Information Processing and Analysis. Cartagena de Indias, Colombia.

[14] Tessler, F. N., et.al. (May 2017). ACR Thyroid Imaging, Reporting and Data System (TIRADS): White paper of the ACR TI-RADS Committee. *Journal of the American College of Radiology*, 14(5): 587–595.

[15] Nguyen, D. T., Pham, T. D., Batchuluun, G., Yoon, H. S., and Park, K. R. (2019 Nov 14). Artificial Intelligence-Based Thyroid Nodule Classification Using Information from Spatial and Frequency Domains. *Journal of Clinical Medicine*, 8(11): 1976 10.3390/jcm8111976. PMID: 31739517; PMCID: PMC6912332.

[16] Fernández Sánchez, J. (2014). TI-RADS Classification of Thyroid Nodules based on a Score Modified Regarding the Ultrasound Criteria for Malignancy. *Revista Argentina de Radiologia*, 78(3): 138–148.

[17] Al-Shweiki, J. (2021). Re: Which Image Resolution Should I use for Training for Deep Neural Network? March 2021. ResearchGate.

[18] De Oliveira, D. C., and Wehrmeister, M. A. (2018). Using Deep Learning and Low-Cost RGB and Thermal Cameras to Detect Pedestrians in Aerial Images Captured by Multirotor UAV. *Sensors*, 18(7): 2244. https://doi.org/10.3390/s18072244.

[19] He, K., Zhang, X., Ren, S., and Sun, J. (2015). Deep Residual Learning for Image Recognition. arXiv, Abs/1512.03385.

[20] Nayak, V., Holla, S. P., Akshayakumar, K. M., and Gururaj, C. (2021 November). Machine learning methodology toward identification of mature citrus fruits. In L. C. Chiranji, A. Mamoun, C. Ankit, H. Saqib, and R. G. Thippa (Eds.), *The Institution of Engineering and Technology*. London: Computer Vision and Recognition Systems using Machine and Deep Learning Approaches IET Computing series, vol. 42, Ch. 16, ISBN: 978-1-83953-323-5, 385–438. 10.1049/PBPC042E_ch16

[21] Srujana, K. S., Kashyap, S. N., Shrividhiya, G., Gururaj, C., and Induja, K. S. (2022 May). *Supply chain based demand analysis of different deep learning methodologies for effective covid-19 detection, innovative supply chain management via digitalization and artificial intelligence*. K. Perumal, C. L. Chowdhary, and L. Chella (Eds.), Springer Publishers, Chapter 9, ISBN: 978-981-19-0239-0. 135–170. 10.1007/978-981-19-0240-6_9.

[22] Gururaj, C. (2018). Proficient Algorithm for Features Mining in Fundus Images through Content Based Image Retrieval. IEEE International Conference on Intelligent and Innovative Computing Applications (ICONIC – 2018), ISBN – 978-1-5386-6476-6, 6th to 7th of December, 108–113, Plaine Magnien, Mauritius. 10.1109/ICONIC.2018.8601259.

[23] Gururaj, C., and Tunga, S. (2020 February). AI based Feature Extraction through Content Based Image Retrieval. *Journal of Computational and Theoretical Nanoscience*, 17(9–10): 4097–4101. ISSN: 1546-1955 (Print): EISSN: 1546-1963 (Online) 10.1166/jctn.2020.9018.

[24] Tan, M., and Le, Q. V. (2019). EfficientNet: Rethinking Model Scaling for Convolutional Neural Networks. ArXiv, Abs/1905.1194.

[25] How to build KNN from scratch in Python (Accessed at https://towardsdatascience.com/how-to-build-knn-from-scratch-in-python-5e22b8920bd2)

[26] Support Vector Machine – Exploring the black box (1/3) (Accessed at https://medium.com/@rahulramesh3321/support-vector-machine-exploring-the-black-box-1-3-93be0af17218)

Vijayarajan Ramanathan, Gnanasankaran Natarajan,
Sundaravadivazhagan Balasubramanian

4 Convergence of AR/VR with IoT in manufacturing and their novel usage in IoT

Abstract: Augmented reality is an interactive experience of a real-world environment, where the objects that reside in the real world are enhanced and superimposed by computer-generated information. Virtual Reality is the use of computer technology to create a simulated immersive environment, which can be explored in 360 degrees. When AR and VR technologies are combined with Internet of Things (IoT), there are endless use cases and industries that benefit from this convergence. AR/VR + IoT can be used to display a machine's performance in real-time; using sensors we can get the real-time data of an object and that data can be superimposed and shown to the user in AR or VR. This helps the manufacturing or service industries know about the status of the machine or a section visually, which helps in finishing the tasks sooner than conventional methods of working.

Data visualization can be done in AR using smartphones, when an object or a machine is scanned using the smartphone. Data visualization on VR can be done using VR head-mounted devices like Oculus Quest, HTC Vive, or Hololens as well as smart glasses. When the user wears the VR device and views the application, he/she gets the feeling as if they are inside the machine or a place. AR + IoT is more suited for fixing day-to-day operations. VR + IoT is suited for training the employees in a given situation or virtual training of machinery.

In this chapter, we will explain how combining AR and IoT can increase the productivity and how VR and IoT can help in reducing the training time of employees. Finally, we touch upon how the data that we extract out of this convergence of ARVR and IoT helps the companies get a meaningful insight into their operations and further optimize their production.

Keywords: Augment reality, Virtual reality, IOT, sensors, real-time data, Manufacturing, Training

Vijayarajan Ramanathan, Hogarth Worldwide, Chennai, Tamil Nadu, India, e-mail: Vijayvrar1@gmail.com
Gnanasankaran Natarajan, Department of Computer Science, Thiagarajar College, Madurai, Tamil Nadu, India, e-mail: sankarn.iisc@gmail.com, Orcid: 0000-0001-9486-6515
Sundaravadivazhagan Balasubramanian, Faculty of IT, Department of Information Technology, University of Technology and Applied Sciences, Al Mussanah, Oman, e-mail: bsundaravadivazhagan@gmail.com, Orcid: 0000-0002-5515-5769

https://doi.org/10.1515/9783110790146-004

1 Introduction

As pioneers investigate the commercial relevance and roles that AR/VR can play in worker enablement, client experience, and interaction, demand for AR and VR) is skyrocketing. This emerging technology is transforming the tectonic shift in increasing the productivity of companies. AR/VR can help boost and enhance consumer engagement as well as deliver an exceptional increase in an organization's bottom line by offering products that purchasers can acquire with only one click. In healthcare, as well as in other industries, AR and VR are rethinking what is possible. This has been proved in [1] The Internet of Things, Augmented Reality And Virtual Reality: A Fruitful Synergy in Industry 4.0 and in [2] AR-Enabled IoT for a Smart and Interactive Environment: A Survey and Future Directions [1, 2], which talks about how AR and IoT can empower manufacturing industries to make use of Industry 4.0 technologies to automate and improve production and remote collaboration, and how it benefits the employers and employees in training, simulation, maintenance operations, and in improving the overall performance.

1.1 What is augmented reality (AR)?

The term "augmented reality" refers to a technology that mixes real and virtual elements. A user may attend a trade show and use her smartphone to activate an AR experience that allows them to view a computer-generated rendition of a product.

A highly visible, engaging method of presenting critical digital information in the context of the physical world, AR bridges the gap between employees and improves corporate outcomes. By adding digital content to actual work surroundings, industrial AR enhances the creation and delivery of easily understandable task instructions.

1.1.1 App-based AR frameworks

Apple's AR platform is known as ARKit. It is made specifically for creating AR experiences for iOS devices. Detecting the ground planes, anchoring, tracking face using face mesh, tracking pictures, and model embedding are all included in ARKit.

ARCore is Google's response to ARKit. It is an open source platform that allows for developing AR on both Android and iOS devices. Face and motion tracking, cloud anchors, light estimation, and plane detection are all included in ARCore, just like they are in Apple's platform.

Vuforia is an app-based framework that is user-friendly. It has enhanced markerless AR technology, which improves the stability of digital objects in physical environments. It also includes the ability to see models from various perspectives in real time, using the "Model Target" feature.

WebAR gives consumers options to use their cellphones to access AR activities. The unique and immersive experience is available to everyone without having to download anything. People may enjoy the immersive and unique experience of AR on demand on most operating systems, mobile devices, and web browsers because no downloading is required. AR that is based on an application is only available for download via an app store. This kind of AR technology makes excellent use of a device's capabilities, memory, and performance, enabling complex experiences. This translates to visually stunning animations and interactive elements.

However, application-based AR provides an additional step to the equation, ultimately reducing the customers' acceptance. Most customers are wary of downloading the app from Play Store or iOS store to observe and interact in AR. When users have to download something, their engagement drops by half. Studies say that roughly 25% of users continue using AR applications after their very first experience. Compatibility across many systems and platforms is particularly expensive when developing apps for AR. On the other hand, a simple Web link, QR code, or NFC tag can all be used to access WebAR. This allows users to enjoy AR view on most mobile devices and browsers, without having to install a whole program.

WebAR also allows makers to give wholesome experience to their intended audience across all platforms. Although the memory-light nature of browser experiences restricts WebAR, its potential market reach more than matches.

1.1.2 Web-based AR frameworks

AR.js is another open source toolkit used to create AR applications on the web. This framework is intended to make creating compelling WebAR experiences that is simple for users.

1.2 What is virtual reality (VR)?

Virtual reality (VR) gives immersive three-dimensional experience to the consumers and makes them feel they are in to the virtual world. VR not only helps in visualization but it also provides tactile feedback of touch sensation using haptic devices; having said that, content availability and computational power are the two limitations to a VR experience.

VR simulations are currently divided into three categories: non-immersive, semi-immersive, and fully immersive.

1.2.1 Non-immersive

Non-immersive virtual experiences are frequently neglected as a VR genre. This technique brings up a system-generated world, by making the consumer manage their real-world view. VR applications that are not head-mounted depend on a laptop or computer for visualization, and interaction is done using keyboard controls. An electronic game is an excellent example of non-immersive VR.

1.2.2 Semi-immersive

Semi-immersive simulation is where consumers watch the three-dimensional experience and they feel as if they are in another dimension; however they remain connected to their real-world view. A more immersive experience is achieved by using more detailed images. Semi-immersive VR is more useful for teaching a concept to the students.

1.2.3 Fully immersive

Fully immersive experience includes vision and hearing. The user will need VR devices like Oculus quest, HTC Vive, or Hololens to visualize and teleport to that environment. The advantage of using hardware devices like Quest or Vive is that it brings ultra realistic visualization to the consumers, and it makes them feel as if they are in a completely new virtual world altogether, making them forget about their real world. This is more useful for training the employees virtually and for remote collaboration.

Just like web-based AR we have web-based VR. Web VR made it easy for anybody to create, consume, and share VR experiences on the web. WebXR, the next step in this progression, will merge Web VR and AR into a single API.

Both virtualization and augmentation are employed in a variety of industries and contexts, including healthcare, retail, manufacturing, and engineering. However, when combined with other technologies like digital twin, Internet of Things, analytics, data science, cloud, and edge computing, AR delivers increased value in a variety of applications in manufacturing. The application of AR and VR in manufacturing and other businesses is not science fiction. Manufacturers are stepping up their digital transformation efforts in order to achieve Industry 4.0 in a deliberate and staged manner. It is happening as we talk (or read), and it is only going to get faster as the benefits become more apparent and infrastructure matures and advances.

1.3 Internet of Things (IoT)

IoT refers to connecting all the physical and electronic objects to the internet. For example, by connecting a water tank with the Internet via sensors, we can get the details, such as whether the water tank contains water or not, to our devices directly. By connecting everyday objects to the Internet, we get massive amounts of data, which can be stored in cloud servers and accessed instantly through devices.

In the manufacturing sector, using IoT in the machineries helps companies get real-time data remotely. Some of the types of IoT are:

- Consumer IoT focuses on everyday use. Examples include lighting systems, voice assistants, and home appliances.
- Most commercial IoT applications are found in the transportation and healthcare industries. For instance, sophisticated monitoring systems.
- Military Things (IoMT) – This term refers to the employment of IoT technologies in the military. For example, surveillance robots and combat biometrics worn by humans.
- Industrial Internet of Things (IIoT) – Used mostly in industrial applications such as manufacturing and energy. For example, industrial big data, smart agriculture, and digital control systems.
- Smart cities primarily use infrastructure IoT for connectivity. For instance, infrastructure management and sensor systems.

1.3.1 How IoT works

Different IoT devices provide different duties, yet they all operate in a similar manner. To begin, IoT gadgets are physical things that track happenings in the actual world. They often have a firmware, integrated CPU, and Dynamic Host Configuration Protocol server connection. It also requires an IP address in order to function across the network.

Most IoT devices are configured and managed by a software program. Think about using a smartphone app to control the lighting in your house. Some devices also come with integrated web servers, removing the need for external software. The lights, for instance, turn on as soon as you enter a room.

2 IoT and VR

VR has been enabled and even enhanced, thanks to the IoT. Law enforcement agencies are using AR and IoT to improve safety and security for the citizens. People use AR smart glasses to interact and fix the issues that appear on their day; even the government uses VR to test the most effective surgical strategy to lower the likelihood of surgical failure. To further improve the immersive experience, a variety of sensors (IoT) can be employed in conjunction with VR. These sensors, like biofeedback sensors, enable the immersive world's narrative to be changed in response to the user's biofeedback, thereby increasing the user's presence.

3 Use cases of IoT in manufacturing

Some of the use cases where IoT is used in the manufacturing industries are as follows:

3.1 Quality control

Quality control is one the key functions of an organization, as compromising on quality will lead to substandard production and, ultimately, loss in customer retention [3]. Quality control processes can be made simpler by employing IoT. Using thermal and video sensors can provide vital information to fix the flaws in the product, which, in turn, meets the quality standard before the product reaches the end customers.

3.1.1 Challenges

1. In terms of business and economics, competition in the IoT industry has a direct impact on the conditions. As a result of the competition, manufacturers of IoT devices must cut their prices. It puts pressure on companies to reduce their time to market. In some IoT applications, such as sensors or camera networks, devices might be placed in locations that make them accessible to others, an intruder; on the other hand, it is difficult for the servicing team to check on a regular basis. These gadgets have the potential to make the entire network insecure.
2. In many IoT devices, customer has little knowhow on the device's underlying mechanisms; also, when the device is updated, the user has little influence over the updates. When combined with GPS, voice recognition, or embedded cameras, major security and privacy risks can arise. Homemade devices that do not adhere to industry standards can be manufactured and integrated with standardized IoT devices. These challenges can be overcome with the help of VR and AR using IoT devices.

3.2 Asset management

The function of IoT is to get relevant and meaningful insights from the huge amount of data collected from sensors. It tracks the location, status, and movement of inventory items and provides users with equivalent results; for instance, an IoT-enabled warehouse management architecture can govern the amount of materials needed for the next cycle, based on available data. This output can be used in many different ways. The system can tell the user when material must be replaced, if just one inventory item is lacking. IoT gives supply chain managers cross-channel insight, enabling them to reduce shared costs throughout the value chain by giving them an accurate assessment of the supplies that are now available, the arrival of new materials, and work-in-progress. By monitoring the flow of traffic and the speed of moving vehicles, manufacturers may prepare better for receiving raw materials. This shortens handling times and makes it possible to process those materials during the manufacturing procedure more effectively.

3.3 Remote monitoring and production

With IoT devices, remote monitoring is feasible. Staff could first remotely gather information about the production process and assess whether the results, or even the process itself, complies with a certain set of rules or specifications.

Second, the device can be fine-tuned and configured remotely, saving significant time and effort.

Furthermore, [4] automated devices simplify device management and control by allowing workers to solve a number of performance issues on virtual networks, without having to be physically present. Employees can also find out about device locations. Example: immunity using virtual device monitoring.

3.3.1 Challenges

Although IoT-driven applications have significant advantages, their implementation might be difficult. The following are major concerns:

- The deployment of an IoT solution necessitates many investments.
- IoT includes spending on administrative labor, technical assistance, and more, in addition to obvious investments in IoT devices, connectivity, and cloud services.
- Many manufacturing plants still employ machines that are more than 20 years old. These devices were not intended to connect to the Internet. Making such machines part of an IoT solution is not easy and takes a lot of time and work.

3.4 Predictive maintenance

Machines with embedded IoT sensors may identify any operational issue (temperature, turning number, pressure, voltage, etc.) and alert the proper personnel, leaving workers to just take corrective action.

Technical support personnel can find and fix issues before they result in major equipment failures, using the technique known as predictive maintenance, which reduces downtime and costs. IoT-connected equipment can be coupled with cutting-edge analytics tools through predictive maintenance in order to foresee when tech support is necessary.

3.4.1 Challenges

One of the key challenges in predictive maintenance is the high probability of false prediction of issues, which leads to time being wasted on a nonissue. However, IoT devices are now smart enough to predict with high accuracy and AR/VR helps in visualizing the issue, so they can be fixed on time.

3.5 Digital twins

Digital twin technology makes use of the IoT, artificial intelligence, machine learning, and cloud computing.

Engineers and management can use digital twins to replicate various processes, conduct tests, identify problems, and get the required outcomes, without endangering or damaging physical assets.

A virtual replica built to accurately reflect a physical object is known as a digital twin [5]. The object under investigation (for example, a machine) possesses a number of sensors that keep track of various elements of its operation. These sensors gather information on the performance of physical items, including their production, temperature, and other features. The processing device then applies this information to the digital version after receiving it.

The virtual model is then able to run simulations, look into performance issues, and make suggestions for changes, all with the goal of gaining important knowledge that can eventually be applied to the actual, physical object [6, 7].

4 Benefits of IoT in manufacturing

1. Cost savings: Using IoT in manufacturing reduces a great deal of possible issues and costs less money.
2. Increased safety: It creates a safe working atmosphere and prevents many accidents.
3. Better decision-making: Employers who use IoT have precise insights and are aware of performance, allowing for faster and improved judgement.
4. Reduction in overall production time: Using IoT throughout the product's lifecycle aids in bringing the product to market faster and providing a better product to consumers.

5 Use cases of AR/VR in manufacturing

Following are some of the use cases where AR and VR are used in the manufacturing industries.

5.1 Real-time training

AR and VR are increasingly being used to teach employees and give directions for repairs and maintenance to service professionals. The technology can be used for employee training and education because it can mimic any procedure or setting. A worker receives virtual instructions on his device or through AR glasses, as he works. The real-time data assists employees in better understanding issues and resolving them faster. By removing the dreaded human element and preventing mishaps, such training allows the company to save weeks, if not months in time. We recently built a VR-based technology solution for one of our clients to boost training efficacy.

5.2 Site operations enhancement

Why should other industries not follow suit and use AR to improve accessibility? Thanks to the IoT's various surveillance integrations and vast bandwidth, the foundations have been laid for distance location workers to monitor actions from faraway places.

Apart from enabling immersive experience of the floor, IoT systems with AR projection may make it easier to view essential operational data in the relevant context. Visual alarm indicators, for instance, could be used by a heads-up display to direct workers to malfunctioning equipment or assist them in navigating situational threats.

5.3 Maintenance and repair work

There are no boundaries between virtual and augmented reality. Any data source can be connected as long as it delivers precise information and contextual mappings to head-mounted VR devices. Consider the case where your employees want assistance with tough activities like maintaining controllers, calibrating sensor devices, or fixing network issues; you can carry out the necessary monitoring with IoT and AR, without ever having to step on the production floor.

Wherever AR is deployed, it encourages human guidance. This feature is particularly beneficial in the field of business-to-business services. AR technology that enables human touchless IoT connections to essential reference sources such as live monitoring feeds or instruction booklets during repairs could be a lifesaver for support centers.

5.4 Robotic production

The fabrication process relies heavily on automation in many manufacturing areas. Each product, such as circuit boards, is made up of hundreds or thousands of extremely distinct, microscopic pieces. To make matters worse, these elements necessitate precision to the micrometer. and clean-room-grade handling choose and place devices are now one of the most prevalent solutions to such production challenges. However, these solutions have drawbacks, such as only being able to hold a limited number of components at any point in time and requiring manual loading.

The IoT and AR help drag and drop activities, humanoid arm manufacturing, and other mechanized processes. AR tags can be used to highlight the actual world elements, allowing computer vision algorithms to detect their presence. After that, the IoT may take over to begin actuator-driven manufacturing process.

5.5 Fleet management

One of fleet management's long-standing issues has been effectively depicting the virtual linkages between the materials. The position of the driver as well as the road conditions can alter at any time, even though contemporary mapping technologies make it a little simpler to maintain elements in place.

With AR eyewear, figures from the IoT become more faithful, interacting with data collection and assets in immersive 3D space; AR panel could provide a realistic sense of crucial trip metrics. Some companies have even recommended that commercial vehicles be fitted with augmented windshields that allow drivers to concentrate on movement, climate, and other relevant data.

5.6 Connecting with subject matter experts

Several companies have shown or sold integrated AR safety helmets, with some garnering tens of millions of dollars in funding. Aside from providing physical protection, these devices can also be used to:

- allow workers to quickly access task lists and job orders
- improve wearer awareness to combat common safety risks
- monitor employees for management purposes
- Warning machinery operators when they are near risky locations or are about to cause accidents involving other people.

5.7 In warehouses

Shipping and handling are time-consuming procedures, but AR has the potential to speed them up. AR and the IoT devices make basic chores more logical, from helping workers discover products to instructing them where to keep inbound deliveries.

In today's warehousing operations, QR codes are everywhere. Many top application enablement platforms already handle activities like computer vision, thus IoT deployments take advantage of the technology, and not much tinkering is required. When these systems are combined with AR, the IoT becomes more tactile and present.

5.8 Training the employees

It is easier for people to recall what they have learnt when they are completely engrossed in the subject. By spreading relevant cues around your range of vision, AR gets you there halfway t. Could the IoT assist learners in completing their corporate knowledge journeys?

AR training is nothing new. It has been used for a long time as a substitute for putting untrained workers in unsafe situations.

The IoT helps by bringing theoretical and continuous staff education closer to real-world corporate concerns and environments. It offers companies to provide training for new recruits and retrain their existing workforce in a immersive way, which benefits both the employer and employee.

5.9 Inventory management

Inventory databases should be managed more efficiently. With the use of AR, give the workers instructions on where certain objects, such as those relevant to the aisle or shelf of their storage are located; this reduces ambiguity, improves precision, and shortens turnaround times.

5.10 How it benefits the companies

Virtualization and digital twin of the environment are used for employee training, which saves time and manpower that would otherwise be required to train them manually. Training the employees using VR instead of the conventional mode of training boosts employee productivity, which directly translates to higher revenues for businesses.

For example, if a company hires 20 new employees, training them manually and getting them ready for production is a huge task for the company. Adding VR training to manual training allows new employees to visualize their work environment and the machines they operate in VR, allowing them to learn more effectively.

6 Challenges in using AR/VR and IoT in manufacturing

Why are some manufacturers still hesitant to adopt IoT technology, despite its tremendous potential benefits?

Regardless of the business, transforming large, established systems needs time and effort. Manufacturing is not any different. The implementation of the IoT in businesses comes with a number of important hurdles [8]. Internet of Things (IoT) for Seamless Virtual Reality Space: Challenges and Perspectives – discusses the challenges that industries face in adopting these technologies in their processes, how to transition from legacy systems to today's technology that powers the industries, and the risks of becoming obsolete if industry 4.0 technologies like VR and the IoT, are not utilized.

6.1 Skills gap

The skills gap among employees is widening as the number of smart factories grows. Modern technologies in production need qualified specialists and data scientists that can comprehend and handle new procedures. Due to their lack of IoT knowledge and skills, industry executives may find it challenging to make judgments. As a result of the skills mismatch, 2.4 million manufacturing positions could become vacant during the next decade [9]. To overcome this issue, as soon as possible, educational training on the IoT and AR/VR technologies should be implemented.

6.2 Security

As data is exchanged between multiple nodes and can be highly vulnerable without effective security measures, security remains the most crucial problem for connected devices.

Manufacturing is now the most common target of cyber-attacks. When traditional factories become digital, they become IP-based systems, making each connected object vulnerable to cybercrime. As a result, manufacturers should place a premium on security and safety by establishing effective protective systems.

To minimize data loss, theft of intellectual property, or other incursions, manufacturers must adapt to modern hacking technologies. Defensive tools should be included in legacy equipment that includes IoT technologies. A security plan should be established from the start, when developing a new linked environment.

6.3 Integrating different technologies

Another issue is how to integrate all devices into a large production system while also providing adequate networking capabilities. To address this issue, businesses should employ intelligent technologies such as artificial intelligence, machine learning, 5 G connection, AR, or digital twins.

7 Technologies for using AR/VR in manufacturing

There are numerous technologies that can be used to construct AR/VR applications for Manufacturing, just as there are numerous use cases for AR/VR.

7.1 Software

Virtual and augmented reality applications are created using software such as Unity and Unreal. VR applications can be seen on Oculus quest, HTC vive, Hololens, and other head-mounted VR devices. Smartphones and tablets can be used to view AR applications.

7.2 Hardware

Enhanced reality smart glasses are being employed in manufacturing facilities, where employees will wear the glasses while performing their duties. For example, if an employee is viewing machinery through smart glasses, vital information such as temperature and how well the machinery has performed on a given day can be displayed on the device. This vital real-time information is possible, thanks to IoT sensors that capture data and display it on the smart glasses. Employees can use smart glasses to contact a subject matter expert to get answers to their questions.

Haptic gloves allow users to feel haptic feedback on their hands. For example, by combining haptic gloves with a VR device, employees will be able to sense the world while being immersed in VR. Haptic gloves provide a realistic touch-and-feel experience of an object or surroundings.

By wearing a full body haptic suit, we can simulate a real life scenario in VR, which helps the employees feel the environment.

8 How AR/VR and IoT are used in manufacturing

8.1.1 Virtual reality

Employees in a car manufacturing factory, for example, receive VR training on how to operate, assemble, and repair equipment [10]. Employers save time and money by providing VR-based training, and employees feel protected while learning in VR. This is a win-win situation for both companies and employees, as employers' revenue increases over time, when employees become more qualified and ready to take on real-world jobs after receiving VR training.

8.1.2 Augmented reality

Companies have begun to use AR for staff training, similar to VR-based training. Using AR has the advantage of not requiring a head-mounted device to watch programs; unlike VR, AR applications can be seen on smartphones and tablets. Employers save time and money by combining AR-based training with manual or traditional training. This enhances productivity [11]. AR, with the IoT, which demonstrates how AR, can use data from IoT devices to assist businesses enhance productivity and lower expenses over time.

8.1.3 Virtual and augmented reality with IoT

Sensors are used to gather real-time data of machinery or a specific operation in a manufacturing unit, which is then saved in a cloud server and shown on virtual and augmented reality applications [10]. Employers can watch the manufacturing unit remotely and oversee the entire operation with fewer personnel, thanks to the convergence of IoT with AR and VR.

8.2 How do we rectify the skill gaps?

Teaching Industry 4.0 technologies to students in their college years will help reducing the skill gap and produce tech savvy professionals who can work on cutting-edge technologies. This allows the country to have a skilled manpower to take on the Industry 4.0 revolution. Another way of reducing the skill gap is training the existing employees on latest technologies such as AR, VR, and IoT, which, in turn, helps the companies increase their productivity. If we lag behind in investing in training the students and employees on Industry 4.0 technologies, western countries will take our place in this domain.

9 Manufacturing industries of the future

With the advent of 5 G technology in the near future, manufacturing industries will become more connected and it will become easier for the companies to produce, reproduce, and even course correct themselves. Some of the trends which will shape the manufacturing industries of the future are:

9.1 Extended reality and the Metaverse

AR technologies such as AR and VR play a big role in manufacturing – from better product design to better production planning, more people on the assembly line, and more immersive training. There are merits. As the Earth expands into the Metaverse, manufacturers will have more options available.

9.2 Automation

Thanks to artificial intelligence, machines are now capable of doing an expanding variety of tasks that were previously reserved for humans [12]. As a result, machines are naturally taking on more and more production work. Automation can benefit manufacturers in a variety of ways, including improved productivity, greater precision, and lower costs [13].

9.3 Robots

One of the key enablers of automation is the use of robots [14]. It is crucial to note, however, that many robots are built to augment human effort rather than completely replace it. For example, we have robots that let assembly line workers lift bigger goods without endangering their safety. We also have "cobots," which are collaborative, intelligent robots designed to work with humans. The utilization of robots and cobots can assist manufacturers [15].

9.4 Smarter and sustainable machines

The emergence of smart, linked IoT devices is transforming not only how, but also what kinds of items are made. There are smart versions of everything these days, and the trend for smart devices shows no signs of abating. As a result, manufacturers will continue to look for new ways to provide customers with the intelligent products they demand [16].

9.5 Industrial internet of things (IIoT)

Networked devices are used in both industrial and manufacturing settings to collect data that is then used to enhance output, according to the Industrial Internet of Things (IIoT) concept. Various sensors are used to get real-time and near real-time data of the production, which are continuously monitored for optimal performance.

10 Case study

10.1 Augmented reality application for a manufacturing company

This AR application lets users know about how the machine operates in real-time. It enables the users to learn the process step by step, with voice over guidance in multiple languages, which supports users from different locations, who speak their own language as shown in Figure 4.1.

10.1.1 Advantages

- Benefit for the company is that users from multiple geographical locations get to know about the machine in detail, which helps them to achieve their tasks sooner.
- Saves development time and it is easier to make changes during design cycles
- Better time to market.

Figure 4.1: Image of AR application.

10.2 Augmented reality application for a manufacturing company

The Figure 4.2 describes the core shooter application. With this AR application, users are able to understand how the core shooter application might work in real-time; using this app, project stakeholders are able to get an immersive experience of how core shooter machinery will look like once it is built. This app allows the stakeholders to visualize and make changes during the design cycle of the development of the machinery.

10.2.1 Advantages

- Stakeholders get to know how the end product would look and how it works.
- It is easier for the development team to make any change in the design phase.
- Better time to market.

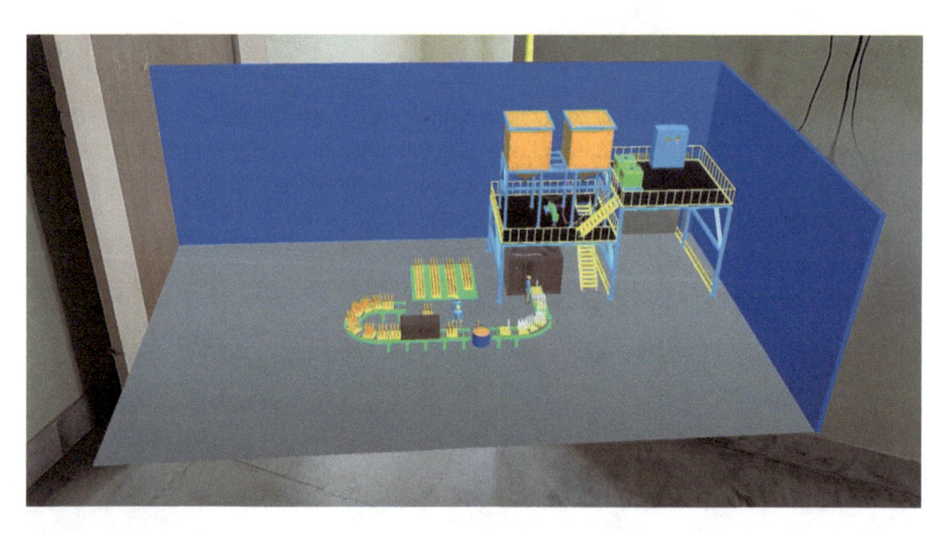

Figure 4.2: Image of core shooter AR application.

10.3 Virtual reality application for inspection

Figure 4.3 shows a VR Application. VR application for inspecting defects saves time as the employers and employees are able to inspect the defects in real-time, since they have already learnt about the inspection using VR application [17].

10.3.1 Advantages

- It allows the new employees to get trained on VR and understand how to inspect a machinery and find its defects [18].
- It saves time and money for the company, compared to training the employees manually.
- VR training allows the users to use it as many times as they want, so that they understand the process effectively.
- When the employees are given real-time tasks, they are able to perform easily, since they have already learnt the concepts using VR applications.

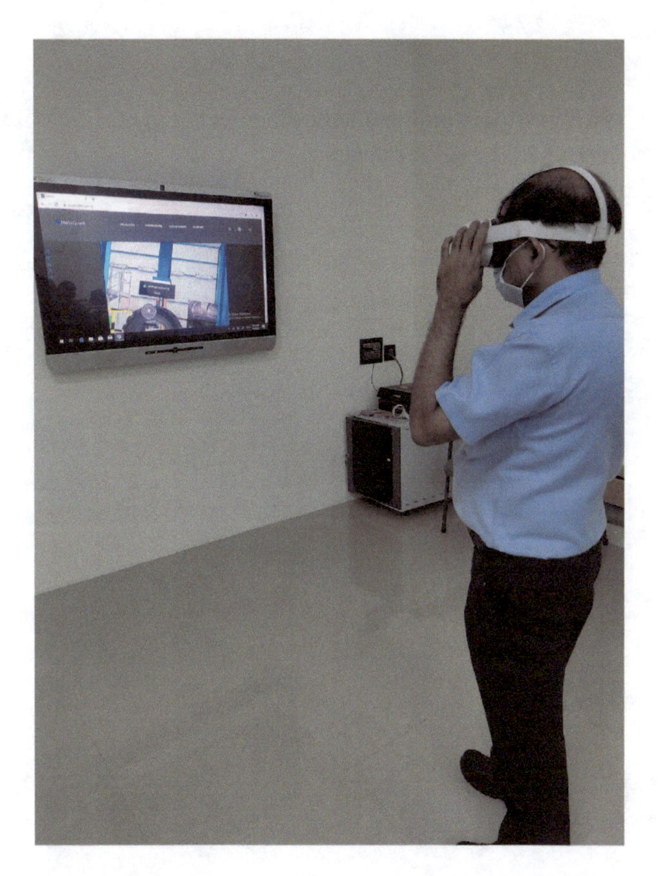

Figure 4.3: Image of user viewing VR application (user experiencing a VR application).

10.4 Virtual reality application for cement manufacturing company

VR application explains how a calciner processes the toxic chemicals and releases non-hazardous chemicals in to the atmosphere, preventing the environment from getting air polluted. This application serves as the reminder for the cement manufacturing companies to construct a calciner to protect the environment from hazardous chemicals that come out as a by-product of cement manufacturing. Users also get to feel how the calciner might look and how it operates using VR, without the need to see the calciner, physically as shown in Figure 4.4 [19].

10.4.1 Advantages

- Visualization of how the machinery functions and how the process works.
- Saves time and money compared to a real-time demo.
- It serves as a marketing tool to market a product to a potential customer.

Figure 4.4: Image of users viewing VR application (user wearing VR device to experience how the calciner works to process the chemicals).

10.5 Augmented reality application for packaging

Using this app as shown in Figure 4.5, the end customer gets to understand how the packaging is done and in what formats packaging is done; the user also gets to know how the packaging machine would look like in their place, how much space it might require, and how packaging is done [20].

10.5.1 Advantages

- Better understanding of how the machinery works
- Understanding of different parameters in the machine
- Easy to market and saves time and money compared to a real-time demo

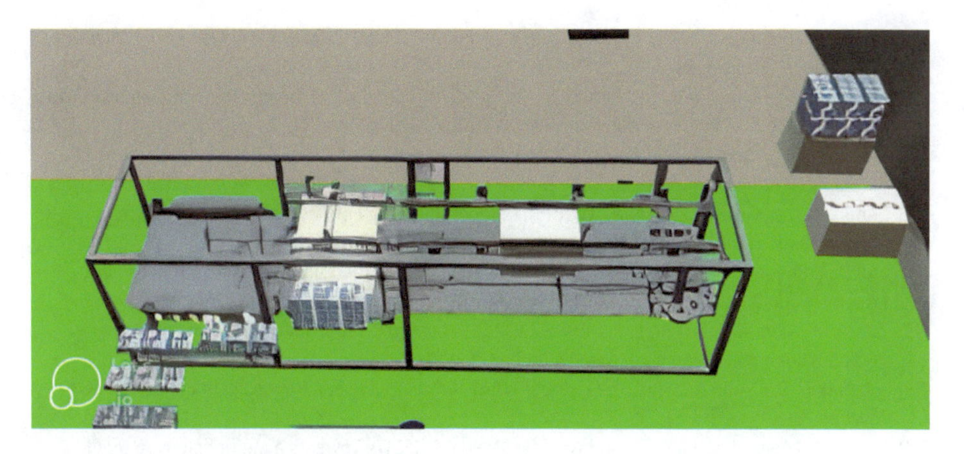

Figure 4.5: Image of the AR application.

11 Conclusion

There are a plethora of opportunities when AR, VR, and IoT are used in manufacturing. To attain their full benefits, it is high time we invest in training the students in these technologies to tackle the next industrial revolution. By having the skilled manpower, India can not only attain self-sustainability domestically, it can also guide the world on smart manufacturing as it did in the software sector.

References

[1] Jo, D., and Kim, G. J. (26 October 2016). ARIoT: Scalable Augmented Reality Framework for Interacting with Internet of Things Appliances Everywhere. *IEEE Transactions on Consumer Electronics*, 62: 334–340.

[2] Lavinga, K., and Tanwar, S. (28 November 2019). *Augmented reality and industry 4.0*. Springer, 143–155.

[3] Bures, M., Cerny, T., and Ahmed, B. S. (24 July 2018). *Internet of things: Current challenges in the quality assurance and testing methods*. Springer, vol. 514, 1–10.

[4] Momeni, K., and Martinsuo, M. (4 October 2018). Remote Monitoring in Industrial Services: Need-to-have Instead of Nice-to-have. *Emerald Insight*, 33: 792–803.

[5] Liu, M., Fang, S., Dong, H., and Xu, C. (9 July 2020). Review of Digital Twin about Concepts, Technologies, and Industrial Applications. *Science Direct*, 58: 346–361.

[6] Negri, E., Fumagalli, L., Cimino, C., and Macchi, M. (2019). FMU-Supported Simulation for CPS Digital Twin. *Sciencedirect*, 28: 201–206.

[7] Kupriyanovsky, V., Klimov, A., Voropaev, Y., Pokusaev, O., Dobrynin, A., Ponkin, I., and Lysogorsky, A. (2020). Digital twins based on the development of BIM technologies, related ontologies, 5G, IoT,

and mixed reality for use in infrastructure projects and IFRABIM. *International Journal of Open Information Technologies*, 8.

[8] You, D., Seo, B., Jeong, E., and Kim, D. H. (23 April 2018). Internet of Things (IoT) for Seamless Virtual Reality Space: Challenges and Perspectives. *IEEE*, 6: 40439–40449.

[9] Mujber, T. S., Szecsi, T., and Hashmi, M. S. J. (30 November 2004). Virtual Reality Applications in Manufacturing Process Simulation. *Science Direct*, 155–156: 1834–1838.

[10] Amin, D., and Govilkar, S. (February 2015). Comparative Study of Augmented Reality SDK's. *International Journal on Computational Science & Applications*, 5: 1–16.

[11] Jo, D., and Kim, G. J. (7 October 2019). AR Enabled IoT for a Smart and Interactive Environment: A Survey and Future Directions. *MDPI*, 19: 1–17.

[12] Dahl, M., Albo, A., Eriksson, J., Pettersson, J., and Falkman, P. (08 January 2018). Virtual Reality Commissioning in Production Systems Preparation. *IEEE*.

[13] Hamid, N. S. S., and Aziz, F. A. (09 October 2014). Virtual reality applications in manufacturing system. *IEEE*.

[14] Calvo, I., López, F., Zulueta, E., and González-Nalda, P. (2017). Towards a Methodology to Build Virtual Reality Manufacturing Systems based on Free Open Software Technologies. *Springer*, 11: 69–580.

[15] Malik, A. A., Masood, T., and Bilberg, A. (26 November 2019). Virtual Reality in Manufacturing: Immersive and Collaborative Artificial-reality in Design of Human-robot Workspace. *International Journal of Computer Integrated Manufacturing*, 33: 22–37.

[16] Howard, M. C. (15 May 2018). Virtual Reality Interventions for Personal Development: A Meta-analysis of Hardware and Software. *Taylor and Francis Online*, 34: 205–239.

[17] Rekimoto, J. (August 1997). Navicam: A magnifying Glass Approach to Augmented Reality. *MIT Press Direct*, 6: 399–412.

[18] Salah, B., Abidi, M. H., Mian, S. H., Krid, M., Alkhalefah, H., and Abdo, A. (11 March 2019). Virtual Reality-based Engineering Education to Enhance Manufacturing Sustainability in Industry 4.0. *MDPI*, 11.

[19] Park, Y., Yun, S., and Kim, K.-H. (June 2019). When IoT met Augmented Reality: Visualizing the Source of the Wireless Signal in AR View. *ACM Digital Library*, 8: 117–129.

[20] Bucsai, S., Kučera, E., Haffner, O., and Drahoš, P. (2020). Control and Monitoring of IoT Devices using Mixed Reality Developed by Unity Engine. *IEEE*, 7: 53–57.

Deepa B. G., Senthil S., Zaiba S., Danish D. S., Kshitiz Tripathi, Alessandro Bruno

5 Proficiency of Metaverse using virtual reality for industry and users perspective

Abstract: Amid the various technologies being used such as AI, Cloud, IoT etc. widespread, the world of the web has seen a gigantic climb within the advancing and digitizing world. Digitalization of every resource and object has fundamentally become a part of our lives. It has helped to make use of resources in an efficient manner, which provides opportunities in various activities. Metaverse is one such technology that has introduced the digital world to our everyday lives. This technology has involuntarily provided an immersive platform to connect people with technology. The paradigm of the digital world is just beginning to take shape and reality as a Metaverse.

It is still emerging, and many key components have started to take shape; it is now a major part of almost every field. MNCs such as Meta (formerly Facebook Inc.), Google, and Electronic Arts have invested huge amounts to set up dedicated infrastructure to improve the existing technology and make it ubiquitous. Several Metaverse worlds have launched their own digital currencies running on Blockchain technology to provide highly secure and seamless trading of digital real estates, avatars, and many more objects. In addition to the subject, it is seen that Immersive Reality (Virtual Reality and Augmented Reality) technologies, such as in the education sector and digital-marketing fields, show interest in this field of Metaverse. When combined with Artificial Intelligence, Metaverse provides a real feel to simulations of aircraft, medical surgeries, and war scenarios to train armed forces.

This chapter presents an effort to showcase the ecosystem and the elements that Metaverse consists of. It examines the capabilities and application of the existing virtual worlds. Every tech giant and field has begun to implement Virtual Reality, coming

Deepa B. G., School of Computer Science and Applications, REVA University, Bangalore, India, e-mail: deepabg03@gmail.com
Senthil S., School of Computer Science and Applications, REVA University, Bangalore, India, e-mail: senthil_udt@rediffmail.com
Zaiba S., School of Computer Science and Applications, REVA University, Bangalore, India, e-mail: zaibaunique786@gmail.com
Danish D. S., School of Computer Science and Applications, REVA University, Bangalore, India, e-mail: danishdeepasoman02@gmail.com
Kshitiz Tripathi, School of Computer Science and Applications, REVA University, Bangalore, India, e-mail: kshitiz.vns94@gmail.com
Alessandro Bruno, Department of Biomedical Sciences, Humanitas University, Via Rita Levi Montalcini, PieveEmanuele (Milan), Italy, e-mail: alessandro.bruno@hunimed.eu

https://doi.org/10.1515/9783110790146-005

up with positive innovative results. Therefore, we discuss the economical and industrial impact, as also the negative aspects of implementing Metaverse.

Keywords: Augmented Reality, Virtual reality, Metaverse, Digitalization, Paradigm, Unity Engine, Technology, Digital-Marketing, Artificial Intelligence, Head-tracking device, Blockchain, Real Estate, Virtual gathering, Simulation

1 Introduction

Marty Resnick, the Research Director at Gartner, suggests that "the Metaverse will encourage people to mimic or improve their physical activities. This can be achieved by transferring or augmenting real-world activities into virtual ones, or by altering the real world."

On October 28, 2021, CEO Mark Zuckerberg publicly announced that his company Facebook had rebranded itself as Meta Platforms, with its ambition to enter "Metaverse" – a technology that can potentially change the way the Internet is being used. The Metaverse word defines itself as "man-made environment augmented to the real world." The company is shifting toward building a new ecosystem to enhance social media experience by implementing virtual reality and augmented reality capabilities.

The Metaverse is believed to be a realistic civilization with more immediate and tangible interactions. Simply put, the Metaverse is a shared virtual environment that is formed by blending digital reality and physically augmented reality. It provides a more enhanced immersive experience and is physically accurate.

Most industries have taken a huge step to adapt to the Metaverse. The Metaverse's main feature is that there isn't one. "isnt one" refers to only Metaverse being the one technology.

There are several virtual worlds that are intended to enhance online social connections. The current activities that occur in isolated environments will start to take part within one Metaverse. One major question though is why Metaverse now? Technology has evolved in such a way that almost all companies have adapted to this change, and almost all the new technologies, especially Augmented reality (AR) and Virtual reality (VR), have been adapted and used to create a better user experience – a number of technologies and trends are required for operation. Internet of Things (IoT), 5G, artificial intelligence (AI), and spatial technologies are some of the contributing technologies.

Our study presents the current phase, development techniques, and challenges of Metaverse. Using the Unity game engine, we developed a mini-interactive model of a museum that defines the qualities of Metaverse. The theme of the museum is inspired by "Greek Civilization," containing related artifacts, monuments, and paintings. One can interact with objects by using a real-time head tracking device. The application allows multi-users to experience virtual reality at the same time through VR headsets and a smartphone.

2 Description of existing techniques for Metaverse

Professor ChunxueBai (2022) [1] has described the Medical Internet of Things, often known as a Metaverse in medicine (MIoT). By solely communicating with virtual and actual cloud professionals, as well as terminal doctors, we learnt how they found it useful. They had the skills necessary to conduct clinical studies, diagnose illnesses, and more. They thought it was noteworthy because it is a safe platform for the medical Metaverse.

Dr. Saraswati from International Journal for Multidisciplinary Research (IJFMR) (2021) [2] has clarified about Virtual Reality (VR) and its applications, which includes education and entertainment; its advantages could be the creation of a realistic world, making education easier and more comfortable, and its disadvantages could be such as (addiction, and health and ethical issues. Simultaneously, it talks about the origin as well as the aim of Metaverse, which is to create a virtual environment where family and friends can gather, learn, play, and work together using VR headsets, AR glasses, and smartphones. According to her, some key features of Metaverse are the Avatars, homespace, teleporting, privacy, and so on. Some companies that are involved in Metaverse are Nvidia, Tencent, Unity, and Snap Inc.

Tomasz Mazuryk and Michael Gervautz (2014) from the Institute of Computer Graphics – Vienna University of Technology, Austria [3]. They gave brief information about the history of VR. One of them is Sensorama, a multi-sensory simulator, which was created by Morton Heilig in the year 1960–62; it had pre-recorded film in color and stereo, and was augmented by binaural sound, scent, wind, and vibration experiences. The Sword of Damocles, created by Ivan Sutherland, was referred to be the first Head Mounted Display (HMD). Virtual wind tunnels were used by NASA to study airflow, and in chemistry, to investigate specific molecules. Additionally, it is employed by advanced flight simulators (training).

Rabindra Ratan, Associate Professor at Michigan State University, and Yiming Lei, a student at the Michigan State University (2021) [4]. According to them, Metaverse is a network of (VE) virtual environments where many people communicate with each other while operating virtual representations. They consider some core elements that consist of standardization, interoperability, and presence. They also say the Metaverse may become successful on the internet and can provide the economy or society with potential benefits. Meta is severely investing in virtual reality. Meta's CEO Mark Zuckerberg says in an interview that they use platforms such as social media as well as 3D immersive technologies, such as virtual reality, to work and play.

StylianosMystakidis from the University of Patras (2022) [5]. This paper describes Metaverse, Augmented Reality, Virtual Reality, Mixed Reality, and Immersive Reality. Metaverse was developed as 3D Internet or Web 3.0. He tells about the limitations of 2D learning and how, because of the pandemic, all had to shift to online learning via web conferencing platforms such as Zoom, Microsoft Teams, and WebEx. Further, he tells about the affordability of VR headsets and the Metaverse.

Anton Nijholt from University of Twente (2017) [6] has made a comparison between human conduct and that of video game avatars. Avatars are digital representations of human participants in video games. In many ways, controlling an avatar in a virtual world also involves influencing how others behave. Players control the avatars, but the AI of the game also has power over them, and the game's narrative, interaction options, and (virtual) physics all place restrictions on what they may do. Environments are dynamic, and the best example of changing environments is Pokemon Go!, which is a game developed with Augmented Reality. Additionally, they have made a distinction between how different games emphasize role-playing, teamwork, environment exploration, puzzle solving, and surviving, in scenarios that mimic real-world chores rather than competition or winning. In short, they have drawn comparisons between video games and smart ecosystems, such as smart cities.

Hannes Kaufmann from the Institute of Software Technology and Interactive Systems (2003) [7]. Construct3D is a three-dimensional geometry construction tool that has been released, and is intended primarily for math and geometry instruction. This is based on the "Studierstube" mobile collaborative augmented reality technology. They have employed head-mounted displays (HMDs) to peer through the computer-generated visuals onto the real world, allowing for organic user interaction. They claim that learning mathematics and geometry in high school and college will benefit from using this program. By using this software, student can see three dimensional objects. Some functions include cubes, spheres, points and lines, and further includes Boolean operations that are based on the OpenCascade tool.

Thippa Reddy Gadekallu, Senior Member, IEEE, Thien Huynh-The, Member, IEEE, Weizheng Wang, GokulYenduri, Pasika Ranaweera, Member, IEEE, Quoc-Viet Pham, Member, IEEE, Daniel Benevides da Costa, Senior Member, IEEE, and Madhusanka-Liyanage, Senior Member, IEEE (2022) [8]. The Metaverse's goal is to provide users with 3D imaginative and personalized experiences by enhancing a variety of relevant technologies. Blockchain is a solution with several unique characteristics like immutability, decentralization, and accessibility that the Metaverse has always needed to secure its users' digital content and data. In this hypothesis, they have brought some applications of blockchain to Metaverse, which includes financial systems, Smart contract deployment, and NFTs. They discuss some technical perspectives of blockchain in the Metaverse (data collection, storage, sharing, interoperability, and privacy). Additionally, they explored holographic telepresence in the Metaverse, multi-sensory XR applications, and how AI may aid in encrypting the data existing in blockchain.

Matthias C. Rillig (2022) [9] says that the Metaverse is tremendously booming, which has broad impacts on our social, technological, and commercial consequences. Here, they have discussed the opportunities and risks of the Metaverse for biodiversity and environment. They claim that the development of the internet is leading to greater coherence between the physical and virtual worlds. Interacting with aspects of biodiversity would be difficult here if there was no Metaverse. If online immersive experiences can replace some visits to sensitive sites, it may be possible to protect certain biodiversity hotspots. It aids the process of environmental science and discusses global environmental changes, such as climate change, among other things. Servers powering this alternative "universe" of experiences will have enormous energy demands, which, unless met by alternative sources, will be associated with increased greenhouse emissions, contributing to climate change and the subsequent effects on ecosystems and biodiversity. Environmental and biodiversity scientists should feel at ease as well. The community of environmental and biodiversity scientists should also feel at ease developing the topic and content for use in the Metaverse. This is a fantastic opportunity for the next generation of scientists.

3 What is Metaverse?

The Metaverse can be called a virtual universe where multi users can meet and experience real-world actions such as communication, gaming, exploration, and learning. It is an interconnected web of social networking and web communities that blends multisensory interaction within Virtual Worlds. With the support of Web 3.0, it enables multiplayers to have seamless experience of interaction in real time. Some features of Metaverse are virtual worlds with massive multiplayer and AR support [10].

According to some, the Metaverse is a realistic civilization with greater in-person interactions. Simply put, the Metaverse is a collaborative virtual environment that is generated by mixing digital reality and physically augmented reality. It offers a more immersive experience and is physically stable. Most industries have taken huge steps to adapt to the Metaverse. The key point about the Metaverse is that there are many worlds that deepen social interactions digitally. A major question though is why Metaverse now? Technology has evolved in such a way that almost all companies have adapted to this change, and almost all new technologies, especially Mixed Reality (MR), have been adapted and have been used to create a better user experience [11].

3.1 Ecosystem of the Metaverse

3.1.1 Experience

The Metaverse is indeed stretching the frontiers of our digital capabilities as the physical and digital worlds collide. In the virtual reality of the Metaverse, everything from human identity, personality, and reputation, to assets, emotions, and history may be interacted with, manipulated, and experienced in whole new ways. As a result, the end-user experience is pure magic! As the Metaverse approaches general use, it is attracting the interest of investors from all around the world, with apparent potential for early adopters. This layer talks about human experience in games like Call of Duty and Battlefield.

The Metaverse is often pictured as a 3D environment that surrounds us. The Metaverse, however, is neither two-dimensional nor three-dimensional. The constant dematerialization of actual location, distances, and materials is the subject of this essay. Alexa in the kitchen and Zoom meetings are also included.

3.1.2 Discovery

This layer discusses the "push and pull" of information that results in experiencing discoveries. This "push and pull" of information introduces consumers to new experiences. While "pull" refers to an inbound system in which users actively seek out information and experiences; "push" refers to processes that notify users about the Metaverse experiences that await them. It introduces people to discover more and more about the technology (Metaverse) on various platforms. Here, people are making friends from all around the world, and are interacting with friends through shared experiences. Discord offers a presence-detection SDK that functions in a variety of gaming scenarios, and boosts in-game "social engagement" in real time.

3.1.3 Creator economy

Creators of content will be crucial in shaping this new universe. They've had a lot of success on social media platforms and will continue to be a big growth engine in the Metaverse's virtual reality. Experts predict that the creator economy will become a multibillion-dollar industry, thanks to the Metaverse. The self-employed creators create digital content like photographs and videos, as well as digital commodities including e-books, webinars, art, blog posts, and more. It contains all the tools and tech that creators use daily to create content and experiences to be consumed by people.

3.1.4 Spatial computing

Spatial computing brings the concept of the "Metaverse" to life by combining AR, VR, and MR. The idea of a parallel, three-dimensional virtual cosmos that interacts with the actual world and never shuts down can be realized with spatial computing. A game that leverages spatial computing, for example, will allow you to play it against the backdrop of your local real-world environment. Spatial computing consists of virtual computation that breaks the barrier between the physical and the ideal worlds, for example, 3D engines (Unity), voice and gesture recognition, and IoT.

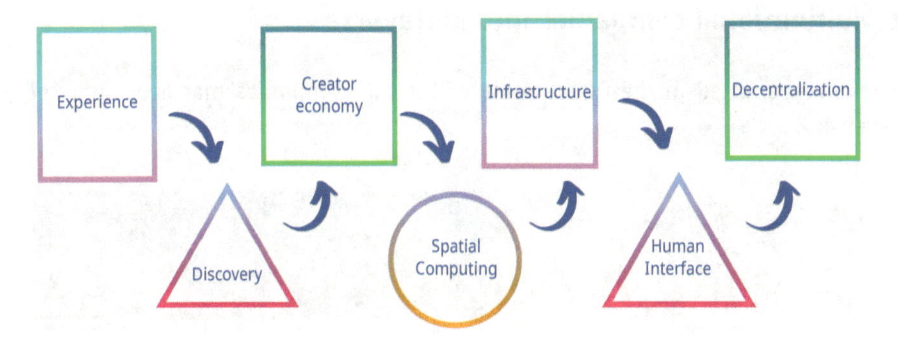

Figure 5.1: Representation of technologies that come together to build the Metaverse ecosystem.

3.1.5 Decentralization

Blockchain is a technology that enables value exchange between software, self-sovereign identity, and new means of diversifying and bundling content and currencies; it is the most basic example of decentralization.

3.1.6 Human interface devices

Human interface devices (HID) is a part computer device through which users input and direct the computer. It is generally portable, plug and play, or always-connected, and ready-to-use device. It can also be found preinstalled on mobile devices. With more time and innovations, these HIDs have become more powerful and accurate. With the right sensors, the distance between input between IoT and Edge Computing has been reduced to a large extent. They learn ever-more Metaverse experiences and applications. HID has mastered an increasing number of applications and abilities in Metaverse. The Oculus Quest is simply a smart gadget that has been reconfigured into a VR device. It has an all-in-one VR system. With a powerful processor with the highest resolution display, it allows unrestricted experience and provides us a sense of where things are going in the future.

3.1.7 Infrastructure

The tools and resources that provide our devices the access to network and content are part of the infrastructure layer. For instance: Bluetooth lets HID connect with VR Headsets and 5G dramatically improves bandwidth and speed by reducing network contention and latency [12].

3.2 Multinational companies into Metaverse

We have represented in the figure some of the top companies that are a part of Metaverse:

Figure 5.2: Representation of popular games and applications used on phones and PCs. From Left: Second Life, Memoji (an AR-based avatar game by Apple Inc.), Roblox, a cross-platform 3D/VR game with over 32.6 million active players.

3.2.1 Meta (formerly Facebook)

Facebook was the first company to adapt to the Metaverse, and had later on renamed itself as Meta Platform. The major focus of the company is to adapt to an interactive virtual world or workspaces. The tech giant has recently launched a VR meeting app called Horizon workrooms. In this app, a person can sit inside a horizon lab by wearing a VR headset and behave as he/ she does in a conference room.

3.2.2 Apple

Apple has not been involved in the race of Metaverse, but it has currently developed a cutting-edge HMD Virtual Meeting application for the Metaverse. Although the company is not in the race for the adaptation of the Metaverse, it has added some equipment with the AR kit, a program that enables the creation of room mapping, as also localization apps for the iPhone using its sensors.

3.2.3 Google

It is not a surprise that Google is on the list, as Metaverse is the next big thing. Google is not new to the field of augmented reality. They have developed various technologies in this field. Sundar Pichai has shared few words during an interview about the virtual world that it is an "evolving Augmented reality that enables immersive computing." The company is planning to integrate popular applications such as Google Maps and YouTube [13].

3.2.4 Roblox

Roblox, a gaming company, has focused on building Metaverse spaces, which provide players with a virtual home experience, and engages them in various tasks. Roblox's focus for the players is not just for gaming, but also to enable them seek adventure, job, avatars, chat with friends, etc. Roblox boasts of organizing world-famous Metaverse events such as the Lil Nas X Concert [13].

3.2.5 Microsoft

Microsoft's approach towards the Metaverse, a technology that enables software developers to create apps that link many devices to the same digital world, aims to create a Metaverse that virtually unites individuals. Participants can utilize avatars to interact in a virtual environment, thanks to the integration of this mesh into Microsoft Teams [13].

3.3 Application of Metaverse

Here, some applications of Metaverse are unveiled. It tells how Metaverse is contributing to various fields, such as gaming, entertainment, marketing, and so on.

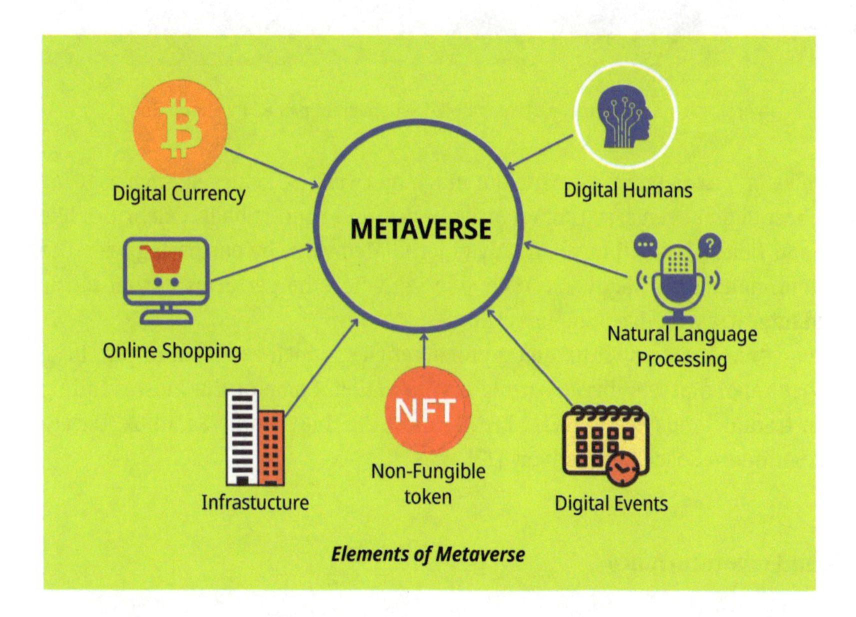

Figure 5.3: Portrayal of the elements of a Metaverse, which narrates how the Metaverse enhances multiple abilities, such as trading NFTs and exchanging digital currencies.

3.3.1 Metaverse land

In the era where we are shifting toward Web 3.0, Metaverse has come up with new means of acquiring assets and properties. Metaverse has made the idea of trading online 3D virtual worlds that are controlled by users of the physical world. In this shared world, one can build avatars, buy land, and socialize with other players [14]. This was first made possible by PlayMining Metaverse, a Singapore-based digital entertainment company [15]. They developed **NFT** (non-fungible token), a digital asset that consists of metadata of the Metaverse land, such as coordinates, records, and owner details. Once acquired, the owner can build, decorate, and monetize the land by making transactions in digital currencies.

Figure 5.4: The image represents the virtual world, named, "Snoopverse" owned by Snoop Dogg [17].

Well, it doesn't stop just as an investment, but much more can be achieved, such as using the space to host digital events, setting up art galleries, and concert halls. Celebrities like Snoop Dogg and Daler Mehendi have already entered Metaverse by partnering with The Sandbox. Snoop named his Metaverse "Snoopverse" where he performs virtual music concerts, and interacts with live audiences [16].

Each Metaverse has its own token/cryptocurrencies, which are used to purchase land and other related assets directly using crypto wallet. Some of the most trending coins used in trading lands are ESTATE, LAND, and CUBES owned by Sandbox, Decentraland, and Somnium Space, respectively [17].

3.3.2 NFTs and cryptocurrency

Non-fungible tokens are unique pieces of artwork that can be produced, published, traded, and transferred using a digital wallet on a desired marketplace. NFT has made possession of digital assets, such as games, virtual lands, avatars, music, and paintings.

NFTs can be generated from any real-world object and sold as digital artwork. One of the most trending NFTs ever sold is the first tweet of Jack Dorsey, former CEO of Twitter [18].

Cryptocurrencies can be defined as money in digital form. Each unit of Crypto is referred to as crypto coins. It can be used for tasks as simple as sending/receiving money. Most commonly, it is used for trading NFTs, buying stocks, and investing in Metaverse.

Table 5.1: Trending cryptocurrency in Metaverse.

Companies	Cryptocurrency	Purposes	Capitalization (in US dollars)
The Ape Foundation	ApeCoin	To get exclusive access to event merch, events, and services.	$49,098,54
Decentraland	MANA	To buy lands and avatars in Metaverse.	$2.149 billion
The Sandbox	SAND	E-sports, trading lands, virtual events like concerts and premiers.	$303,827652.67
Facebook	libra	Decentralized crypto-based payment platform.	$1,168,882
Sky Mavis		Mobile games, to trade and battle with NFTs called Pets.	$302,240,664

3.3.3 Medical science

Various organizations and hospitals already use VR, AR, MR, and AI to train medical staff and doctors by simulating real-world procedures and details about the human body. Majorly, it is used for diagnosis, and pre-surgical training and preparation. Secondly, it is used to train medical students, surgeons, nurses, and so on.

Metaverse provides a 360° view of ailments in the body, which is like real-world procedures. It provides security to patient's data so that people don't have to worry about

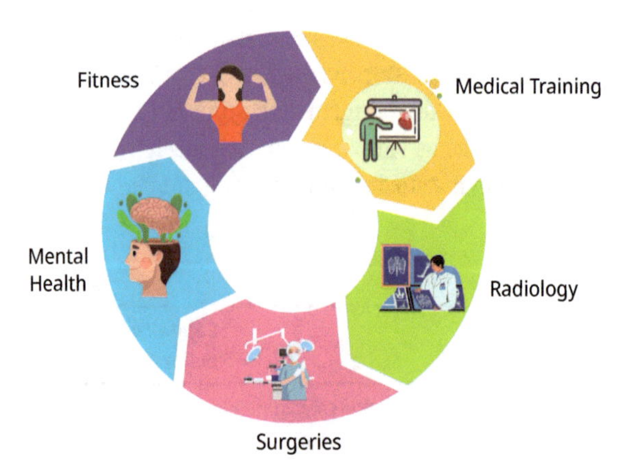

Figure 5.5: The application of Metaverse in the medical or healthcare segment.

security breaches anymore. Metaverse is gaining vast acceptance in the healthcare industry as it needs high-tech hardware, such as gloves, VR headsets (glasses), sensors, and so on; these things can impact people to not experience the immersive space, which may cost more.

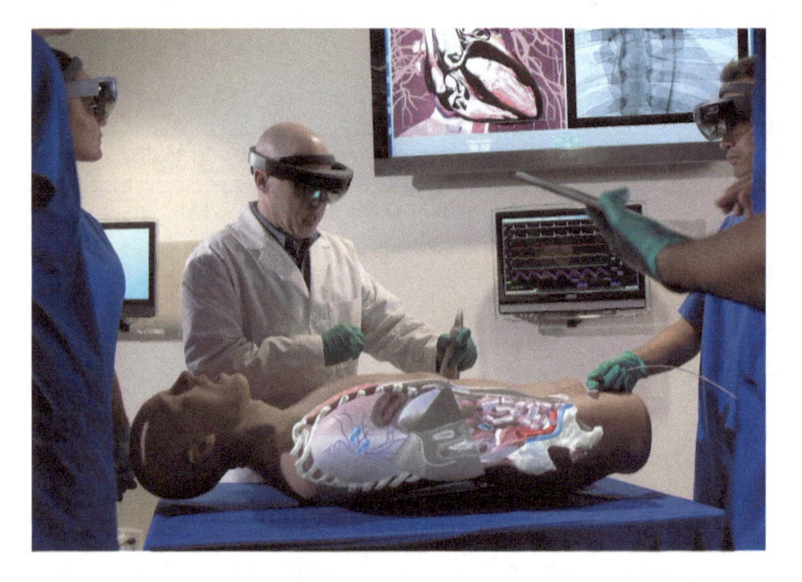

Figure 5.6: Healthcare teams using Microsoft HoloLens to perform virtual cardiac surgery [19].

Artificial Intelligence with Mixed Reality to enhance medicine [20]. MR has now made it possible to simulate and learn realistic anatomy and other essential processes of the body. Through this, they can train before performing actual surgeries [21].

3.3.4 Tourism and hospitality

During the pandemic, we have becoming increasingly aware about climate change, which influences individuals while making decisions regarding destinations and hotels. This aids in educating public about climate change, shifting consumer demands, and other outside variables that are likely to encourage hospitality businesses to practice sustainability by travelling to local rather than to foreign destinations. Customers' attitudes may change as a result, and they may be inspired to partake in certain hospitality and tourist activities like attending events, gatherings, museums, and other things [22], without having to spend any extra time or money; these can be made available in the Metaverse. These potential shifts in consumer behavior underline the significance of industry efforts to create and provide authentic hospitality and travel experiences in the Metaverse, particularly as the required technology develops and gains wider user acceptance. Metaverse applications are projected to become more important to the sector as customers and technologies advance, improving marketing, customer connections, communication, customer decision-making processes, and visitor experiences. Thus, Metaverse may also ruin experiences in future industries.

3.3.5 Metaverse in military

Immersive reality is one's viewpoint of being virtually present in an imitation of the real world. This environment is produced by surrounding the user with Virtual Reality systems, sound effects, and other peripheral devices like motion controllers and haptic vests. Immersive technologies have also made it possible to hold, crash, and interact with objects using motion controllers and gestures. The demand for Immersive Reality is so high in industries like First-Person gaming, Virtual Parks like Jurassic Worlds, and for simulating war-like situations to train armed soldiers [23]. It represents the next step of reality, which creates an immersive virtual reality experience in a specific surrounding to, in which camera angles are toward three or more displays using gyroscope sensors to look around. In this case, a smartphone or PC device connected to a special 3D goggle allows multi-sensorial effects.

3.3.6 Shopping marketing and analysis

Metaverse helps in growing various industries in an advanced way. There are reports that provide some details or analysis of several leading market vendors such as Roblox Corp, Epic Games Inc., Niantic Inc., and Tetavi Ltd. in entertainment in Metaverse. These kinds of reports and analyses present studies from various sources, which focus on parameters like profit, promotions, pricing, etc. and present to us how the Metaverse is taking the lead or drastically dominating the market [24].

Figure 5.7: Illustration of virtual shopping on Walmart Metaverse [25].

3.3.7 Entertainment industry

The Metaverse is adding three core functionalities to entertainment businesses, such as immersive storytelling, world-building, and fan's unlocking their creativity. Here, it has interoperable virtual spaces where it is easy for the users to choose between various digital entertainment events, venues, and services of their preferences. Therefore, these virtual events and digital media are promoting the Metaverse, a continuous, unified virtual realm where users can freely create, interact, and explore with varied environments and with other users (participants), with the help of internet connection. Users get an immersive experience by a combination of elements of social media, online games, video communication tools, and so on [25].

3.3.8 Metaverse in real estate

The real estate business in the Metaverse is already becoming a huge business; many companies have already purchased virtual land that they want to develop for a number of purposes. Those who have already bought land in the early stages have seen significant gains. The average price of a parcel of land available to buy on decentral land or the sandbox – these are two of the largest Metaverse platforms – was less than $1000 a year ago, and now its current worth is roughly around $13,000. There are two main reasons to buy land in the Metaverse – one, when the user wants to build something (a house to live and premises from which to do business) and two, for investment purposes [26].

3.4 Pros and cons of Metaverse

While we can say that the Metaverse is one of the coolest things, in that it gives the user an endless experience in the virtual world where they can interact with the objects using the avatar, it is said that the Metaverse is likely to offer that mysterious aura of an alien world, which makes everything more enjoyable and more interesting, and improves work from home and homeschooling. While we talk about the brighter side, we also have the darker side. The major disadvantage can be said to be that it separates people from reality, and makes the agent or user lose track of time.

Imagine a world where humans are dependent on the virtual world for completion of most of the tasks; they will soon become lazy and drowsy. Over time, their physical strength will also reduce. Unless one is physically challenged and incapable of functioning, Metaverse is nothing less than fortune.

Advantages

A. Geographic factor
The geographical factor can be one of the leading advantages of Metaverse as it removes all the geographic barriers that are irrelevant. In the Metaverse world, geography does not matter anymore as it acts as a neutral space where everyone can meet as equal to one other, since it offers the user with a platform to be comfortable where it is easy to share interests and ideas.

B. Improvement in education
During the time of Covid, education was a major hindrance to students; this was when online learning became a normal form, and students could continue their education through Zoom or Teams. In the same way, Metaverse offers real-time support for on educational environment, and people can share their information and study together.

C. Improvement in gaming sector
Gaminggaming has played a major role in AR and VR; it is now is riding piggyback on the development of the Metaverse and is also implementing games into Metaverse. One of the leaders is EPIC, which has invested almost $1 billion into Metaverse in 2021.

D. Office environment from home
During the Metaverse meet, Mark Zuckerberg introduced the concept called Infinite office, which is to make working from home the goal – offer convenience to users, which enables the user to work efficiently and achieve overall productivity.

Disadvantages

A. Privacy issues
We can fully immerse ourselves in the Metaverse, which is the next iteration of the internet, thanks to AR and VR technologies. With all of this digitization, privacy is an issue. We must consider privacy concerns when using the internet. In the Metaverse, the technology that monitors our online activity will also be in use, and it will probably become even more intrusive, intense, and powerful.

B. Health concern
People also experience post-VR sadness. As soon as we have gained experience in the amazing immerse world and when we come back to the real world, it can make us depressed. Internet and online gaming are already a big problem for kids, adults, and the addicted, and with the Metaverse, it could even be worse [27].

C. Metaverse laws
Metaverse will bring regulatory challenges and will call for the introduction of new laws. For example, you are wearing a haptic suit, and someone in the virtual world touches you without your permission. We will need to face these regulatory challenges as technology advances and thorny legal issues emerge.

D. Child safety
The Metaverse will make it even more difficult for parents to monitor what their children are doing online, which is already difficult. Because we can't see what our children are seeing via their VR headsets and there is no mechanism in place to monitor their screens on tablets or phones, it will be even more difficult for us to understand what they are doing in the Metaverse.

E. Privacy and implication issues
Today's digital solutions are widely used, yet many of them pose privacy and security issues. Digital solutions' ability to acquire user data is their main point of contention. Such data can be used for intrusive web marketing and identity theft. Additionally, companies have not been able to entirely resolve these problems. As a result, security and privacy issues could be drawbacks of the Metaverse. As an online environment, the Metaverse may raise new security and privacy issues for both people and organizations.

4 What is VR?

Virtual Reality (VR) is a scene created by a computer with life-like objects and visuals that gives the viewer the impression that they are completely immersed in their surroundings. A virtual reality headset is used to view this world. VR is mainly used to give the user an immersive experience – immerse themselves into video games as if they were a character or avatar. It was not just in the gaming industry, but also in the educational sector where they could do heart operations and enhance the standard of athletic training to increase performance. Virtual reality (VR) seems quite futuristic, yet it's not what we might think. Applications outside of leisure, tourism, marketing, and education are in high demand in the market. To prevent flaws, virtual interfaces must also be upgraded. Major technology firms have already made progress toward virtual reality [28].

Virtual reality (VR) is a computer-generated experience that immerses the viewer in their environment by using realistic, third-dimensional scenes and objects. A VR headset, often known as a head-mounted device, is used to create this experience. We can also use compatible motion controllers and haptic devices to be able to interact around that world. Virtual reality technology can be conceptualized and researched in a variety of ways. On the other hand, industries characterize VR as a type of immersive simulation that exposes our senses to an explorable, computer-generated virtual environment. Ivan Sutherland and Myron Krueger, among the first wave of VR researchers, described VR as an application of technological equipment systems. Sutherland, the inventor of the first computer-connected head-mounted display, was laser-focused [29].

4.1 Types of VR?

A. Desktop Based VR

Due to its 3D compatibility on a standard computer screen, desktop-based VR can be observed in old-school computer games. In VR games that are played on a PC, characters can easily change the environment [30].

B. Avatar Based VR

These virtual reality gadgets with avatar-based user interfaces are easily accessible online or for purchase. These can be created by using a responsive model and interacting with the surroundings through Avatar. Second Life is an appropriate example [30].

C. Simulation-Based VR

This form of VR is used in training environments like aviation schools and pilot virtual reality training to control a real-world incident in an existing device [30].

D. Projector-Based VR

Projector-based VR is when a user is completely submerged in a virtual world via massive volume displays that project the virtual setting. Soarin' Over, one of California Adventure's virtual rides, takes passengers through an actual environment using projection technology [30].

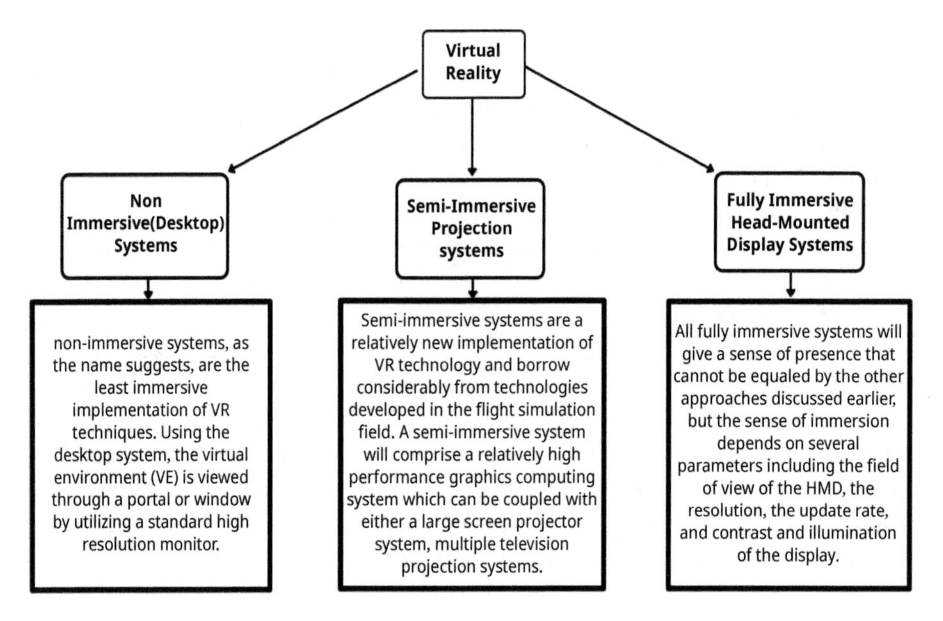

Figure 5.8: Different forms of VR and their distinguishing features.

4.2 Importance of virtual reality in the economy

VR is one of the fastest-growing markets that uses blockchain technology to maintain security, and makes an open-to-all platform, which eventually boosts its development. In 2021, Fortune Business Insight reports suggest that the estimated number of VR users would exceed a population of 500 million, making it an industry of more than USD 57.55 billion by 2027 [31], bigger than the size of the current PC gaming market. However, its use is not just limited to gaming; it extends to several areas like education, film-industry, real estate, tourism, and medical science.

Table 5.2: Distribution of virtual platform across number of users.

Virtual platforms	No. of users
Gaming/e-sports/digital events	250,000,000
Non-fungible tokens	412,578
Blockchain gaming	2,364,576
Web 3.0 Metaverse worlds	50.000
Trading of digital currencies	3,450,000
Meta (Formerly Facebook)	2.970.000.000
Global crypto cap	220,000,000

Currently, MNCs like Sony, Apple, Google, Meta (formerly Facebook), and Microsoft are researching and developing new ways to implement the technology. Video game companies like Electronic Arts and Activision, have invested over $125–$150 million to set up a whole new ecosystem for the future of the industry. Several video games franchises like *Red Dead Redemption* (Rockstar Games), *Tomb Raider* (Activision) and *Fortnite* were launched with dedicated VR modes. Within a year, Sony has sold over 2 million copies of PS VR and 12.2 million Headset games [32].

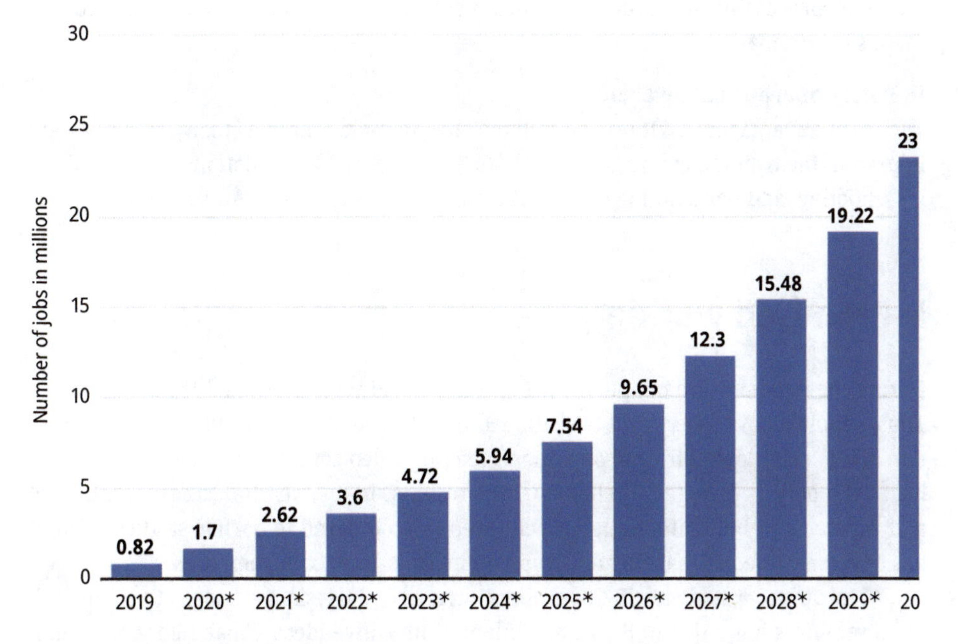

Figure 5.9: Number of jobs added worldwide by augmented reality (AR) and virtual reality (VR) from 2019 to 2030 (in millions) [33].

According to a report presented by Thomas Alsop, Virtual Reality will create more than 20 million jobs by 2030. The world has seen an effective growth in the marketing of AR/VR devices in the recent past. It is projected that it will boost the global economy by $60 billion [33].

4.3 Benefits of VR

A. Advanced training

Virtual reality creates realistic environments to train service members, for example, simulation-based VR for as aviation school and pilot virtual reality training where the person in the immersive environment can experience, from beginners' level till expert (professional) level.

B. Better operational awareness
In the army, to accurately pinpoint their location in the region on a map at the headquarters, so that the commander may make decisions based on accurate real-time information from the battlefield, commanders use virtual workbenches to coordinate the system [34].

C. Cost savings
The current state of VR makes it difficult to adapt and operate; however, improvements in processing speed and graphics display will continue to reduce the cost of these systems [34].

D. Better operational awareness
The main advantages of VR for the military are lifesaving. Earlier, many soldiers were injured in the training environment [35]. Virtual reality (VR) simulations of urban fighting, shooting, and parachuting can significantly lower the hazards for soldiers.

5 Proposed working

During the pandemic, many people around the world were restricted from socializing. Thus, people all over the world transitioned to online mode. Every industry and institution had to shut down physical operations, leaving video conferencing and online learning as the only options to be relied on. This resulted in an overall decrease in physical and mental well-being in adults and children, as compared to earlier studies. Also, it has adversely affected learning and innovation skills, impacting their jobs.

According to a study by SB Goyal, City University, Malaysia, the usage of virtual reality in seminars has aided in the development of inventive ideas – making the presentation more understandable to the audience. As a result, establishing such an interactive platform may assist us in addressing the issues that traditional virtual interactions have generated.

Our vision is to make Virtual Reality common to users of all ages. Through our application, we have contributed to the study and promotion of Metaverse. It allows the user to learn about Greek civilization through 3D models of the Greek statues, paintings, horses, and monuments. By using only a gyroscope-enabled smartphone and a VR headset, one can easily move and interact in the scene, making it compatible with most devices available in the market. It can be classified as an educational VR application that can be used by children or by educational institutions to provide “virtual tours” to their students.

Figure 5.10: A glimpse of the Greek VR museum model based on ancient Greek civilization.

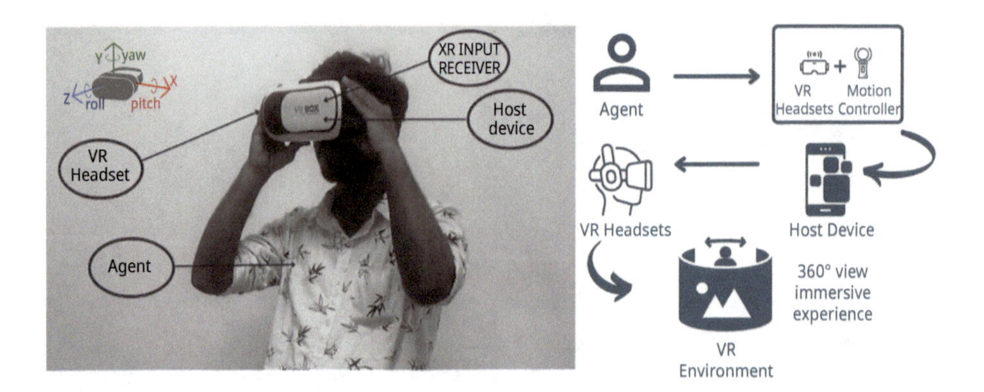

Figure 5.11: Representation of a person experiencing Metaverse through the VR system.

5.1 Methods and materials

A. Unity 3D 2020.3.30 (v)

Unity 3D is a game development software kit that provides cross-platform IDE and tools to develop 2D, 3D, and VR games for any platform. It can be integrated with Unreal Engine 5 and Vuforia Engine to develop traditional First-Person and Virtual Reality games like Counter Strike, Fall Guys, and Among Us.

B. Google Cardboard SDK

The open-source Cardboard SDK focuses on Android and iOS platforms to have an immersive cross-platform VR experience, with some important VR features such as Motion tracking, user interaction, etc. Cardboard enables a two-eyed view on the application running on the gyroscope-enabled device, making VR easy-to-use on HMD devices. Two-eyes-view makes the user run a VR application directly on his phone and attach it with the VR Headset.

C. Unity Asset Store

Unity Asset Store, owned by Unity, is an online platform where developers can publish, trade, and download thousands of free and paid assets – assets like characters, environment, audio, and many multimedia add-ons. These assets can be downloaded and imported directly to the project using the Unity account.

There are hundreds of extensions available, which makes cross-platform development possible with various controller supports.

D. XR interaction tool

The XR interaction toolkit is a unique system for interfacing that offers a framework for making 3D and UI interactions accessible via Unity input controls. This toolkit includes the following set of components:

1. XR controller input for various platforms
2. Basic object hovering movement
3. Haptic feedback like throwing, grabbing, and pushing
4. Visual feedback to highlight possible interactions and a VR camera module rig for handling static and close-scale VR experiences
5. Users need to have motion controllers to interact in VR when the XR plugin is enabled

E. Gyroscope

A gyroscope is a sensor built in mobile devices or VR headsets that works on the principles of gravity to specify the direction of the item. It comprises a disc that may freely rotate, which is placed in the middle of bigger wheels. The disk's free axis of rotation adopts a certain orientation. Any tilting or shifting of the outer wheels' orientation has no impact on the disk's orientation while it is rotating. Pitch, yaw, and roll rotational movements are tracked.

Algorithm 1: Algorithm for enabling connection between Host and VR input devices.

Input: To connect the VR headsets and get it running using the gyroscope.

Output: To get connected to the VR headsets and navigate to the environment.

Step 1: Start

Step 2: If Gyroscope is enabled

If yes,
Then install the app build
Else,
Exit

Step 3: If yes, Connect VR headsets with Motion Controllers

Step 4: Test whether the VR environment is running

If yes,
the user must navigate to the environment
exit (Step 5)
Else,
Start from Step 2

Step 5: Exit

With Unity 3D and Google Cardboard, we can create decent VR applications at an exceptionally low cost. Inspired by Metaverse, we have generated our own VR Museum model that represents artifacts, paintings, and weapons based on Greek and Egyptian civilizations. Our project is to provide an educational tour-like experience to the user where he can move, interact, and connect with other users.

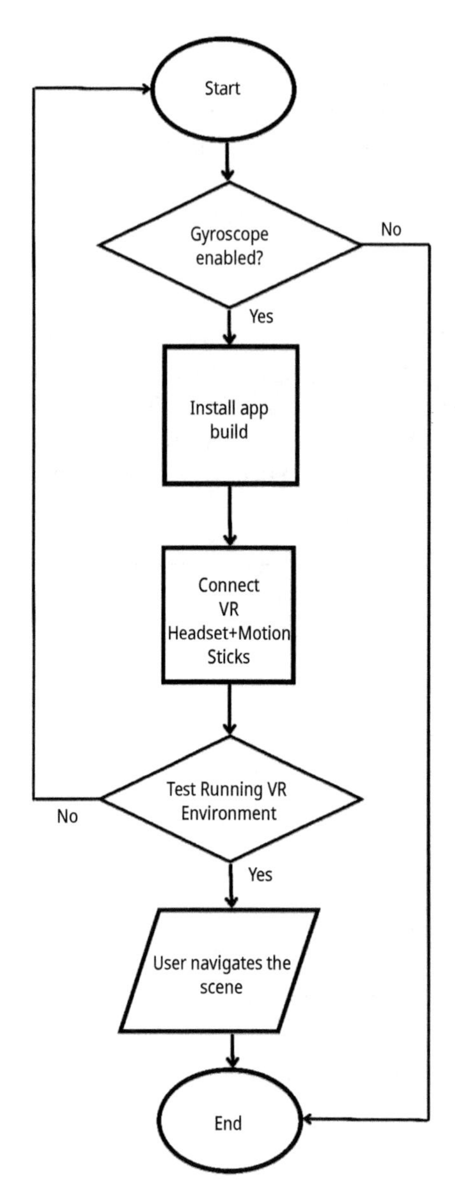

Figure 5.12: Flowchart to represent the working principle of gyroscope.

The major drawback which was also the reason behind Google's previous failed attempt to capture VR market is that not all the experiences made for VR could work on Google Cardboard – a user requires a motion controller for interaction. VR tools that company like 'Oculus' provided were expensive. A solution to this is Google's open-source VR library "Cardboard SDK," which made interaction through head tracking as the main controller of VR in devices.

To generate and process 3D models, we used Sketchfab.com. It is a website that provides advanced development tools to draw 3D models for various extensions. Some artifacts, statues, and historical monuments used in our museum were created on Sketchfab, whereas many textures and materials were imported from other sources.

To give the user a more realistic experience, we added human-like avatars to the scene. The characters could move and react. In the enhanced version of the application, users will be able to control those characters by using motion controllers or HMD, giving the user a first-person perspective view. Once the modeling is done, a test of the scene is initialized using the "game-view" in Unity. It allows a developer to access scenes from a user perspective by setting up one or more VR camera angles. Its "character controller" script allows it to use original controls, like touch and motion tracking.

Table 5.3: Challenges faced during application development.

	Challenge of application development
1	Selection of best available software development kit.
2	Configuring XR-Plugins according to available VR enabled devices.
3	Importing necessary plugins, toolkits, backend-supporting.
4	It is difficult to generate and place 3D-Models.
5	Only one platform at a time the project may be exported for.
6	No support for older versions of OS mobile and web.
7	Modifications and testing take a lot of time.
8	Maintaining the small size of the application.

6 Conclusion

We can conclude by saying that various companies and giant technology organizations have plans toward bringing in new and ambitious materialistic plans to adapt to the Metaverse. In this paper, a study of Metaverse has been made, considering the new era of socialization, where the basic aim is to make a virtual society where people can interact and get live experience in a real space. It is said to be a new interactive way to work and live in the Metaverse. As there is a drastic growth in the field of Virtual Reality, we can witness it creating a space where user interactions are made more multidimensional than with any other technology.

Rather than viewing the digital content, we will be able to immerse them into the Metaverse. Mark Zuckerberg's version of the Metaverse is where the users can take on the role as an avatar, and attend meetings using the VR headsets and motion controllers. The activities and tasks that take place in the physical world can be made possible in the Metaverse worlds too. One of the major sectors that might have a positive approach is the social gathering / events and concerts, and buying real estate. Metaverse helps build a centralized environment that will smoothen the transition to remote working, and makes idea-sharing and collaboration more successful and available than ever. It is simple to connect virtual gatherings and take part effectively in talks by picking up a near-physical encounter.

Metaverse's Virtual World has the potential to be the next big technological breakthrough. It appears to be yet another science fiction used in various fields; to name a few, we can say blockchain and cryptocurrency. Metaverse has already attracted the interest of some of the world's technological giants due to its multifaceted characteristics. Metaverse has arrived, and with it comes the opportunity for marketers to participate – a virtualized version of the Internet in which most aspects of human life are virtualized.

Be it entertainment, economy, or education, it not only brings people closer together, but it also improves their experience – create interactive content, including quizzes, chatbots, and eCommerce recommendations, as a first step to provide users with a real-time engaging experience. We are still at the point where we can readily use Metaverse for digital marketing. The Metaverse has a lot of potential to become the next "reality." However, we may already participate in Metaverse-like projects and continue to integrate blockchain into our daily lives in the interim.

References

[1] Yang, D., Zhou, J., Chen, R., Song, Y., Song, Z., Zhang, X., Wang, Q., et al. (2022). Expert consensus on the metaverse in medicine. *Clinical eHealth*, 5: 1–9.

[2] Sarswati. (2021). Metaverse: The next big technology. *International Journal For Multidisciplinary Research*, 3(6), November-December 2021. DOI 10.36948/ijfmr.2021.v03i06.001.

[3] Mazuryk, T., and Gervautz, M. (1996). History, applications, technology and future. *Virtual Reality*, 72.

[4] Kensinger, R. (2021). Metaverse. 10.13140/RG.2.2.19971.22564.

[5] Mystakidis, S. (2022). Metaverse. *Encyclopedia*, 2(1): 486–497. https://doi.org/10.3390/encyclopedia2010031.

[6] Nijholt, A. (2017). Humans as avatars in smart and playable cities. 10.1109/CW.2017.23.

[7] Kaufmann, H. (2003). Construct3D: An augmented reality application for mathematics and geometry education. Proceedings of the ACM International Multimedia Conference and Exhibition. 10.1145/641007.641140.

[8] Gadekallu, T. R., Huynh-The, T., Wang, W., Yenduri, G., Ranaweera, P., Pham, Q.-V., da Costa, D. B., and Liyanage, M. (2022). Blockchain for the metaverse: A review. *arXiv Preprint arXiv:2203.09738*.

[9] Rillig, M. C., Gould, K. A., Maeder, M., Kim, S. W., Dueñas, J. F., Pinek, L., Lehmann, A., and Bielcik, M. (2022). Opportunities and risks of the "Metaverse" for biodiversity and the environment. *Environmental Science & Technology*, 56(8): 4721–4723.

[10] Mystakidis, S. (2022). Metaverse. *Encyclopedia*, 2: 486–497. 10.3390/encyclopedia2010031.

[11] Team, C. (2021, November 8). *Metaverse – Explained in layman's terms*. ClearIAS. https://www.clearias.com/metaverse/#:%7E:text=A%20metaverse%20is%20a%20form%20of%20mixed%20reality,ways%20that%20go%20beyond%20what%20we%20can%20imagine.

[12] Radoff, J. (2021, November 24). Market map of the metaverse. Medium. Retrieved May 19, 2022, from https://medium.com/building-the-metaverse/market-map-of-the-metaverse-8ae0cde89696

[13] Traqq Team. (2023, March 1). 10 metaverse companies Shaping Virtual reality. Traqq Blog. Retrieved March 9, 2023, from https://traqq.com/blog/metaverse-companies/

[14] Scientya.com. (2022, January 4). *Why fintechs will play a key role in the metaverse – Scientya.com – The digital world publication*. Medium. https://scientya.com/why-fintechs-will-play-akey-role-in-the-metaverse-3b1daa3ed6

[15] Kamui. (n.d.). PlayMining Metaverse To Launch First Land NFTs For Creator Nations. Cryptopolitan. Retrieved March 10, 2023, from https://www.cryptopolitan.com/playmining-metaverse-to-launch-first-land-nfts-for-creator-nations/

[16] *Daler Mehndi Buys Land InMetaverse, Names it BalleBalle Land*. (2022, March 24). MSN. https://www.msn.com/en-in/news/india/daler-mehndi-buys-land-in-metaverse-names-it-balle-balle-land/ar-AAVqZZb

[17] Joseph, M. (2021, December 9). *Snoop Dogg Metaverse: What is Snoop Dogg Doing in the Metaverse and What is Snoop Dogg NFT Collection?* Stealth Optional. https://stealthoptional.com/metaverse/snoop-dogg-metaverse/#:%7E:text=As%20the%20name%20suggests%2C%20Snoop%20Dogg%E2%80%99s%20universe%20in,already%20been%20sold%20in%20the%20initial%20land%20sale.

[18] Okonkwo, I. E., and Okonkwo, I. E., NFT, Copyright; and Intellectual Property Commercialisation (May 29, 2021). *International Journal of Law and Information Technology*, 2021, eaab010, https://doi.org/10.1093/ijlit/eaab010, Available at SSRN: https://ssrn.com/abstract=3856154 or

[19] CAE Healthcare. (2017, January 27). *CAE Healthcare announces first mixed reality ultrasound simulation solution with Microsoft HoloLens*. Medgadget. https://www.medgadget.com/2017/01/cae-healthcare-announces-first-mixed-reality-ultrasound-simulation-solution-with-microsoft-hololens.html

[20] Yang, D., et al. (2022).Expert consensus on the metaverse in medicine. *Clinical eHealth*, 5: 1–9.

[21] Koo, H. (2021). Training in lung cancer surgery through the metaverse, including extended reality, in the smart operating room of Seoul National University Bundang Hospital, Korea. *Journal of Educational Evaluation for Health Professions*, 2021 Dec 31, 18.

[22] Gursoy, D., Malodia, S., and Dhir, A. (2022). The metaverse in the hospitality and tourism industry: An overview of current trends and future research directions. *Journal of Hospitality Marketing & Management*, 1–8. 10.1080/19368623.2022.2072504.

[23] Donovan, F. 2019. U.S. Military Employs AI, AR to Boost Medical Modeling, Simulation. [online] Hit Infrastructure. Available at: https://hitinfrastructure.com/news/u.s.-military-employs-ai-ar-to-boost-medical-modeling-simulation [Accessed 31 March 2022].

[24] ReportLinker. (n.d.). The Global Metaverse in Entertainment Market is expected to grow by $ 28.92 bn during 2022–2026, accelerating at a CAGR of 8.55% during the forecast period. Globenewswire. Retrieved March 10, 2023, from https://www.globenewswire.com/news-release/2022/05/17/2444679/0/en/The-Global-Metaverse-in-Entertainment-Market-is-expected-to-grow-by-28-92-bn-during-2022-2026-accelerating-at-a-CAGR-of-8-55-during-the-forecast-period.html

[25] White, L. (2022, January 5). *Walmart Metaverse Shopping envisions a more complicated online shopping experience*. Stealth Optional. https://stealthoptional.com/news/walmart-metaverse-shopping-envisions-a-more-complicated-online-shopping-experience/

[26] Lee, J. Y. (2021). A study on metaverse hype for sustainable growth. *International Journal of Advanced Smart Convergence*, 10(3): 72–80.

[27] Baus, O., and Bouchard, S. (2014). Moving from virtual reality exposure-based therapy to augmented reality exposure-based therapy: A review. *Frontiers in Human Neuroscience*, 8: 10.3389/fnhum.2014.00112.

[28] Lee, L.-H., Braud, T., Zhou, P., Wang, L., Xu, D., Lin, Z., Kumar, A., Bermejo, C., and Hui, P. (2021). All one needs to know about metaverse: A complete survey on technological singularity, virtual ecosystem, and research agenda. *arXiv preprint arXiv:2110.05352*. 2021 Oct 6.

[29] Blanchard, C., Burgess, S., Harvill, Y., Lanier, J., Lasko, A., Oberman, M., and Teitel, M. (1990). Reality built for two: A virtual reality tool. In *Proceedings of the 1990 symposium on Interactive 3D graphics*, 35–36.

[30] SoftwareandVirtualReality. [http://edel518spring2011.wikispaces.com/Software+%26+Virtual+Reality]

[31] Insight, F. B. (2021, August). *Virtual Reality (VR) Market Size, Share & Covid-19 Impact Analysis* (No. FBI101378). https://www.fortunebusinessinsights.com/industry-reports/virtual-reality-market-101378#:~:text=The%20global%20virtual%20reality%20%28VR%29%20market%20size%20was, manufacturing%2C%20entertainment%20and%20more%20is%20driving%20the%20market.

[32] Mahmoud, M. (2020, January 7). *CES 2020: Sony announces updated PlayStation statistics – PS4 sales reach 106 million*. KitGuru. https://www.kitguru.net/tech-news/mustafa-mahmoud/ces-2020-sony-announces-updated-playstation-statistics-ps4-sales-reach-106-million/

[33] Alsop, T. (2021, February 23). *Number of jobs enhanced by AR/VR worldwide 2019–2030*. Statista. https://www.statista.com/statistics/1121601/number-of-jobs-enhanced-globally-by-vr-and-ar/#:%7E:text=As%20of%20a%202019%20report%2C%20it%20is%20forecast,where%20enhanced%20by%20VR%20and%20AR%20in%202019.

[34] Mathew, S. Importance of virtual reality in current world. Last accessed at www.ijcsmc.com on June 19th (2014).

[35] Rillig, M. C., Gould, K. A., Maeder, M., Kim, S. W., Dueñas, J. F., Pinek, L., Lehmann, A., and Bielcik, M. (2022). Opportunities and risks of the "Metaverse" for biodiversity and the environment. *Environmental Science & Technology*, 56(8): 4721–4723. 10.1021/acs.est.2c01562.

Vijay Joshi, Sukanta Kumar Baral, Manish Pitke, Rocky J. Dwyer

6 Retailing and e-commerce riding on technology: augmented reality and virtual reality

Abstract: Today's world and economy are driven by technology and technology-savvy people across all sectors. Moreover, since industrial revolutions have a long-term impact on businesses and society, it is also associated with the progression and recognition of new technologies. Therefore, emerging technologies such as augmented reality (AR) and virtual reality (VR) will have immense potential worldwide in the dynamic retail and e-commerce industry. This chapter gives the reader an idea about the use of AR and VR in the e-commerce and retail sectors.

The retail industry has grown considerably over the last two decades with the advent of e-commerce. It is no longer just a "purchasing" activity-based industry. Instead, it has become an "interactive" industry that provides customers with a "shopping experience." Very soon, AR and VR could be critical factors in establishing customer engagement and product promotion.

The Indian retail business is close to $1 trillion in size. It is undergoing various noteworthy changes, including premiumization, a movement toward freshness, a resurgence of vintage habits, increased customer trust on the internet, and a need for 10 min delivery. Some were secular developments, while the Covid-19 tailwind helped others. As a result, the old-world order of organized retail, the value of location and scale, and the power of shelf space are all up for grabs.

Some upcoming advances (supported by tech innovations) are virtual reality, augmented reality, robots, facial recognition, 3D printing, etc. They are making their presence felt. These show that they are gradually being accepted. They provide retailers with new ways to engage with their customers and retain them. They are also managing their stores more efficiently while reducing costs, where possible. As a result, retail has become more transformative.

Corresponding author: Manish Pitke, Freelance Faculty, Prin. L. N. Welingkar Institute of Management Development and Research (WeSchool), Matunga, Mumbai, Maharashtra, India, e-mail: manishpitke@gmail.com

Vijay Joshi, Dr. Ambedkar Institute of Management Studies and Research (DAIMSR), Deekshabhoomi, Nagpur, Maharashtra, India, e-mail: vijayjoshi62@gmail.com

Sukanta Kumar Baral, Department of Commerce, Faculty of Commerce and Management, Indira Gandhi National Tribal University – IGNTU (a central university), Amarkantak, Madhya Pradesh, India, e-mail: drskbinfo@gmail.com

Rocky J. Dwyer, D. B. A. Contributing Faculty Member, College of Management and Human Potential (CMHP), Walden University, Canada, e-mail: rocky.dwyer@mail.waldenu.edu

https://doi.org/10.1515/9783110790146-006

Some of the changes are:
- Use of AR/VR
- Social commerce
- Giving importance to deploying machine learning (ML) and artificial intelligence (AI)
- Video commerce
- Digitization of distribution
- Hyperlocal commerce
- Omni-channel retailing

Other enabling technologies include dynamic demand forecasting and computer vision to understand buyer behavior in store.

Objective

This chapter gives the reader an idea about the use of AR and VR in the e-commerce and retail sectors.

The following topics are covered in this chapter:
- AR/VR: understanding the terms
- AR/VR: advantages and disadvantages
- The changing nature of the e-commerce and retail industry (across the globe and in India)
- AR/VR: benefits to e-commerce and retail industry
- Factors responsible for the limited use
- A few practical examples (in the e-commerce and retail industry)

Keywords: emerging technologies in retail and e-commerce, virtual showrooms, interactive purchasing, immersive shopping experience, real-life simulation, augmented reality, virtual reality

1 Introduction

AR/VR: understanding the terms [1, 2]

Today's world and economy are driven by technology and technology-savvy people across all sectors. AR and VR are seen as emerging technologies benefiting the changing retail and e-commerce sectors.

AR is increased or augmented reality. AR appends computerized components to a real-time view. It mainly uses a camera (from a smartphone). Some examples are the game – Pokemon Go – and Snapchat. AR can be seen in terms of older and newer versions of AR. The older may be referred to as "marker" type AR. This is because a marker is needed in this case. It could be a specially marked poster or any poster. The modern type may be called "markerless" AR. In this case, it is not required (for the camera) to look at any unique object (in the real world). Pokemon Go is considered an example of this type of AR.

VR is virtual reality. It is a different experience outside the real world. It may utilize special gadgets like Oculus Rift, Google Cardboard, HTC Vive, or similar. Here, viewers are immersed into other worlds and with different envisioned conditions.

The real world and the modernized objects work together in mixed reality (MR). It is an experience that combines parts of AR and VR. Mixed reality development has recently gained momentum with Microsoft's HoloLens, perhaps the most famous early mixed reality device.

The collective term used for each of these various advances is extended reality (XR). It encompasses resources in two ways. One is by providing more details of the current situation/scenario. Another is by creating a stunning simulated universe that can be viewed and experienced. It is the consolidation of all three realities.

AR and VR features are shown in Table 6.1.

Table 6.1: AR and VR features [1, 2].

Parameter	Augmented reality (AR)	Virtual reality (VR)
What does it do?	In addition to the actual physical world, there are superimposed digital elements.	It is an artificial scenario. Instead, it uses simulation to provide an immersive experience.
Equipment in use	Personal devices such as smartphone	Special VR headset.
Transforming the shopping experience by providing	Enhanced shopping occasions online Guided journey in store Contact-less (Product) trials Online marketing Individual product offerings One-to-one (individualized) marketing and product search	Creating virtual happenings. V-commerce. In-store VR. Easing the task of employee training.
Hardware needed	A mobile device is enough.	Bulky, and sometimes ineffective.
Social Environment	One can interact with the people in the same room.	They are not allowed. You and the rest of the world are separate from each other.
Access and Affordability	Most people (billions of users) are using smart mobile devices.	Hardware costs are high, and could be costly and alarming.

As indicated in Table 6.1, there is a trend toward using these technologies to make an organization a prospective asset. In the near future, AR and VR technology will drive the "e-commerce and retailing" sector. This is a response to tackling the COVID-19 pandemic and the broader economic slowdown.

The arrival of AR technology is a good sign for the end users of the retail industry. This allows them to see the natural environment overlaid with digital data. Thus, it can enhance user capability and thereby have the potential to transform the retail industry. But unfortunately, augmented reality consumers have been slow to adopt it. However, continued investment in technology and Metaverse's ideas show that emerging technologies still attract significant investment value in this information technology domain [3].

Mr. Rajeev Dwivedi, CEO and Founder, LivePixel Technologies quips: "When augmented reality is widely used for commercial purposes and consumer engagement, people will likely get the "Metaverse" environment and the experience in the retail industry."

AR/VR: Advantages and disadvantages [4]

Augmented reality
- Pros
  - It provides valuable and interactive data when appropriately implemented (on the scene).
  - AR apps will become mainstream with the widespread use of these new phones. As a result, it will create more (business) opportunities for marketers. Some recent examples are Apple's latest Phone/Tablet and AR Developer Kit, the AR-enabled new Google Android smartphone, and the IKEA app, allowing users a "see-before-buy" feature before ordering.
  - A simple AR application equipped with a smartphone is well-established in the market.
- Cons
  - Hardware may be of awkward shape and is only for specific applications. Therefore, it is not immediately used daily.
  - Users have to incur substantial costs toward hardware (e.g., HoloLens: USD 3500, Magic Leap: USD 2300). The current format may have certain (severe) restrictions. Both have limited field choices, and sometimes the motion controls are not easy to use.

Virtual reality
- Pros
  - When carried out appropriately, VR can be an unbelievably captivating tangible ride. Utilizing PC-created symbolism (Computer Graphics Interface), there are no restrictions besides the capital expenses and the creative mind while making different universes, item demos, or spaces in novel and intriguing ways.
  - In the education/learning sector, VR is more agreeable and is seen as a straightforward tool.
  - Computer-generated reality clients can do trial and error, utilizing imaginary situations.

- Cons
  - Content creation will generally be altered and, in many cases, costly. Best practices for a compelling as well as drawing in a happy design are also being worked out.
  - To provide a demonstration, VR can be said to be a bit slow. This takes a lot of time to manage the entire setup, organize the headset, put on/replace the headset, understand the controls, and let the customer see what's happening. Fortunately, one demonstrator, experienced or new, can guide 15 to 20 people through this experience in an hour.
  - Currently, VR may be in its nascent stage. Also, there is a possibility that the content made for one location will usually exhibit incompatibility with another.
  - You are entirely inaccessible in VR. It is a lonely environment that brings you individual experiences, but it takes you away from the authentic atmosphere, resulting in loneliness.
  - In terms of hardware costs, VR is fragmented with different price ranges. Headset valuing goes from USD 15 (Google Cardboard utilizing a cell phone) to USD 1500 (HTC Vive Pro). Abroad, the range of choices is varied (with changing abilities in between).

2 Literature review

Augmented reality is a spearheading innovation incorporating computerized components with a genuine view. Hence, AR is an endeavor to improve reality with carefully produced pictures and elements. Frequently considered to be like VR (Virtual Reality), AR doesn't supplant reality with computerized or virtual space. However, expanded reality can adjust the ongoing climate to give a fascinating and improved insight. In 2020, approx. 37% of the total populace (of the world) was using social media [5, 69]. It has grown impressively in the year 2022. At present, 58.4% of the total populace (of the world) is engaged with social media. These included diverse ways of social networking (using various online entertainment formats). Thus, advertisers are giving more significance to the online entertainment advancement that depends on AR. This permits likely clients to attempt various items before making a buy, prompting more educated choices, faithful clients, expanded deals, and so forth.

Smink et al. [6] and their study explored the benefits of shopping with Augmented Reality (AR). The authors believed it offers a "try before you buy" experience. Augmented Reality improves the extent of providing information (to the consumer). Hence, there is a considerable enhancement in enjoyment. Purchase intentions and willingness (to buy) increase because of more shared data. Consumers' pleasure leads to an interactive, practical, and decisive process. This, in turn, yields a more positive brand attitude while AR raises perceived intrusiveness.

2.1 The changing nature of the e-commerce and the retail industry (across the globe and in India)

Individuals have been involved in web-based buying for more than twenty years. It has grown considerably due to the Smartphone and Internet transformation over the last ten years. Today, internet shopping has progressively become common. Purchasers can purchase anything from online business sites to their homes. Ease of buying (i.e., convenience), and the variety of low-priced products, which can't be found customarily, make an alternate way to deal with web-based buying for shoppers [5, 70–71]. Numerous experts opine that the size of internet shopping is enormous. The justification behind the massive buying or shopping evolution is the improvement in the capacity to purchase items whenever and products being delivered to the doorstep. Conventional shopping – It is perhaps the most seasoned approach to shopping; it requires the customer's physical presence at each shop. In this case, you must visit (each) shop and select the one meeting your needs. Then, pick an appropriate item from that point and bring it home. Conventional shopping empowers buyers. They can touch and feel the article before paying the cash. The shopper can be aware of the quality, estimate its value, and address a lot more different issues on the spot. For example, purchasing design stuff on the web requires clients to be specific – garments or shoes be in the correct tone or size. The younger age is leaning toward web-based shopping because of the cutting-edge occupied way of life. Additionally, the excellence of connecting business with innovation blends the universe of creation and interaction, which is how AR is doing socially.

In the past two decades, e-commerce and retail have changed considerably. Now, people are hooked on to e-commerce websites to fulfill their buying needs. This ranges from daily needs purchases to gifts for special occasions. Increasing internet penetration, easy availability of smartphones, and development in telecommunication technologies have resulted in the growth of e-commerce. Another strange reason for the recent change is the advent of the pandemic Covid-19. Due to the pandemic, people are compelled to follow restrictions such as social distancing and wearing masks. Due to the lockdown imposed, the physical operations of the retail industries are disturbed, but at the same time, are well-supported by their online presence. This has resulted in the growth of online retailing. This is mainly because of the precautions taken by the people (due to fear of the pandemic) to stay at home and avoid any kind of social interaction. This also impacted their physical visits to stores and shops. As a result, they opted to stay at home and preferred online shopping. This has resulted in the growth of e-commerce in most of the developed and developing countries in the world.

A report by MGI – McKinsey Global Institute (https://www.mckinsey.com/featured-insights/future-of-work/the-future-of-work-after-covid-19) has the subject of interest, "the future of work after COVID-19." The report investigates different aspects of the post-pandemic economy in eight countries. This document presents an evaluation of the impact of the pandemic on various factors of the economy. Some of the aspects of interest to this study are workforce skills (required in the future), supply and demand scenario of human resources, the changing nature of the jobs (occupations), and accordingly, the resulting matrix of the job profiles, and so on. These eight selected countries have diverse economic scenarios and different workforce market models. These include China, France, Germany, India, Japan, Spain, the United Kingdom, and the United States. It is estimated that they will collectively have about 62% of the world GDP, and all these eight countries (together) will have almost half of the global population [7, 6–7]. The share of e-commerce (as % of total retail sales) in these countries (in 2020) is shown in Table 6.2.

Table 6.2: E-commerce in select countries in 2020 [7, 6–7].

Country	E-commerce sales as % of total retail sales (2020)
India	7%
Spain	9%
France	9%
Japan	10%
Germany	14%
United States	20%
United Kingdom	24%
China	27%

2.2 AR commerce is driving the AR/VR market growth

AR commerce, enabled by both AR and VR, is a very promising idea for the world of retail. Therefore, many companies are already pioneering the technology. Given the many advantages of both AR and VR (with their respective limitations), online retailers are making efforts to understand their importance (to their business) and started investing in them. According to Samta Arora, who cites a GlobalLogic report, there is good potential for a wide range of applications that may use AR/VR soon [8]. The diverse potential of AR and VR applications (in 2025) is shown in Table 6.3.

Table 6.3: AR and VR applications: estimated market size in 2025 [8].

Sector	Value (in US $ billion)
Enterprise and public sector (A)	16.1
This includes various segments such as healthcare, engineering, real estate, retail, military, and education.	
Consumer (B)	18.9
This includes various segments such as videogames, live events, and video entertainment.	
Total (A + B)	**35.0**

2.3 Changing habits of consumers (rise of hybrid shopping)

Apart from the demand, the changing nature of shoppers compels retail companies to use new technologies to attract and retain their customers. Luq Niazi, IBM Global Managing Director of Consumer Industries, mentioned that they had completed a significant study that has 19,000 people surveyed around the world [9]. What they saw from that – and of course, all demographic groups and all categories – is the rise of hybrid shopping. This is what's happening right now. Niazi said people are not just visiting the shops but exploring different methods. They are going online and buying the goods. They are using a combination of sources (for their shopping). Accordingly, there will be a combination of delivery models. A survey response indicating the rise of hybrid shopping is shown in Table 6.4.

Table 6.4: The rise of hybrid shopping: IBM study of global shoppers [9].

Goods	Primary method of buying by category (as % of respondents)		
	In-store	**Hybrid**	**Online**
Groceries	62%	20%	19%
Personal care and beauty	50%	22%	29%
Apparel and footwear	40%	25%	35%
Home goods	28%	40%	31%
Total	**45%**	**27%**	**28%**

The above fact is also supported by the GlobalLogic Report [8]. According to the said report, the following are seen as significant segments that can implement AR/VR technology for better performance:

- Apparel and footwear
- Personal accessories and cosmetics
- Home furniture
- Home kitchen appliances

In the Indian context, the country is increasingly embracing and adopting advanced technology to support its business. Guided by technological advances, multiple industries in different industries are seeking additional impetus for growth, opportunity, and success. Indian retailing and e-commerce sectors are also not far behind in finding new advances and strengthening their roots to secure the country's future. AR is anticipated to accelerate profitability and make e-commerce and retailing more competitive, according to the YourStory article [10].

The article from YourStory [10] believes that the country has been perceptive of advancements in various technologies. Thus, it is exploring different avenues regarding virtual and increased reality for some time to advance further. Combining heightened reality (AR) in organizations assists clients with having an intuitive encounter, which is additionally upgraded by visual, wearable, and haptic-tactile modalities.

3 Some of the emerging technologies used in the retail industry

Emerging technologies, as identified by some of the works of literature, are presented in Table 6.5.

Table 6.5: Some of the literature reviewed for the emerging trends in retailing.

Title of the article	Article presented by	Trends identified
Retail Innovation	– StartUs Insights [11]	– Mobile Technology – Augmented Reality (AR) – Big Data – Virtual Reality (AR) – Artificial Intelligence (AI) – Internet of Things (IoT)
Technological Advances in the Retail Industry	– Acquire [12]	– Augmented Reality (AR) – Facial Recognition – Cashier-less Stores – Robots as Store Assistants – Internet of Things (IoT)
Trends in Indian Retail Beyond 2021	– Salesforce Blog [13]	– Virtual Reality (AR) – Omni-Channel Interaction and Presence – Contact-less Engagement – Digital Payments
Leveraging the Retail Technology Trends	– Vend Retail Blog [14]	– Virtual Customer Engagement – Automated Commerce (A-Commerce) – Contactless Payments – Emphasis on Pre-store Research – Artificial Intelligence (AI)

As said by earlier studies, the use of technology is evident in the retailing industry. It is required, given the intense competition in the global retail sector. Also, technology ensures that retailers seamlessly deliver their goods and services. It is also possible for them to gather information from their users/consumers regarding their views and opinions, likes and dislikes, inclinations, and experiences.

4 Some of the areas in the retail industry addressed by the AR/VR

When we are talking about the retail industry (and that too in the online environment), the extent of VR/AR technology use varies. There are numerous future possibilities, and they are analyzed in various contexts. These may range from personal things like cosmetics to clothing to real estate [15, 11–13].

Compared with daily web-based shopping highlights, this innovation will upgrade the shopping experience in different folds. For example, the LensKart application offers an online selection of frames. This choice is based on the appropriateness of the client's face. Another example is the Nykaa application in the area of personal care products. This offers the option or the selection of make-up items by considering the user's complexion. It uses the virtual environment supported by AR innovation [16].

The clients can receive constant feedback or good input about the clothing that looks great on them by utilizing the virtual fitting option. This (virtual try-out or fitting) empowers them to go with the buy or no-buy choice. Specific research studies have explored the equipment associated with VR/AR advancements, while others investigated the guidance methods of internet-based stores [17, 635–637].

According to [18, 147], AR/VR technology may address several areas in the retail industry. This is presented in Table 6.6.

Table 6.6: Some of the areas in the retail industry addressed by AR/VR technology.

Area	Article presented by	Article outcome
Personal navigation	– Doctoral Dissertation [19]	– Familiarity with the user interface enhances user interaction and usage duration.
Using simulation (to sense the touch)	– Bonetti et al. [20, 3–16]	– The surface, weight, and sample of the item might be felt. There may be the use of haptic gloves. It may involve electric pulses while doing online buying.
Data visualization	– Hariharakrishnan and Bhalaji [21, 69–75]	– Furnishes an advanced shopping experience with constant information perception.
Apparel fitting	– Edriss and Babikir [22, 16–30]	– The users can pick garments that fit them best.
Interaction in 3-D	– Xue et al. [23, 27–31]	– They are providing a better shopping experience. The immersive settings enable users to create virtually.

As said by earlier studies, the use of technology is evident in the retailing industry. It is required, given the intense competition in the global retail sector. Also, technology ensures that retailers seamlessly deliver their goods and services. It is also possible for them to gather information from their users/consumers regarding their views and opinions, likes and dislikes, inclinations, and experiences.

AR/VR: benefits to e-commerce and retail industry [24]
Some benefits of AR and VR in retail (and e-commerce) are elaborated herewith.

4.1 Customer interaction

Changes and innovations in VR/AR primarily impact the online business sector. This will change how you search, and buy and sell things on the Internet. Customers can use virtual and physical advances to get more information about an item and see if they have been to the store. Computer-generated reality headsets provide a rich shopping experience that enhances the reality, and brings objects into the actual environment. Customer loyalty and (shopping) participation is supported and extended by VR/AR. You can use augmented reality to browse stores and quickly observe what you want. Home comfort allows users to move around the company and achieve unprecedented customer contributions virtually.

4.2 Create a unique shopping experience (for your existing and potential customers)

VR/AR gives a particular point of view on the clients' tendencies and helps retailers make organized shopping decisions for their clients. Tweaked and assigned exhibiting has proactively demonstrated to be profitable to retailers. For flexible workspaces and retail web organizations, they convey modified advancing strategies. Clients can likewise utilize AR techniques to make things custom-made to their needs. This guides in making a custom-fitted shopping experience in which every buyer gets unique and one-of-a-kind item, thus expanding the retail location's purchase rate. Customers will experience a better way of "shopping" as an activity using AR and VR. It is an advantage to retailers when aligned with their business strategy. Instead of a conventional two-dimensional display, traders can provide three-dimensional graphics-rich content to deliver. This creates an immersive environment and enhances user participation and interaction. Here, it is likely that the trader may witness a growth in revenues. This may be due to more customer demands. Users can also use AR tools to create personalized items tailored to their preferences.

4.3 Increase client satisfaction and the number of customers who return (for purchasing)

In the most competitive and dynamic sectors (like retailing), VR and AR will be handy for retailers. They enable them to manage the customers and satisfy them. This will increase profits. Organizations that enter the VR/AR markets first have an advantage over organizations that enter later. The "customized services" feature of VR/AR will treat each customer individually. Hence, it will provide an affirmative involvement and (buying) knowledge. This will ensure that the customers will come again and again (to the store). Also, it is unlikely that the goods purchased will be returned. Almost certainly, a fulfilled client will get back to shop in the future.

4.4 Using innovative ways to market and promote the offerings

Increased reality and computer-generated reality can end up being incredible instruments for showcasing. These give convincing and instinctive substance to computerized advertising. Expanded and augmented reality advances can be utilized for advertising via online entertainment sites. Current advertising systems just give clients data about the item. The advertisements you see are non-intuitive and may not necessarily hold any importance to you. Expanded and augmented reality can provide clear promoting choices that draw clients' focus, and enhance their expectations and usage of the item. The online entertainment showcasing this standpoint has become broadly utilized by numerous organizations nowadays, and retailers can likewise profit from it.

4.5 Worker preparing

Using VR and AR in retail stores can minimize the risks involved in the training process. It is possible to train and prepare the workers for actual situations. These situations may be created artificially (in a simulated environment). With this, it is possible to manage different scenarios such as a lathe machine operation or guiding a customer to locate a suitable product (in the store). These examples show that employee training is a personal activity and is not an easy task. It is a time-consuming process and needs active human participation. Training may be capital-intensive if not appropriately planned or implemented systematically. This may result in a loss of revenue for the organization.

5 Analysis and interpretation

Factors responsible for the limited use [25]
In the case of such emerging technologies, their use is not that widespread across all the industry sectors. This is due to low awareness, high costs, complexities involved in the implementation, etc.

Some of the factors indicated by the Statista survey are shown in Table 6.7.

Table 6.7: Limiting factors for AR and VR (in the US in 2020) (as % of respondents) [25].

Factors	AR	VR
Review and opinion of the users (e.g., bulky hardware, technical issues, etc.)	32%	19%
Limited product and service offerings (in terms of the magnitude of the content, its sources, reliability, etc.)	18%	27%
Unwillingness to shift to consider AR and VR technology for their business	15%	19%
Legal frameworks such as regulations and legal risks involved therein	14%	12%
Capital investment requirements	11%	9%
Cost to consumers	7%	11%
Government authority is not interested in this aspect	4%	3%

Though these factors covered are in the context of the US, these could be generically applicable at any other location.

5.1 A few practical examples (in the e-commerce and retail industry)

5.1.1 Elimination of product returns (or reducing them)

3DLOOK has provided a product YourFit. It is an integrated offering that deploys a “try-before-you-buy” feature for the buyers. It takes the help of a virtual try-on scene and provides suggestions based on the body size and the selected fittings.

YourFit is the first solution on the market to remove any uncertainties around shopping for clothes online. This new B2B SaaS (Software-as-a-Service) product will create a win-win situation for both the fashion brands and their consumers. Furthermore, this solution will enable the fashion brand and retailers to achieve minimum product returns (from their consumers) [26].

The CEO and co-founder of 3DLOOK, Vadim Rogovskiy, believes that their product (YourFit) will enable the progression of e-commerce to the advanced stage, including personalization [27]. Also, this will improve the "shopping experience."

5.2 Product promotion on the virtual platform

Rivaah by Tanishq, a wedding adornment focused sub-brand from Tanishq has revealed its amazing Polki assortment. This was done in a virtual public interview facilitated on the Metaverse stage to make a profoundly vivid and customized 3D virtual experience for its watchers. This is the first time any Indian adornment's brand has sent a gems assortment through an experiential drove occasion on the Metaverse [28, 29].

5.3 AR fitting rooms/virtual try it out

Currently, people don't need to enter the store to perceive how they will look in a specific outfit. Lily's, a Chinese clothing brand, has placed an AR show on the window of a store situated at the metro subway of Shanghai's West Nanjing Road Station [30] (https://www.youtube.com/watch?v=zBKtZvy5r1A). Travelers can give garments a shot as they are waiting (for their trains to arrive) [31].

5.4 Large products shopping/cost savings/space visualization

Ikea came up with the Ikea Place ARKit app (https://youtu.be/UudV1VdFtuQ). This app lets users see a three-dimensional preview of the selected furniture at the desired spot in their homes/offices [30]. Users can see how it will fit into their spaces and how it will look. Accordingly, they can make informed purchasing decisions [32].

5.5 In-store guidance makes product finding and purchasing easy

Lowe's AR-enabled app helps (their) customers to locate the product quickly [33]. In addition, the application provides in-store navigation [30]. Thus, it directs people through the brand's stores, selects better routes, and quickly locates what they need (https://www.lowesinnovationlabs.com/projects/lowes-vision-navigation).

5.6 Virtual try-out/real-life simulation

Now, people can virtually try products before they buy. This is the requirement of the personal accessories segment that includes cosmetics. French cosmetic brand Sephora has developed an AR technology app [34]. It allows the user to display her picture and try out different cosmetics (https://www.youtube.com/watch?v=NFApcSocFDM). The 'Virtual Artist' app scans the face and determines the lips and face position. Then, as the customer selects the product, it will show the changes in the customer's appearance [35].

A similar experiment was tried by another French cosmetics and perfumery company – L'Oréal. The company provides an augmented reality make-up experience (by using virtual assistants). This exercise was done in association with Facebook. As stated in the above case, it also facilitates the users in product selection (of their interest). This does not require physical visits to their offline stores [36].

5.7 Assistance in luxury product buying (virtual try-out)

Forbes magazine quotes, "visitors to BMW showrooms can also put on virtual reality goggles and experience what it's like to drive cars. Hence, they understand their capabilities and can make the perfect choice for their new vehicle."

Imagine entering a car dealership and viewing the car that you want. It sounds crazy, but yes, this is happening. BMW, a reputed luxury car maker, deploys AR to lure potential buyers. By using AR, they are enabling viewers (or the car buyers) to walk into a car dealership (place). Upon their entry, they can customize based on specific parameters (such as different colors or styles) using their personal devices (such as tablets or phones) [36].

5.8 When the customers can't visit the shop, the shop comes to them (virtually)

Earlier, we indicated the importance of e-commerce and its growth (due to the recent pandemic and spread across different countries). This has created the situation that the customers cannot visit any shops but can do it virtually. Leading brand, Apple, has taken advantage of this AR/VR technology to have customer engagement during the pandemic [36]. It is learned that Apple has used the "pandemic time" to create an environment for their customers, through which they can visit Apple stores using AR technology. This enabled the company to showcase its products. Within this environment, now customers are using AR Quick Look for new iPhone or iMac models and experiencing how they will look in their space or their hand.

5.9 Other aspects like training and industry expert views

Earlier, we talked about the training. Of course, this aspect is of much importance in the retail sector. However, some IT companies have taken a cue by creating virtual environments and using the same for training. For instance, Accenture has created a virtual workplace, Nth floor, to train new joiners [37]. This trend will be followed by the retail industry soon.

Earlier, we talked about a marketing campaign by Titan on the metaverse for the product launch of Tanishq. The company combined VR and AR to make the project come to life. Thus, there are different ways of creating this experience. Numerous platforms are available for launching a Metaverse [38]. These may range from browser-based platforms to native applications. Each of these has some unique functionalities. For example, Tanishq has used a web VR-based platform to create an immersive three-dimensional experience. In this context, some industry experts think that these new platforms (such as Metaverse) are changing the marketing, buying, and selling process [38].

6 Conclusion

To sum up and finish up, we can express that there are a few other empowering advancements like the utilization of AR/VR, active interest anticipation, utilization of PC vision to comprehend in-store customer conduct, and so on.

Concerning the utilization of AR/VR, it is accepted that few proposal-based buys for the touch-and-feel sorts of classes will see expanded penetration of AR in their purchasing cycle. This could be an individual consideration item (skincare or magnificence care brand) requesting that you share an image of your face to figure out what things will suit you best. A design brand offers virtual fitting assistance or a virtual vehicle sales center, where you can wander around and connect with various vehicle models, much like in the disconnected world. There are a few fundamental justifications for why this will become more famous. In the first place, online retailers need to give a detached sort of involvement to change over clients. Second, innovation can sometimes provide preferable choices over a human, or answer item questions better than a human can. Third, a store with countless SKUs, helps clients pursue faster and better decisions.

Considering how quickly the (retail and e-commerce) business is transforming, one should rest assured that there will be another rearrangement of scene by 2023.

Statista believes that about one-third of US shoppers will use augmented reality technologies by 2025 when shopping online.

Considering the intensely competitive retail (and e-commerce) sector, AR and VR will sooner or later become fault-finding tools for organizations. This will be valid for their marketing operations that decide (their) accomplishments (in the market) or their business. Any pitfalls in the marketing campaigns may be identified and rectified to achieve the desired objectives and results that the organization aims to achieve.

Another advantage of a brick-and-mortar store is leveraging technology that benefits the store and its customers. In this case, the store is following a technology-based approach to increase the number of retail purchases therein.

References

[1] Butler, S. (2022). What Is Extended Reality (XR), and Will it Replace VR? How-To Geek, May 2, 2022. Accessed May 4, 2022. https://www.howtogeek.com/794396/what-is-xr/.

[2] Kiger, J. P. (2020). What is the Difference Between AR, VR, and MR? Updated on Jan. 6, 2020. Accessed April 7, 2022. https://www.fi.edu/difference-between-ar-vr-and-mr.

[3] Rajeev, D. (2022). How AR, VR, and Metaverse Helping the Retail Industry Transform. Indian Retailer, March 25, 2022. Accessed May 10, 2022. https://www.indianretailer.com/article/technology/digital-trends/how-ar-vr-and-metaverse-helping-the-retail-industry-transform.a7858/.

[4] Azurdy, J. (2020). Virtual and Augmented Reality: Pros and Cons. Encora, June 2020. Accessed April 7, 2022. https://www.encora.com/insights/virtual-and-augmented-reality-pros-and-cons.

[5] Anuradha, Y. (2020). Digital Shopping Behaviour: Influence of Augmented Reality in Social Media for Online Shopping. *Journal of Multidimensional Research & Review (JMRR)*, 1(3), Sept. 2020, 68–80.

[6] Smink, A. R., Frowijn, S., van Reijmersdal, E. A., van Noort, G., and Neijens, P. C. (2019). Try Online Before you Buy: How Shopping with Augmented Reality Affects Brand Responses and Personal Data Disclosure. *Electronic Commerce Research and Applications*, 35: 100854. https://doi.org/10.1016/j.elerap.2019.100854.

[7] Susan, L., Anu, M., James, M., Sven, S., Kweilin, E., and Olivia, R. (2021). McKinsey Global Institute. The future of work after COVID-19, Feb. 2021. Accessed April 21, 2022. https://www.mckinsey.com/featured-insights/future-of-work/the-future-of-work-after-covid-19.

[8] Arora, S. (2021). Impact of Augmented and Virtual Reality on Retail and E-Commerce Industry. A GlobalLogic Report. Accessed April 11, 2022. https://www.globallogic.com/wp-content/uploads/2021/05/Impact-of-Augmented-and-Virtual-Reality.pdf.

[9] IBM Study. (2022). Grocery Shortages Abound Amid Supply Chain Issues and Omicron. Yahoo Finance. Accessed Jan. 28, 2022. https://finance.yahoo.com/video/grocery-shortages-abound-amid-supply-191515320.html.

[10] Sethi, R. (2021). Role of Augmented Reality in Making Retail Future-Ready. YourStory. November 15, 2021. Accessed April 7, 2022. https://yourstory.com/2021/06/role-of-augmented-reality-in-making-retail-future-/amp.

[11] Startus. (2021). Retail Innovation Map Reveals Emerging Technologies and Start-ups. A blog by Startus Insights. 2021. Accessed Jan. 28, 2022. https://www.startusinsights.com/.

[12] Acquire. (2021). Technological Advancements in the Retail Industry to Make You Say Wow! October 2021.Accessed Jan. 28, 2022. https://acquire.io/blog/technological-advancements-retail-industry/.

[13] Salesforce (2021). Indian Retail: Trends and Opportunities in 2021 and Beyond. Salesforce Blog. July 2021. Accessed Jan. 28, 2022. https://www.salesforce.com/in/blog/2021/07/indian-retail-trends-and-opportunities-in-2021-and-beyond.html.

[14] Nicasio, F. (2021). Seven Retail Technology TrendsMaking Waves – and How to capitalize on them. April 2021.Accessed Jan. 28, 2022. https://www.vendhq.com/blog/retail-technology-trends-2/.

[15] Chen, J., Zong, I., and Hengjinda, P. (2021). Early Prediction of Coronary Artery Disease (CAD) by Machine Learning Method-A Comparative Study. *Journal of Artificial Intelligence*, 3(01): 7–33.

[16] Jung, T., and Dieck, T. (2018). Augmented reality and virtual reality. In *Empowering human, place, and business*. Springer International Publishing: Cham.

[17] Yadav, N., Omkar, K., Aditi, R., Srishti, G., and Shitole, A. (2021). Twitter Sentiment Analysis Using Supervised Machine Learning. In Intelligent Data Communication Technologies and Internet of Things: Proceedings of ICICI 2020, pp. 631–642, Singapore. Springer.

[18] Kumar, S. T. (2021). Study of Retail Applications with Virtual and Augmented Reality Technologies. *Journal of Innovative Image Processing (JIIP)*, 03(02): 144–156. https://doi.org/10.36548/jiip.2021.2.006.

[19] Zhang, J. (2020). A Systematic Review of the Use of Augmented Reality (AR) and Virtual Reality (VR) in Online Retailing. Doctoral dissertation, Auckland University of Technology.

[20] Bonetti, F., Pantano, E., Warnaby, G., Quinn, L., and Perry, P. (2019). Augmented reality in real stores: Empirical evidence from consumers' interaction with AR in a retail format. In *Augmented reality and virtual reality*. Springer, Cham, 3–16. DOI: 10.1007/978-3-030-06246-0_1.

[21] Jayaram, H., and Bhalaji, N. (2021). Adaptability Analysis of 6LoWPAN and RPL for Healthcare applications of Internet-of-Things. *Journal of ISMAC*, 3(02): 69–81.

[22] Edriss, A., and Babikir, E. (2021). Evaluation of Fingerprint Liveness Detection by Machine Learning Approach-A Systematic View. *Journal of ISMAC*, 3(01): 16–30.

[23] Xue, L., Parker, C. J., and McCormick, H. (2019). A Virtual Reality and Retailing Literature Review: Current Focus, Underlying Themes, and Future Directions. *Augmented Reality and Virtual Reality*, 27–41. DOI: 10.17028/rd.lboro.6177005.

[24] Joshi, N. (2019). Retailers Have a Lot to Gain from AR and VR. Oct. 2019. Forbes. Accessed April 7, 2022. https://www.forbes.com/sites/cognitiveworld/2019/10/01/retailers-have-a-lot-to-gain-from-ar-and-vr/?sh=4b808b727a1c.

[25] Statista. (2021). Obstacles to Mass Adoption of AR/VR Technologies as per U.S. XR Experts 2020. Published by Thomas Alsop, May 2021. Accessed April 11, 2022. https://www.statista.com/statistics/1098558/obstacles-to-mass-adoption-of-ar-technologies/.

[26] Apparel Resources. (2022). 3DLOOK launches 'YourFit 2.0' – The First Shareable Omnichannel Virtual Fitting Room. *Apparel Resources News Desk*. March 30, 2022. Accessed April 7, 2022. https://apparelresources.com/technology-news/retail-tech/3dlook-launches-yourfit-2-0-first-shareable-omnichannel-virtual-fitting-room/.

[27] Fit, Y. (2022). YourFit 2.0: A Virtual Trial Place from 3DLOOK. Accessed April 17, 2022. https://3dlook.me/yourfit/.

[28] AdGully. (2022). Titan is Launching a Specific Collection on Metaverse. Accessed May 10, 2022. https://www.adgully.com/img/800/202204/untitled-design-2022-04-07t181455-470.png.jpg/.

[29] Sakal. (2022). Romance of Polki' Collection on the Metaverse. *Sakal, Pune edition*, April 8, 2022.

[30] Chalimov, A. (2021). Bringing Augmented Reality to Your Retail App. Oct. 2021. Easternpeak. Accessed April 5, 2022. https://easternpeak.com/blog/bringing-augmented-reality-to-your-retail-app/.

[31] Lily. (n.d.). A Video Showing AR Fitting Rooms at Shanghai Metro. Accessed April 5, 2022. https://www.youtube.com/watch?v=zBKtZvy5r1A.

[32] Ikea. (n.d.). A Video Showing IKEA Place App. Accessed April 5, 2022. https://www.youtube.com/watch?v=UudV1VdFtuQ.

[33] Lowe. (n.d.). A Video on the Company Website Shows In-Store Navigation at Lowe's. Accessed April 5, 2022. https://www.lowesinnovationlabs.com/projects/lowes-vision-navigation.

[34] Thakkar, M. (2018). How AR and VR are Transforming E-commerce. December 20, 2018. Accessed April 5, 2022. https://synoptek.com/insights/it-blogs/ar-vr-ecommerce/.

[35] Sephora. (n.d.). A Video Showing a Demo of Sephora's "Virtual Artist". App. Accessed April 5, 2022. https://www.youtube.com/watch?v=NFApcSocFDM.

[36] DNA India. (2022). Some of the Best Examples of Augmented and Virtual Reality in Retail. April 15, 2022. Accessed May 10, 2022. https://www.dnaindia.com/business/report-5-best-examples-of-augmented-and-virtual-reality-in-retail-2946389.

[37] Shilpa, P. (2022). Accenture's Our 1.5 Lakh New Joinees to Train, Meet in the Metaverse. *Times of India, Pune edition*, April 18, 2022.

[38] Shilpa, P., and Akhil, G. (2022). Metaverse is Already Changing How We Market, Buy, and Sell. *Times of India, Pune edition*, May 18, 2022.Accessed May 18, 2022. https://timesofindia.indiatimes.com/business/startups/trend-tracking/metaverse-is-already-changing-the-way-we-market-buy-sell/articleshow/91631492.cms.

Ipseeta Satpathy, B. C. M. Patnaik, S. K. Baral, Majidul Islam

7 Inclusive education through augmented reality (AR) and virtual reality (VR) in India

Abstract: With the advancement of technology, the education-imparting scenario also changed drastically. The present study lays the framework to understand how far augmented reality (AR) and virtual reality (VR) have contributed to inclusive edification in India. For the study, various leading engineering colleges in the different parts of Odisha were considered, and initially 17 variables were identified with the help of 5 core group discussions and review of literature; however, after the pilot study, the number of variables was restricted to 11 only. The sample for the current research was collected through non-probabilities sampling, precisely, through convenient sampling technique. The findings of the study conclude that with the help of AR and VR, there were transformations in the education system, leading to inclusive education, evidenced by improved learning, improved motivation for study, enhanced students engagement, authentic learning opportunity, better communication, learning opportunity, etc.

Keywords: Augmented Reality (AR), Virtual Reality (VR), Ultrasound, Inclusive education and B.Tech.

1 Background of the study

Information and communication technology (ICT) transformed the education scenario in recent times with the change in the process of education [1, 2]. The initiative of integrating ICT and with education began in the twentieth century [3]. The implementation of AR and VR in education has made its significant presence in the last decade [4, 5]. These AR and VR technologies include technologies, such as VR head-mounted displays, simulated environment replications, collective cybernetic environment, hypnotic simulated environment, and practical ecospheres [6–10]. Most of these technologies, meanwhile, include 3D immersive visuals with avatars that have their control mechanisms and communication systems [11]. With the passage of time and with the increased use of

Ipseeta Satpathy, D.Litt, KIIT School of Management, KIIT University, Odisha, India, e-mail: ipseeta@ksom.ac.in
B. C. M. Patnaik, KIIT School of Management, KIIT University, Odisha, India, e-mail: bcmpatnaik@gmail.com
S. K. Baral, Department of Commerce, Indira Gandhi National Tribal University, MP, India, e-mail: drskbinfo@gmail.com
Majidul Islam, John Molson School of Business, Concordia University, Montreal, Canada, e-mail: majidul.islam@concordia.ca

https://doi.org/10.1515/9783110790146-007

smartphones by students, it has been discovered that AR and VR apps make technology adoption quite straightforward. Digital facets are more accessible than ever, thanks to affordable devices such as Samsung Gear VR, Google Cardboard, and Google Daydream [12], and the utility of AR with the help of smart devices has made education more accessible than earlier to a wide range of people [13]. AR and VR have caught the considerable interest of various teachers and researchers [14]. VR and AR implementation in education has transformed the students understanding to a large extent; they can connect with the subject with more clarity, and learning outcomes have also improved a lot [15].

"AR is a new technology that merges digital information with data from real-world surroundings, allowing users to interact virtually while concurrently viewing the real world (typically through a digital camera on a smartphone or tablet) [16]". VR, on the other hand, uses digital visuals to create real-time realistic experiences [17]. As a result, AR incorporates virtual items into a physical location, while VR isolates content from the actual environment and takes people to a completely virtual environment. To put it in another way, virtual reality gives users the sensation of becoming mentally involved in a virtual world [18]. In the Figure 7.1 the images of augmented reality which clarifies the concept very simplified manner. In the Figure 7.2 it tries to explain the concept of virtual liability with more practical approach.

Figure 7.1: Image of augmented reality.
Source: iStock

2 Review of previous studies

It will be imperative for academicians to implement VR and immersion in a meaningful way, so as to create a sense of presence [19]. One of the obstacles to VR adoption is its economic viability [20]. Applied use of pedagogical settings also plays an important role in implementing VR [21]. With VR, pupils can practice complex and repetitive

Figure 7.2: Image of virtual reality.
Source: Unsplash.com

tasks in a safe environment, which has been a significant contribution to education. Surgical operations and dental treatments are procedural jobs that require a specific level of expertise in order to be performed in real life [22]. "Students have been able to improve their cognitive abilities through experiential learning, such as being exposed to settings that would be too logistically difficult to visit in real life [23]." By providing students with a way to directly experience environments or scenarios that are impossible to duplicate in old-fashioned methods of teaching, such as lectures, slideshows, or 2D movies, virtual reality has made a significant contribution to education [24]. "Another recent study focuses on three sub-categories of virtual reality: games, simulations, and virtual worlds. Immersion allows the actor to roam freely throughout the virtual environment, testing hypotheses, fulfilling goals, and evoking motivation and learning [25]". This leads to improvement of students' learning outcomes through AR and VR [26], enhancement of motivation or learning interest with the help of AR and VR [27], and creating a positive perception in the direction of employing AR and VR [28]. AR and VR lead to improved students engagement [29]. AR and VR facilitate authentic learning opportunities [30]. AR and VR facilitate communication and interaction [31]. Promoting self-learning is possible through AR and VR [32]. In addition, there is increased satisfaction with the help of AR and VR [33], better learning experience by VR and AR [34], and more research focuses are possible through AR and VR [35]. Students may practice complicated and challenging skills in a secure setting using AR and VR [36].

3 Objectives of the study

To understand the relevance of AR and VR for imparting inclusive education and adding to the literature of the field

4 Scope of the study

This research focuses on Odisha's numerous engineering institutions. The study subjects include students from the following B. Tech courses: Computer Science, Electronics and Communication, Electrical Engineering, Mechanical Engineering, and Biotechnology.

5 Methodology of the study

We used both secondary and primary data in our research. The data were collected by visiting various libraries and online mode by visiting various websites. For collecting primary data, initially 17 variables were identified from the review of literature and 5 core group discussions consisting of 6 members each. Forty-eight students from different branches were surveyed using a questionnaire. Among the initial variables, 11 variables were finalized after the pilot study. Five-point Likert type scale method was used for the computation of data, and for this purpose, score 5 is for completely agree (CA), score 4 is for agree (A), score 3 for neutral (N), score 2 for disagreeing (DA), and score 1 for completely disagree (CDA). For the collection of the desired data, 512

Table 7.1: Schedule of research work.

Progress of research work	**Month 1**				**Month 2**				**Month 3**				**Month 4**			
	Weeks				**Weeks**				**Weeks**				**Weeks**			
	1	**2**	**3**	**4**	**1**	**2**	**3**	**4**	**1**	**2**	**3**	**4**	**1**	**2**	**3**	**4**
"Conceptualizing and drawing outline of the research work"	■	■	■	■												
"Extensive literature review"					■	■	■	■								
"Gap analysis and finalization of the research questions"									■	■						
"Data collection phase"											■	■	■	■		
"Data analysis"															■	■
"Conclusion and finalization of the research paper work"															■	■

Source: Author's own compilation

questionnaires were distributed and, of that, 402 responses were received, which was 78.51%. However, 387 responses were received in proper form, and all were being considered for the computation of data, even though 384 responses were targeted as per sample size determination of unknown population. The total period of study was 4 months i.e., November 2021 to February 2022. The sample for the current research was collected through non-probabilities sampling, precisely, through the convenient sampling technique.

Table 7.1 explains that the schedule of the first one month focused on conceptualizing and drawing the outline of research the work; in the next one month, the concentration was on extensive literature review. The third month and half of the fourth month were dedicated to gap analysis and finalization of research questions, along with data collection phase; and finally, the remaining period was for data analysis and conclusion.

6 Sample size for unknown population

$$N = Z2\,(P)\ (1 - P)/\,C2$$

$$(1.96)2(0.5)\ (1 - 0.5)/\ (0.05)2$$

$$= 384$$

where

Z = Standard normal deviation, set at 95% confidence level, which is 1.96
P = Percentage picking choice or response, which is 0.5
C = Confidence interval, which is 0.05

Table 7.2: Sampling frame.

City	B. Tech (CS)	B. Tech (EC)	B. Tech (EEE)	B. Tech (Mech)	B. Tech (Biotech)	Total
Bhubaneswar	21	18	19	22	12	**92**
Cuttack	23	16	19	17	14	**89**
Sambalpur	17	15	14	16	6	**68**
Rourkela	15	13	14	16	5	**63**
Berhampur	10	10	3	10	42	**75**
Total	**86**	**72**	**69**	**81**	**79**	**387**

Source: Authors own compilation

With reference to Table 7.2, from Bhubaneswar, 92 respondents were included; from Cuttack, 89 respondents; from Sambalpur, the number of participants included was 68; from Rourkela, it was 63; and from Berhampur, the response was 75 out of the total responses of 387.

7 Data inclusion and exclusion criteria

Respondents were only those students having a basic idea about AR and VR and had experience in using them. Those who did not have any clarity about these realities were ignored for the current study. It was a very challenging job to classify the exclusion and inclusion criteria; however, the researchers managed to collect the desired sample size.

Table 7.3: Socio-demographic profile of the respondents under the study (N = 387).

Items	Category	Frequency	Percentage (%)
Gender	Male	219	56.59
	Female	168	43.41
Age in years	18–22	254	65.63
	23–26	78	20.16
	Above 26	55	14.21
Specialization in B.Tech	Computer Science (CS)	86	22.22
	Electronics and Communication (EC)	72	18.60
	Electrical and Electronic Engineering (EEE)	69	17.83
	Mechanical (Mech)	81	20.93
	Biotechnology (Biotech)	79	20.42
Nature of institution	Private university	104	26.87
	Autonomous	76	19.64
	Private institution	113	29.20
	Government	94	24.29
Religion	Hindu	117	30.23
	Muslim	68	17.57
	Christian	56	14.47
	Others	146	37.73
Residential status	Rural	114	29.46
	Semi urban	84	21.71
	Urban	189	48.83
Knowledge of AR and VR	Good	234	60.46
	Acceptable	153	39.54

Source: Primary data.

With reference to Table 7.3, the percentage of respondents joining the survey: gender-wise 56.59% were male and 43.41% were female. Regarding the composition of age, 65.63% were in the 18–22 range, 22.16% were in the age group 23–26, and the rest were above 26 years of age. In the composition of various branches of B. Tech, 22.22% were from the CS stream, 18.60% were from EC, 17.83% were from EEE, 20.93% consisted of Mechanical engineering students, and the remaining were from Biotechnology. The composition of the nature of institution included 26.87% from Private universities, 19.64% from Autonomous institutions, 29.20% from Private institutions, and the rest were from Government organizations. Regarding religion, 30.23% were Hindus, 17.57% were Muslims, 14.47% were Christians, and the rest were others. In the case of residential status, 29.46% were from rural areas, 21.71% were from semi-urban regions, and the rest were from urban areas. About the knowledge of AR and VR, 60.46% were good and the rest were acceptable knowledge.

Table 7.4: Computation of maximum possible score and least possible score.

Items	B. Tech (CS)	B. Tech (EC)	B. Tech (EEE)	B. Tech (Mech)	B. Tech (Biotech)
Maximum possible score (MPS)	4730	3960	3795	4455	4345
Equation	86 × 11 × 5	72 × 11 × 5	69 × 11 × 5	81 × 11 × 5	79 × 11 × 5
Least possible score (LPS)	946	792	759	891	869
Equation	86 × 11 × 1	72 × 11 × 1	69 × 11 × 1	81 × 11 × 1	79 × 11 × 1

Whereas
Maximum Possible Score = Number of items × Number of respondents × Maximum allotted score
Least Possible Score = Number of items × Number of respondents × Least allotted score

Table 7.5: Analysis of data based on the actual score.

Variables	Aggregate score				
	B. Tech (CS)	B. Tech (EC)	B. Tech (EEE)	B. Tech (Mech)	B. Tech (Biotech)
Improvement of students' learning outcomes through AR and VR	386	328	310	377	369
Enhancement of motivation or learning interest with the help of AR and VR	369	319	308	386	357
Positive perception towards using AR and VR	374	335	311	368	359
AR and VR leads to improved students engagement	367	339	308	382	374

Table 7.5 (continued)

Variables	Aggregate score				
	B. Tech (CS)	**B. Tech (EC)**	**B. Tech (EEE)**	**B. Tech (Mech)**	**B. Tech (Biotech)**
AR and VR facilitates authentic learning opportunities	368	325	314	372	363
AR and VR facilitates communication and interaction	354	333	314	353	362
Promoting self-learning possible through AR and VR	383	331	332	367	372
Increased satisfaction with the help of AR and VR	379	317	336	370	365
Better learning experience by VR and AR	398	328	315	391	381
More research focuses possible through AR and VR	370	319	303	364	371
AR and VR allows students to practice complicated and tough assignments continuously in a safe environment	366	332	326	384	367
Total Score	4114	3606	3477	4114	4040
Maximum Possible Score	4730	3960	3795	4455	4345
Least Possible Score	946	792	759	891	869
% of total score to maximum possible score	86.98	91.06	91.62	92.35	92.98

Source: Annexure- A, B, C, D and E.

Table 7.6: Analysis of data based on percentage component score to total actual score.

Variables	Aggregate score					
	B. Tech (CS)	**B. Tech (EC)**	**B. Tech (EEE)**	**B. Tech (Mech)**	**B. Tech (Biotech)**	**Average**
Improvement of students' learning outcomes through AR and VR	9.38	9.10	8.92	9.16	9.13	9.14
Enhancement of motivation or learning interest with the help of AR and VR	8.97	8.85	8.86	9.38	8.64	8.94

Table 7.6 (continued)

Variables	Aggregate score					
	B. Tech (CS)	B. Tech (EC)	B. Tech (EEE)	B. Tech (Mech)	B. Tech (Biotech)	Average
Positive perception towards using AR and VR	9.09	9.29	8.94	8.95	8.89	9.03
AR and VR leads to improved students engagement	8.92	9.40	8.86	9.29	9.25	9.14
AR and VR facilitates authentic learning opportunities	8.95	9.01	9.04	9.09	8.99	9.02
AR and VR facilitates communication and interaction	8.60	9.23	9.04	8.58	8.96	8.88
Promoting self-learning possible through AR and VR	9.31	9.18	9.55	8.92	9.20	9.23
Increased satisfaction with the help of AR and VR	9.21	8.79	9.66	8.99	9.03	9.14
Better learning experience by VR and AR	9.67	9.10	9.06	9.50	9.43	9.35
More research focus possible through AR and VR	8.99	8.84	8.71	8.85	9.18	8.91
AR and VR allows students to practice a complicated and tough assignment continuously in a safe environment	8.91	9.21	9.36	9.29	9.10	9.17
Total Score	100	100	100	100	100	

Source: Table 7.3.

8 Result and discussion

Table 7.5 analyses the data based on the actual score for the various groups of respondents, which includes students from B. Tech (CS), B. Tech (EC), B. Tech (EEE), B. Tech (Mechanical), and B. Tech (Biotechnology). Their scores were 4114, 3606, 3477, 4114, and 4040, respectively, as against the maximum possible score of 4730, 3960, 3795, 4455, and 4345, respectively. The percentage of actual scores to the total maximum possible score were 86.98%, 91.06%, 91.62%, 92.35%, and 92.98%, respectively. In no case, the actual score was near the least possible score.

Table 7.6 shows data computed similarly, based on data in Table 7.5, Answering the question, improvement of students' learning outcomes through AR and VR; the average opinion of all the groups was 9.14% and the opinion shared by the B. Tech (CS) was 9.38% of the total score in the group, followed by B. Tech (Mechanical) with 9.16%, B. Tech (Biotech) with 9.13%, and B. Tech (EEE) with 8.92%. For the variable, enhancement of motivation or learning interest with the help of AR and VR, the opinion of B. Tech (Mechanical) 9.38%, followed by B. Tech (CS) with 8.97%, B. Tech (EEE) with 8.86%, B. Tech (EC) with 8.85%, B.Tech (Biotech) with 8.64%, and the average score was 8.94%. Answering the question, positive perception towards using AR and VR, B. Tech (EC) leads the table with 9.29%, followed by B. Tech (CS) with 9.09%, B. Tech (Mechanical) with 8.95%, B. Tech (EEE) with 8.94%, B. Tech (Biotechnology) with 8.89%, and the average score was 9.03%.

Answering the question, AR and VR leads to improved students engagement, B. Tech (EC) leads with 9.40%, followed by 9.29% of B. Tech (Mechanical), 9.25% of B. Tech (Biotechnology), 8.92% of B. Tech (CS), 8.86% of B. Tech (EEE), and the average score was 9.14%. For the question, AR and VR facilitates authentic learning opportunities, B. Tech (Mechanical) leads the table with 9.09%, followed by 9.04% by B. Tech (EEE), 9.01% by B. Tech (EC), 8.99% by B. Tech (Biotech), 8.95% by B. Tech (CS),and the average score was 9.02%. Responding to the question related to AR and VR facilitates communication and interaction, B. Tech (EC) leads with 9.23%, followed by B. Tech (EEE) with 9.04%, B. Tech (Biotech) with 8.96%, B. Tech (CS) with 8.60%, B. Tech (Mechanical) with 8.58%, and the average score was 8.88%.

Answering the question related to promoting self-learning possible through AR and VR; the average score was 9.23%, B. Tech (EEE) leads with 9.55%, followed by B. Tech (CS) with 9.31%, B. Tech (Biotechnology) with 9.20%, B. Tech (EC) with 9.18%, and B. Tech (Mechanical) with 8,92%. Answering the question related to AR and VR allowing students to practice a complicated and tough assignment continuously in a safe environment, B. Tech (EEE) leads the table with 9.36%, followed by B. Tech (Mechanical) with 9.29%, B. Tech (EC) with 9.21%, B. Tech (Biotech) with 9.10%, B. Tech (CS) with 8.91%, and the average score was 9.17%.

Answering the question about more research focus possible through AR and VR, B. Tech (Biotech) leads the table with 9.18%, followed by B. Tech (CS) with 8.99%, B. Tech (Mechanical) with 8.85%, B. Tech (EC) with 8.84%, B. Tech (EEE) with 8.71%, and the average score was 8.91%. With respect to the question related to better learning experience by VR and AR, B. Tech (CS) leads the table with 9.67%, followed by B. Tech (Mechanical), B. Tech (Biotech), B. Tech (EC), B. Tech (EEE), and the average score was 9.35%. And, with respect to the query about augmented satisfaction with the help of AR and VR, B. Tech (EEE) leads with 9.66%, followed by B. Tech (CS) with 9.21%, B. Tech (Biotech) with 9.03%, and later by B. Tech (Mechanical) and B. Tech (EC), and the average score was 9.14%.

9 Concluding note

In recent times, with the invention of new technologies, the lives of people have positively changed a lot. This has not only saved time but has also helped to reach the remote corners of the universe. Now, there is no need for physical presence for discharging duties or delivering any work; it can be done through the digital mode also. With augmented reality (AR) and virtual reality (VR) technology, the education system has been transformed from a conventional model to the most sophisticated digital model. The present study was undertaken to understand the relevance of augmented reality and virtual reality in imparting inclusive education, and the research was conducted taking the inputs of B. Tech students of various branches, who have been the direct beneficiaries of AR and VR, leading to their improved learning, improved motivation for study, enhanced students engagement, authentic learning opportunity, better communication, learning opportunity, etc. in addition to other constructive benefits.

10 Ethical consideration

The present work considered all the parameters related to ethical issues. Equal opportunity was given to respondents during the collection of data. There was no forceful collection of data and each participant was explained the various attributes before participating in the survey.

Annexure

Annexure A: Opinion of B. Tech (CS) respondents – 86

Variables	**CA**	**A**	**N**	**DA**	**CDA**	**Score**
	5	**4**	**3**	**2**	**1**	
Improvement of students' learning outcomes through AR and VR	59	14	10	2	1	386
Enhancement of motivation or learning interest with the help of AR and VR	57	11	8	6	4	369
Positive perception towards using AR and VR	55	17	6	5	3	374
AR and VR leads to improved students engagement	53	16	7	7	3	367
AR and VR facilitates in authentic learning opportunities	56	12	9	4	5	368

(continued)

Annexure A: Opinion of B. Tech (CS) respondents – 86						
Variables	**CA**	**A**	**N**	**DA**	**CDA**	**Score**
	5	**4**	**3**	**2**	**1**	
AR and VR facilitates communication and interaction	54	11	5	9	7	354
Promoting self learning possible through AR and VR	59	15	6	4	2	383
Increased satisfaction with the help of AR and VR	58	14	7	5	2	379
Better learning experience by VR and AR	63	16	5	2	0	398
More research focus possible through AR and VR	52	19	8	3	4	370
AR and VR allows students to practice a complicated and toughassignment continuously in a safe environment	55	13	7	7	4	366

Source: Primary data

Annexure B: Opinion of B. Tech (EC) respondents – 72						
Variables	**CA**	**A**	**N**	**DA**	**CDA**	**Score**
	5	**4**	**3**	**2**	**1**	
Improvement of students' learning outcomes through AR and VR	53	12	3	2	2	328
Enhancement of motivation or learning interest with the help of AR and VR	50	11	5	4	2	319
Positive perception towards using AR and VR	53	9	6	6	4	335
AR and VR leads to improved students engagement	57	10	4	1	0	339
AR and VR facilitates in authentic learning opportunities	52	12	3	3	2	325
AR and VR facilitates communication and interaction	55	13	2	2	0	333
Promoting self learning possible through AR and VR	58	4	6	3	1	331
Increased satisfaction with the help of AR and VR	51	5	7	8	5	317
Better learning experience by VR and AR	58	4	3	6	1	328
More research focus possible through AR and VR	49	12	6	3	2	319
AR and VR allows students to practice a complicated and toughassignment continuously in a safe environment	52	14	5	0	1	332

Source: Primary data

Annexure C: Opinion of B. Tech (EEE) respondents – 69

Variables	CA	A	N	DA	CDA	Score
	5	4	3	2	1	
Improvement of students' learning outcomes through AR and VR	50	9	6	2	2	310
Enhancement of motivation or learning interest with the help of AR and VR	52	8	2	4	2	308
Positive perception towards using AR and VR	54	6	3	2	4	311
AR and VR leads to improved students engagement	51	8	4	3	3	308
AR and VR facilitates in authentic learning opportunities	55	4	6	1	3	314
AR and VR facilitates communication and interaction	49	12	5	3	0	314
Promoting self learning possible through AR and VR	57	11	1	0	0	332
Increased satisfaction with the help of AR and VR	58	5	4	2	0	336
Better learning experience by VR and AR	52	10	2	4	1	315
More research focus possible through AR and VR	48	11	3	3	4	303
AR and VR allows students to practice a complicated and toughassignment continuously in a safe environment	54	12	2	1	0	326

Source: Primary data

Annexure D: Opinion of B. Tech (Mechanical) respondents – 81

Variables	CA	A	N	DA	CDA	Score
	5	4	3	2	1	
Improvement of students' learning outcomes through AR and VR	63	11	4	3	0	377
Enhancement of motivation or learning interest with the help of AR and VR	67	9	5	0	0	386
Positive perception towards using AR and VR	62	8	6	3	2	368
AR and VR leads to improved students engagement	70	5	3	0	3	382
AR and VR facilitates in authentic learning opportunities	66	6	2	5	2	372
AR and VR facilitates communication and interaction	61	4	6	4	6	353
Promoting self learning possible through AR and VR	65	3	6	5	2	367

(continued)

Annexure D: Opinion of B. Tech (Mechanical) respondents – 81						
Variables	**CA**	**A**	**N**	**DA**	**CDA**	**Score**
	5	**4**	**3**	**2**	**1**	
Increased satisfaction with the help of AR and VR	66	5	4	2	4	370
Better learning experience by VR and AR	73	4	2	2	0	391
More research focus possible through AR and VR	59	12	4	3	3	364
AR and VR allows students to practice a complicated and toughassignment continuously in a safe environment	66	11	3	0	1	384

Source: Primary data

Annexure E: Opinion of B. Tech (Biotechnology) respondents – 79						
Variables	**CA**	**A**	**N**	**DA**	**CDA**	**Score**
	5	**4**	**3**	**2**	**1**	
Improvement of students' learning outcomes through AR and VR	64	4	11	0	0	369
Enhancement of motivation or learning interest with the help of AR and VR	62	7	2	5	3	357
Positive perception towards using AR and VR	65	4	3	2	5	359
AR and VR leads to improved students engagement	68	6	2	1	2	374
AR and VR facilitates in authentic learning opportunities	62	9	2	5	1	363
AR and VR facilitates communication and interaction	65	3	5	4	2	362
Promoting self learning possible through AR and VR	66	7	2	4	0	372
Increased satisfaction with the help of AR and VR	64	6	3	6	0	365
Better learning experience by VR and AR	70	4	5	0	0	381
More research focus possible through AR and VR	65	5	4	4	1	371
AR and VR allows students to practice a complicated and toughassignment continuously in a safe environment	69	5	0	3	2	367

Source: Primary data

References

[1] Jonassen, D. H. *Computers as mindtools for school: Engaging critical thinking*, NJ: Prentice-Hall, 1999.

[2] Smeets, E. (2005). Does ICT contribute to powerful leaning environments in primary education? *Computers & Education*, 44 (3): 343–355.

[3] Saettler P. (1968). *A history of instructional technology*, New York: McGrawHill.

[4] Wang, F., Lockee, B., and Burton, J. (2011). Computer game-based learning: Chinese older adults' perceptions and experiences. *Journal of Educational Technology Systems*, 40(1): 45–58.

[5] Fernandez, M. (2017). Augmented virtual reality: How to improve education systems. *Higher Learning Research Communications*, 7(1): 1–15.

[6] Strickland, D. C. et al. Brief report: Two case studies using virtual reality as a learning tool for autistic children. *Journal of Autism and Development Disorders*, 26 (6): 651–659, 199.

[7] Mitchell, P., Parsons S., and Leonard, A. (2007). Using virtual environments for teaching social understanding to 6 adolescents with autistic spectrum disorders. *Journal of Autism and Development Disorders*, 37(3): 589–600.

[8] Fabri, M., Moore, D., and Hobbs, D. (2004). Mediating the expression of emotion in educational collaborative virtual enviornments: An experimental study. *Virtual Reality*, 7(2): 66–81.

[9] Wallace, S. et al. (2010). Sense of presence and atypical social judgements in immesive virtual environments: Responses of adolescents with autism spectrum disorders. *Autism*, 14 (3): 199–213.

[10] Kandalaft, M. R. et al. (2013). Virtual reality social cognition training for young adults with high-functioning autism. *Journal of Autism and Development Disorders*, 43(1): 34–44.

[11] Newbutt, N. et al. (2016). Brief report: A pilot study of the use of a virtual reality headset in autism populations. *Journal of Autism and Development Disorders*, 46: 3166–3176.

[12] Riva, G., Wiederhold, B. K., Gaggioli, A. (2016). Being different. The transfomative potential of virtual reality. *Annual Review of Cybertherapy and Telemedicine*, 14:1–4.

[13] Das, P., Zhu, M., McLaughlin, L., et al. (2017). Augmented reality video games: New possibilities and implications for children and adolescents. *Multimodal Technologies and Interaction*, 1: 8.

[14] Bower, M., Howe, C., McCredie, N., et al. (2014). Augmented reality in education-cases, places and potentials. *Educational Media International*, 51: 1–15.

[15] Merchant, Z., Goetz, E. T., Cifuentes, L., et al. (2014). Effectiveness of virtual reality-based instruction on students' learning outcomes in K-12 and higher education: A meta-analysis. *Computers and Education*, 70: 29–40.

[16] Milgram, P, Takemura, H, Utsumi, A, et al. (1995). Augmented reality: A class of displays on the reality-virtuality continuum. *Telemanipulator and Telepresence Technologies: International Society for Optics and Photonics*, 282–293.

[17] Burdea, GC, and Coiffet, P. (2003) *Virtual reality technology*. London, United Kingdom: Wiley-Interscience.

[18] Steuer, J. (1992). Defining virtual reality: Dimensions determining telepresence. *Journal of Communication*, 42: 73–93.

[19] Häfner, P., Dücker, J., Schlatt, C., and Ovtcharova, J. (2018). Decision support method for using virtual reality in education based on a cost-beneft-analysis. In *4th International Conference of the Virtual and Augmented Reality in Education*, VARE 2018, 103–112.

[20] Olmos-Raya, E., Ferreira-Cavalcanti, J., Contero, M., Castellanos, M. C., Giglioli, I. A. C., and Alcañiz, M. (2018). Mobile virtual reality as an educational platform: A pilot study on the impact of immersion and positive emotion induction in the learning process. *Eurasia Journal of Mathematics, Science and Technology Education*, 14: 2045–2057.

[21] Hodgson, P., Lee, V. W. Y., Chan, J. C. S., Fong, A., Tang, C. S. Y., Chan, L., et al. (2019). Immersive virtual reality (IVR) in higher education: Development and implementation. In M. Dieck & T. Jung (Eds.), *Augmented reality and virtual reality*. New York: Springer, 161–173.

[22] Larsen, C. R., Oestergaard, J., Ottesen, B. S., and Soerensen, J. L. (2012). The efcacy of virtual reality simulation training in laparoscopy: A systematic review of randomized trials. *Acta Obstetricia Et Gynecologica Scandinavica*, 91: 1015–1028.
[23] Çalişkan, O. (2011). Virtual feld trips in education of earth and environmental sciences. *Procedia-Social and Behavioral Sciences*, 15: 3239–3243.
[24] Bailenson, J. N., Markowitz, D. M., Pea, R. D., Perone, B. P., and Laha, R. (2018). Immersive virtual reality feld trips facilitates learning about climate change. *Frontiers in Psychology*, 9: 2364.
[25] Gee, J. P. (2004). What video games have to teach us about learning and literacy. *Education Training*, 46: 20.
[26] Abad-Segura, E., González-Zamar, M. D., Rosa, A. L. D. L., and Cevallos, M. B. M. (2020). Sustainability of educational technologies: An approach to augmented reality research. *Sustainability*, 12: 4091.
[27] Redondo, B., Cózar-Gutiérrez, R., González-Calero, J. A. , and Ruiz, R. S. (2020). Integration of augmented reality in the teaching of English as a foreign language in early childhood education. *Early Childhood Education Journal*, 48: 147–155.
[28] Radu, I. (2014). Augmented reality in education: A meta-review and cross-media analysis. *Personal and Ubiquitous Computing*, 18: 1533–1543.
[29] Chen, M. P. , Wang, L. C., Zou, D., Lin, S. Y., Xie, H., and Tsai, C. C. (2020). Effects of captions and English proficiency on learning effectiveness, motivation and attitude in augmented-reality-enhanced theme-based contextualized EFL learning. *Computer Assisted Language Learning*, 1–31.
[30] Chang, S. C., Hsu, T. C., Chen, Y. N., and Jong, M. S. Y. (2020). The effects of spherical video-based virtual reality implementation on students' natural science learning effectiveness. *Interactive Learning Environments*, 28: 915–929.
[31] Kami ́nska, D., Sapi ́nski, T., Wiak, S., Tikk, T., Haamer, R. E., Avots, E., Helmi, A., Ozcinar, C., and Anbarjafari, G. (2019). Virtual reality and its applications in education: Survey. *Information*, 10: 318.
[32] Shadiev, R., and Yang, M. (2020). Review of studies on technology-enhanced language learning and teaching. *Sustainability*, 12: 524.
[33] Bensetti-Benbader, H., Brown, D. Language acquisition with augmented and virtual reality. In *Proceedings of the Society for Information Technology & Teacher Education International Conference*, Las Vegas, NV, USA, 18–22 March 2019, 1476–1480.
[34] Bacca Acosta, J. L., Baldiris Navarro, S. M., FabregatGesa, R., and Graf, S. (2014). Augmented reality trends in education: a systematic review of research and applications. *Educational Technology & Society*, 17: 133–149.
[35] Akçayır, M., and Akçayır, G. (2017). Advantages and challenges associated with augmented reality for education: A systematic review of the literature. *Educational Research Review*, 20: 1–11.
[36] Garzón, J., Pavón, J., Baldiris, S. (2019). Systematic review and meta-analysis of augmented reality in educational settings. *Virtual Reality*, 23: 447–459.

Aditya Singh, Siddharth Mishra, Shubham Jain, Sandeep Dogra, Anubhav Awasthi, Nihar Ranjan Roy, Kunwar Sodhi

8 Exploring practical use-cases of augmented reality using photogrammetry and other 3D reconstruction tools in the Metaverse

Abstract: In today's world, people increasingly rely on mobile apps to do their day-to-day activities, like checking their Instagram feed and online shopping from websites like Amazon and Flipkart. People are depending on WhatsApp and Instagram stories to communicate with local businesses and to leverage the said platforms for online advertising. Using Google Maps to find their way when they travel and finding out the immediate road and traffic conditions with digital banners around the road, has obviously led to a boom in advertising and marketing. Recently, internet users have increasingly desired to immerse themselves in a Metaverse-like platform where they can interact and socialize. Meta's Metaverse is a tightly connected network of 3D digital spaces that allow users to escape into a virtual world. It is designed to change the way you socialize, work, shop, and connect with the real and virtual world around you. These platforms are not fully submerged in the real world; they are inclined toward virtual spaces only, making it obvious to fill this gap. Thus, the proposed framework in this chapter would be a new kind of system that may develop a socio-meta platform, powered by augmented reality and other technologies like photogrammetry and LiDAR. Augmented reality provides an interactive way of experiencing the real world, where the objects of the natural world are enhanced by computer-generated perceptual vision. One of the significant problems with augmented reality is the process of building virtual 3D objects that can be augmented into real spaces, which could be solved with photogrammetry and LiDAR. Photogrammetry is the technique of producing 3D objects using 2D images of a physical object taken from different angles and orientations. LiDAR, on the other hand, is another 3D reconstruction technology used widely by Apple's eco-system. The functioning of LIDAR is very similar to sonar and radar, and the detection and ranging part are where it stands out from the others. The idea behind this platform is to open tons of virtual dimensions in the real world using the principles of mixed reality and geographic mapping tools such as Google Maps, MapBox, and GeoJSON; it would consequently transition the way people spend their time on social media by opening a portal for generating 3D objects that can be augmented to the real-world location using photogrammetry and cloud anchors by just a few 2D digital photographs taken from their camera.

Aditya Singh, Siddharth Mishra, Shubham Jain, Sandeep Dogra, Anubhav Awasthi, Nihar Ranjan Roy, Sharda University, School of Engineering and Technology, Greater Noida, Uttar Pradesh, India, 201310
Kunwar Sodhi, Dominus Labs LLC 5900 Balcones Dr. STE 100, Austin, Texas , 73301, USA

https://doi.org/10.1515/9783110790146-008

Keywords: Entertainment, Social-networking, Navigation, Mixed Reality, Maps, Cloud Anchor, LiDAR

1 Introduction

With increased internet participation and need for additional facilities and developments, there is a desire for innovation. Apps like Instagram and Facebook tend to advertise in one-dimension using either vertical scroll views or horizontal scroll views [1]. On the other hand, the proposed framework is trying to revolutionize the way people interact with each other on the internet in their daily social lives using the latest technologies like augmented reality and photogrammetry [2] by transitioning the two-dimensional scroll views to an augmented reality view of the real world. With it, the proposed framework aims to provide services where users can amalgamate via their digital selves into a single map-based user interface where users may purchase, sell, and socialize on a Metaverse-like platform, also known as the "Reality-virtuality continuum" [3], as shown in Figure 8.1.

This chapter mainly talks about the use-cases of augmented reality in social media, digital marketing, and tourism. The chapter is organized as follows: in Section 2, a discussion of the products in the same domain, such as OLX, Oyo, and Dubizzle, and a comparison of the revenue, user base, and all the different aspects of these companies have been done; Section 3 discusses the methodology; followed by experiments in Section 4. The conclusion of this chapter is given in Section 5.

Figure 8.1: Reality-virtuality continuum.

2 Related work

Virtual environment (VE) [4] is a subset of augmented reality (AR). The VE technology immerses a person in an artificial internal environment. The person cannot see the actual world around them, while engaged. On the other hand, AR allows people to interact with the natural world by superimposing or compositing digital items onto or into the natural environment [5].

2.1 Virtual reality on decentralized networks

Before talking about the practical use-cases of Metaverse with augmented reality, some discussion about its related work and the platforms that are deployed on a decentralized network [6], commonly known as the blockchain network, is required.

Decentraland (stands for decentralized virtual lands) [7] is the first fully decentralized world, powered by technologies like web3 and JavaScript. It's a user-driven platform that aims to build a virtual marketplace where users can explore lands owned by different users and immerse themselves in a virtual world, right from their web browser. Decentraland consists of various spaces, like virtual villages, dungeon mazes, and more. Decentraland also allows its users to create scenes and artworks using their simple yet effective builder tool and a complete SDK for intermediate or advanced level users. Every user in Decentraland has their virtual own avatar. That avatar acts as the user in this virtual world, and the whole experience feels like playing an RPG game while using this platform.

Combining the concept of the Decentraland with an indoor navigation system could bring up more use cases as discussed below:

Current outdoor guidance systems can assist us with where to go and how to get there on the streets [8] since there is no such thing as a versatile navigation system. One can make use of Indoor Navigation System to navigate inside building structures with augmented reality [9]. If a user requires a navigation software to get to a specific location, they consider using indoor navigation. This technology can assist the user while navigating confusing areas, streets, and buildings. A user with a home navigation system can find their current location on a map. The route they want to take will appear on the map after they select the place or area that interests them. This is a client setting because the device calculates the site area.

Use case: In case someone gets lost in a mall or airport, augmented reality can assist the user with navigation, such as locating elevators, stores, and other amenities. Instead, the system will guide user where to go with arrows on the screen, similar to how Google Maps and other map applications work on city streets.

2.2 Social media, e-commerce, and augmented reality

Coming to a centralized network, there are a bunch of very mature platforms that span the domain of e-commerce [10], social media, and management systems:

Urban Company is a cross-platform application that offers a variety of services including home installation, maintenance, and repair. It uses precise location of user to provide the best services available at that time. It also provides a history of the work that user has posted. This app employs a sophisticated search filter with home repair and beauty & wellness services, which accounted for 45 percent of total revenue in the previous fiscal year. Urban Company had a network of around 25,000 professionals and was serving over 22 locations worldwide, as of March 2020.

OLX is an online marketplace where people can buy and sell various products and services. It had over 50 million monthly active users who accessed the platform every month across 4,000+ cities in India and over 8 million monthly transactions, as of September 2019. Dubizzle.com is a fantastic website that allows users to buy and sell anything in their neighborhood. OYO is a leading hotel booking service that connects people with excellent hotels in various cities. It creates about a million job openings in India and South Asia every year.

Instagram and Reddit are two of the most appreciated social media platforms [11] of this generation, with more than a billion monthly active users on Instagram and more than 65 million monthly active users on Reddit.

Unlike the above-mentioned platforms, Flipkart is slowly adopting the AR concept [12] on its e-commerce platform by adding AR views for a group of products, out of their entire product spectrum. Although applications like Flipkart and Decentraland are adopting the AR/VR concepts at their own pace, a significant problem still exists: they cannot build a pure customer-centric platform. Hence, a need for a new conglomerate system is arising.

Table 8.1: Comparison of different platforms offering similar services.

Platform	Year released	Monthly active users	Supported platforms	Domain
Dubizzle	2005	2 million	iOS, Android, Web	E-commerce
Instagram	2010	1 billion	iOS, Android, Web	Social media
OLX	2006	15 million	iOS, Android, Web	E-commerce
OYO	2013	~1 million	iOS, Android, Web	Management services
Reddit	2005	62 million	iOS, Android, Web	Social media
Urban Company	2014	~6 hundred thousand	iOS, Android, Web	Household services

3 Methodology

The primary aim is to build a conglomerate platform that is powered by augmented reality and Google Maps, with a centralized as well as a decentralized network. Keeping this in mind, the proposed framework offers the following services on a single map-based UI, so that every object on the platform can be pinned to a real-world geolocation:

- E-commerce – purchase, sell and rent items with augmented reality by building 3D objects with 2D images utilizing technologies like photogrammetry and LiDAR [13].
- Use case: Consider selling an old LED TV on a marketplace like OLX; the product page on OLX only shows some low-quality images, and other details like the

contact details of the seller and price. Having a system that lets the user take some images of the LED TV as the seller, which then creates an augmenTable 3D object in the cloud will surely change the whole experience of e-commerce.

- Social media – socializing in a reality-virtuality continuum, customizing the virtual avatars, and participating in virtual events tethered to real-world geo-locations.

Use case: Users can organize different social events around the most expensive real-world locations that people can attend, right from their phone or any other compatible device, without leaving their place. This opens tons of possibilities for marketing as well. Interested companies can pre-book the billboard for advertisements in this virtual world, which will be more convenient for them.

This framework provides everything that one needs on a single map-based UI, where users may purchase and sell by categories, receive or deliver services like babysitting, tutoring, elder care, beauty care, dog walking, cleaning, and much more. Users can also find entertainment, compete in sports and game tournaments, get a cheap ride, find a place to stay, meet new friends, and socialize in mixed reality that runs on a blockchain network [14].

The technical hierarchy of the proposed framework (as shown in Figure 8.2) essentially combines all the services, and targets all the audiences, as shown in Figure 8.4.

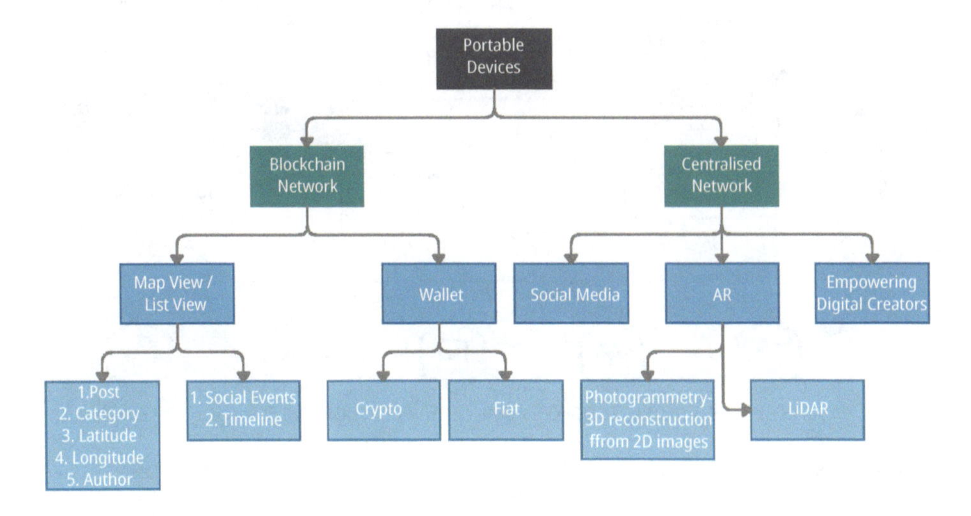

Figure 8.2: Technical hierarchy of the proposed framework.

A wallet system helps store private keys, keeping users' crypto safe and accessible. Crypto wallets [15] technically do not store crypto. Instead, users' holdings live on the blockchain but can only be accessed using a private key. Keys prove users' ownership of digital money and allow the users to make transactions. Crypto wallets have use cases ranging from simple-to-use apps to more complex security solutions.

The same can also be said for fiat currencies, as it minimizes the process of adding money from online methods. However, unlike crypto wallets, these wallets store digital money, which get their approval from the central bank. Therefore, transactions from these wallets are not sent to the bank, instead they are sent from one wallet to another.

With AR as a medium, content creators can showcase their stuff more creatively and innovatively than traditional videos and images.

Augmenting their work into 3D objects will bring more depth and transparency to their content and ideas.

The system architecture of the proposed framework (as shown in Figure 8.3) demonstrates a high-level proof of concept. This demonstrates the whole concept thoroughly.

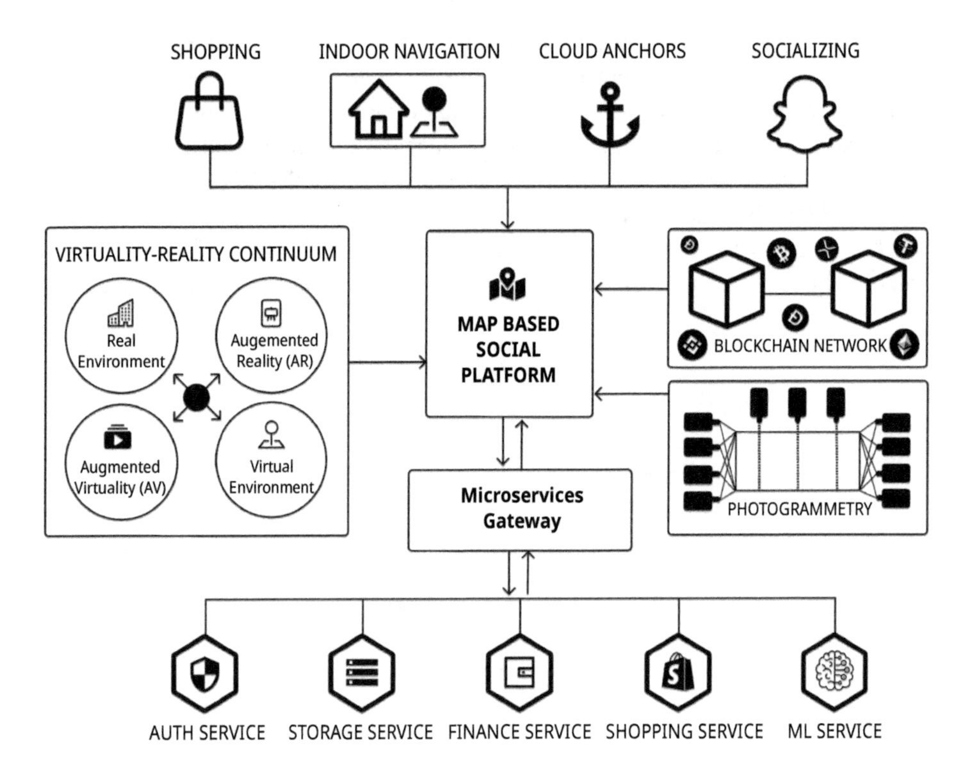

Figure 8.3: Proposed framework.

This proposed framework demonstrates how all the atomic pieces of the framework are stitched together to produce the overall outcome of a Metaverse-based social media platform. Each component in this framework has its own vital role, which is irreplaceable with old technologies and the traditional social media platforms like Facebook, Instagram, and TikTok. A comparison of MAU (monthly active users) of traditional social media platforms is show in Figure 8.4.

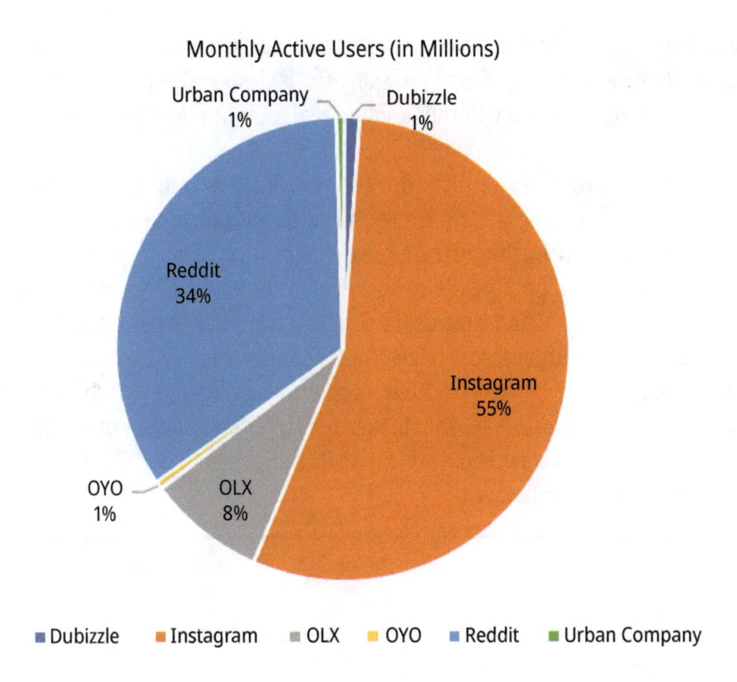

Figure 8.4: Comparing similar products based on user adoption.

Let's talk about all the architectural components of the framework one by one (in bottom-to-top order) and see how each component has its own vital role.

3.1 Proposed framework components

A. Microservices

With the increase in internet participation, a typical online platform nowadays receives over a million requests to their API and databases, which may lead to downtime due to heavy traffic. This leads to exploration and experimentation of different architectures available while designing a system architecture that best suits the use cases and the features offered by the platform.

Among the different kinds of system architecture designs, microservice architecture is most compatible with our platform. A microservice architecture comprises of multiple levels of accessibility and a collection of resources like databases, load balancers, daemon sets, deployments, services, ingress, cluster IPs, and replica-sets.The proposed model offers three different levels of the same microservice architecture:

- Development
- Staging
- Production

Development and staging are non-production levels that are designed for internal debugging and testing, while production, as the name suggests, is designed for production track only. For the sake of simplicity, we will talk about development and staging levels only in this chapter.

The development and staging tracks utilize docker and docker-compose for containerization and internal networking, which are required to pass data to each other over a TCP channel, which is used in this architecture. Each microservice, shown in Figure 8.2, is running on a unique port inside a docker container that has specified resources for RAM usage, CPU usage, and disk usage. Docker-compose is used to stitch these different containers containing microservices together so that they can communicate with each other using their ports and their respective internal IP addresses (this IP address is given to the containers by the docker itself). This mapping of different containers in docker-compose is achieved by a YAML file. A small section of the YAML file of our framework is given below.

```
version: "3.9"
services:
    finance:
        build: ./mappe-finance-service
        expose:
          - "8003"
        links:
          - "auth"
        environment:
          - LEVEL=staging
        email:
             build: ./mappe-email-service
             links:
          - "auth"
        expose:
          - "8004"
        environment:
             - LEVEL=staging
    ---
    ---
    ---
    EOF
```

Each microservice is written in a NodeJS framework, named NestJS, which is a very mature and acknowledged framework for building microservice backend applications in the industry. The microservice pattern of NestJS framework offers tons of options for the communication channel, such as AMQP, Kafka, TCP, etc. Our choice of communication channel is TCP, as discussed above.

Below is a code snippet that shows how a TCP microservice is made using NestJS.

```
async function bootstrap() {
  const app = await NestFactory.createMicroservice(AuthenticationModule, {
    transport: Transport.TCP,
    options: {
      host: config.get('authenticationService').ip,
      port: config.get('authenticationService').port,
    },
  } as TcpOptions);
  app.useGlobalPipes(new ValidationPipe())
  await app.listen();
}
```

A microservice architecture has at least one gateway service. A gateway service has an ingress attached to it, which is mostly open on port 80 for http and port 443 for https, which makes the whole API exposed to external machines. Microservices, other than gateway, are not exposed to external traffic, as only the gateway can communicate with those microservices. In some exceptional cases, microservices other than gateway can also be exposed to external traffic, if required.

The same architecture is used in production, with some changes in ingress and load balancer, since different technologies are used in production that are more optimized for production-ready deployment requirements. Kubernetes is widely used in place of docker-compose and docker-swarm in production environments. Kubernetes is a good choice for production environments since almost all cloud providers like GCP, AWS, and Azure have developed an abstraction layer over it, which is easy to use and also acts as "platform as a service," often abbreviated as PaaS.

GKE (Google Kubernetes Engine) used in the proposed framework, offers services like HPAs (Horizontal Pod Auto-scalers), which can scale the microservices, based on defined metrics. This metric is known as scaling metric, which decides when to scale-up or scale-down the deployments running in the GKE cluster, based on defined parameters like number of requests, CPU usage, or memory usage. If primitive parameters are not enough to decide when to scale-down or scale-up, then custom metric providers can also be written and added to the cluster, which can periodically monitor the cluster health and resources using Google's cloud logging service.

Augmented reality requires a fault-tolerant and robust system, which can be achieved by using the microservices pattern and orchestration tools like Kubernetes and docker-swarm.

Hence, the proposed framework uses docker-compose in the development and staging environments, while Kubernetes is being used in the production environment.

B. Blockchain

Apart from our own network with load-balancers and other optimizations, a fair share part of the platform also depends on the blockchain network. Blockchain is being used to implement the wallet system, and augmented NFTs pinned to a geolocation in the proposed framework can be generalized as a Metaverse-based social media platform. The system runs on Ethereum network using solidity smart contracts. Solidity is a programming language that is used by developers to write smart contracts, which can run on the EVM (Ethereum Virtual Machine); it works similar to how JVM (Java Virtual Machine) works. Each transaction on the Ethereum network requires a gas fee to run, which is paid in ETH coin by the user who is creating the transaction on the network. The reason we are using a blockchain network for handling the wallet system of this platform is to prevent any charge backs made by buyers, which can lead to financial frauds. Non-blockchain services like PayPal, Stripe, Payoneer, etc. do not have a good charge back protection system, hence blockchain is needed.

C. Indoor navigation

Outdoor navigation can be achieved by geo-positioning satellites, using services like Google Maps, Apple Maps and Microsoft Maps, etc., but there are very limited services that can offer a full-fledged indoor navigation system. The proposed framework aims to develop an indoor navigation system, which would be powered by augmented reality and GeoJSON points. Augmented reality would be used to place virtual objects like arrows or turns on the real scenes to guide the user to their destination. Sophisticated places like malls, universities, and railway stations can use this application to replace their directional boards and stickers with an indoor navigation system powered by augmented reality.

Of course, this system may get tons of requests every minute, but the microservices pattern discussed above will be able to handle all the load by utilizing the HPA (Horizontal Pod Auto-scalers) metric of the production environment Kubernetes cluster, powered by the Google Cloud Platform.

D. E-commerce with augmented reality

The proposed framework aims to combine e-commerce services with a Metaverse-powered social media platform. The e-commerce platform that is being developed is not limited to only physical goods but also supports digital products like yoga classes, video tutorials, NFTs, and paid game-spaces powered by augmented reality, where users can spend their free time and form a better internet community. This concept is similar to Decentraland that has developed a decentralized world on the web where users can interact, advertise, and carry out transactions of digital goods among themselves. Decentraland lacks the fundamental concepts of augmented reality and mixed reality on their platform, which can be filled by the proposed framework, which has great possibilities in e-commerce.

Advertising and marketing spots in the real world are often very expensive and hard to get due to low availability of the spots. Brands can be advertised in this e-commerce space, which is running in a reality-virtuality continuum.

Along with big brands, this platform will be open for small content-creators as well, where they can advertise their content and products in a more intuitive way, using augmented reality, and get a global exposure. This can lead to tremendous conversion of their users and the audience spectrum, as people will be able to try out the products offered by brands and content creators from their own place using augmented reality or mixed reality headsets.

Artists are also increasingly combining their art and technology to create NFT face filters. Marc-O-Matic from Melbourne is a good example – his artwork incorporates traditional graphic art, 3D animation, and augmented reality.

3.2 Photogrammetry and LiDAR

The major problem in the existing platforms is that they cannot develop a user-driven system. In order to make a user-driven AR-powered conglomerate platform where users can purchase, sell, rent, and socialize, a 3D construction architecture is needed. This 3D reconstruction architecture [16] can be used to create augmentable 3D objects by a bunch of 2D images in the cloud, taken from the user's camera. This type of system is not offered by any e-commerce or social media platform, to date.

This idea can be made possible using Photogrammetry and LiDAR. The basic principle of Photogrammetry is building 3D objects using 2D images of a physical object, taken from different angles and orientations (shown in Figure 8.4). The outcome of the Photogrammetry majorly depends on the following constraints:

- Visibility constraints: Fog, rain, or dense vegetation cover.
- Lightning conditions: A light temperature of around ~6,500 Kelvin (D65) – similar to daylight; light the object with an illuminance of 250 to 400 lux [17].
- Capturing angles play a vital role; sequential images should have 70% or more overlapping angles.
- Background: A matte, middle gray background works best.

LiDAR (Light Detection and Ranging) is another 3D reconstruction technology used widely by Apple's eco-system. The functioning of LiDAR is very similar to sonar and radar, and the detection and ranging part is where it stands out from the others.

Sonar uses sound waves to detect the position of an object, whereas radar uses radio waves for the same. However, in LiDAR, the medium is ultraviolet, visible, or near-infrared light.

Apple has added a new feature to its Pro series with the introduction of the LiDAR sensor. The LiDAR sensor [18] makes it possible to seamlessly create 3D objects from 2D images by a portable device such as iPhone or an iPad. People mostly use social media from their phones; this opens many possibilities in the augmented reality world.

The scene created in Figure 8.5 can be posted on an e-commerce platform that supports an AR viewer, making it possible for other users to augment the object at their place to view, explore, and get a better idea of the whole object before purchasing or renting it.

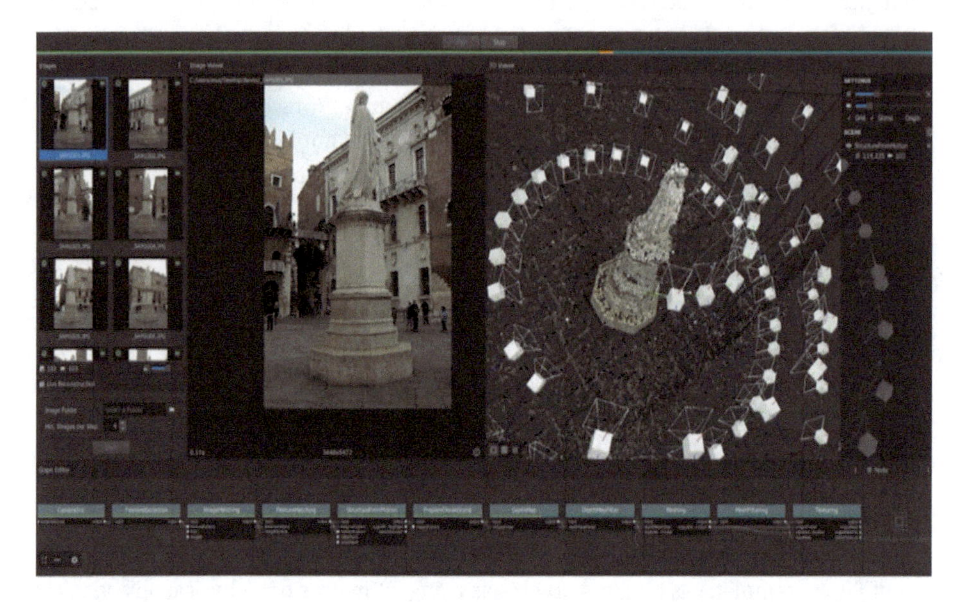

Figure 8.5: A user interface depicting a Photogrammetry tool "Meshroom." Left view is the 2D images navigator, the right view is the 3D viewer, and the bottom view is the blueprint of the reconstruction setup.

Photogrammetry competes with LiDAR in a variety of domains, as shown in the table below, which makes Photogrammetry an optimal technology for the features that this proposed framework aims to offer.

Table 8.2: Comparing LiDAR with Photogrammetry.

Technology	Field of view	Initial expense	Image overlap	Platform
LiDAR	60–90 degrees	More	Less	Apple only
Photogrammetry	60 degrees	Less	Medium	Cross-platform (open-source)

A LiDAR-enabled device emits multiple beams of lights that get bombarded to the surface of an object (as shown in Figure 8.6a). Those beams bounce back to the LiDAR sensor, and the returned beams are then processed by the device (in this case, an iPhone), which calculates the total time taken for the round trip.

Figure 8.6: Image depicting the 3D reconstruction of a scene using the LiDAR sensor of an iPhone 13 Pro. LiDAR scanner of iPhone, intermediate screen and final reconstruction of scene (from left to right).

Since each beam returns at a different time (as the further the object from the sensor is, the more time it takes to produce) to the sensor, the device can calculate the subject's size, shape, and position with the help of this time difference.

The result obtained from the process helps create a 3D picture of the environment, which includes multiple variables such as length, depth, height, and position, relative to nearby objects, as shown in Figure 8.6b. The accuracy of the generated 3D object competes with traditional photogrammetry in terms of details (as shown in Figure 8.6c).

The small footprint of LiDAR sensors is very suitable and feasible for cameras. With the 3D map created, low-light photography is drastically improved, which helps take low light shots.

3.3 Meshroom, 3DF Zephyr, and Apple Reality Composer

The three most promising Photogrammetry-based 3D reconstruction tools are Meshroom [19], 3DF Zephyr [20], and Apple Reality Composer [21]. Meshroom is an open source software. 3DF Zephyr has a free version with limited features and also offers additional paid features, while Apple Reality Composer is a proprietary software by Apple that is free to use.

Apple Reality Composer is supported on macOS and iPadOS, whereas 3DF Zephyr and Meshroom are supported on Windows and Linux but not on macOS.

Table 8.3: Comparing different methods used for photogrammetry, based on their memory usage, GPU usage, CPU usage, and time taken to produce a 3D object from a 2D image.

Method	Memory usage	CPU usage	GPU usage	Time taken
Apple Reality Composer	~75%	~60%	~48%	~8 min
Meshroom	~90%	~80%	~90%	~40 min
3DF Zephyr	~85%	~84%	~73%	~26 min

3.4 Augmented reality and the Metaverse

Metaverse is a virtual world powered by the blockchain network. It allows users to explore the realm between the real and the virtual world by using different technologies like 3D graphics, blockchain, mixed reality, and web technologies such as WebXR and Web3 [22].

WebXR and other similar technologies such as ArKit and ARCore provide freedom using hardware devices like head-mounted displays, commonly known as VR headsets. Mobile devices can also be used for AR, as they have a multitude of different sensors, such as an accelerometer, gyroscope, and GPS. Augmented reality focuses on making this augmented realm as real as possible, which is known as degree of freedom in scientific terms. The degree of freedom enabled by WebXR and similar technologies offers 3DoF or 6DoF, based on the number of sensors available on the device while using an application.

Augmented reality with Metaverse is an exciting concept because both technologies are new and are ready to run at their full potential, thanks to the latest advancements in portable devices. Since the Metaverse consists of heavy graphical objects, mobile devices like the newest iPhone lineup [23] and other flagship Android devices provide high-speed GPUs (Graphical Processing Units) that are perfectly capable of rendering sophisticated 3D and virtual views in no time.

Combining things like organizing social events, spending time with friends, live converts, sightseeing, non-fungible tokens (NFTs – ERC721), and standard fungible tokens (FTs – ERC20) with augmented reality or mixed reality can fill the gap that exists at the moment in the Metaverse. The Metaverse aims to focus on a single concept, allowing people to do in a virtual space, the things they do in real life. This dream of a virtual world can only be possible with mixed reality technologies.

Some applications of augmented reality offered by the proposed framework are:

A. Interactive playgrounds powered by augmented reality

- In-built AR playgrounds/virtual labs for students to interact with augmented reality objects like human anatomy and chemical components.
- Ability to create custom AR playgrounds with 3D objects and discover all the things that are impossible to visualize with traditional learning techniques, which are possible with education games in augmented reality.

B. Promoting sales with augmented reality

Shopkeepers can utilize AR-powered billboards to layout their best-sellers and featured items outside their store. These billboards would co-exist in the real-world using cloud anchors, so users may interact with the augmented objects before visiting the store. This will increase the chances of a store visit by customers who are interested in purchasing the item, but are too busy to visit each and every store.

C. 3D product NFT

Art NFTs are not limited to works of art and face filters. 3D models of unique products are also growing. NFT fashion sounds like a dress you can wear, which is made possible by AR filters.

The collaboration between the crypto brand, RTFKT Studios, and Atari has created a line of digital sneakers that consumers will be able to show off at all multiplayer blockchain games, including the upcoming Atari Metaverse. Interestingly, users can try these sneakers at home before using Snapchat or the Metaverse ME app.

D. Augmented reality powered games

In today's world, the gaming industry is increasing day by day; be it population-wise or technology-wise, the graph of AR technology in the gaming industry is rising continuously. AR gaming integrates the game's visual and audio with the user's environments in real life. In contrast, Virtual Reality (VR) needs a separate area to create such an environment to play games using a VR headset. On the other side, AR games are played on devices like smartphones, tablets, and other gaming systems. In AR gaming, an environment is created based on the user's surroundings in real time. Users can immerse themselves completely in the game as they experience the activities performed by other game characters, combined with the real world at the user's location. The main attraction of any game is its scenery and environment, which attracts the users to be involved in the game and enjoy it. Still, after some time, when users have completely explored the game and its environment, they may get bored and may want to move to a different scenery or environment. Environment creation is one of the most time-consuming parts of game development. By considering the constant demands for new scenery, AR gaming expands the game field by taking advantage of a real-world environment to keep users interested in a game for a long period.

For example, Pokémon Go is an example of an AR app for gaming that uses a smartphone's camera, gyroscope, clock, and GPS to enable a location-based augmented reality environment. In this game, there is a display of map of the current environment and the movement of the grass indicates the presence of a Pokémon. In AR mode, after touching the screen, it displays the Pokémon in the user's real-world environment using augmented reality

E. Impact of AR in advertisements

Consider the case of soccer; the televised advertisement will be different on the American television screen compared to that of Germany. The same can also be seen in other countries where brands in that region can pay to advertise their product on TV.

Figure 8.7: The English FA used the technology to deliver modified feeds for England's World Cup warm-up against Costa Rica [24].

The Bundesliga is one of Europe's most significant big five leagues to thoroughly test and approve the latest version of virtual advertising that substitutes ads on the perimeter board, which has been made possible by augmenting the ads provided by the brands, which means that one international match can be advertised to people living in different regions.

The augmented reality software has been created by a company called Supponor, and their software is built into virtual hybrid LED systems. With this software, brands can speak to their local audience more effectively.

4 Experiments

The reconstruction process with the same 2D images on three different tools has been done. The tools of interest here are Meshroom, Apple Reality Composer, and 3DF

Zephyr. After the reconstruction of the 3D models, the system augments them in the real-world locations with cloud anchors, so they can stay pinned to a geo-location point in our cloud servers.

4.1 Apple reality composer

To begin with, Apple Reality Composer, or macOS 12.0 or above is required along with XCode. A command-line application is used to make these experiments quick and intuitive. The command-line application takes the following parameters while recording the 2D images shown in Figure 8.6 as inputs:

- d (detail): preview, reduced, medium, full or raw
- o (sample-ordering): sequential, unordered
- f (feature-sensitivity): normal, high

Based on the complexity of the input images, it outputs a reconstructed 3D object in 10–60 min (these results are based on our experiments).

In this experiment, 40 images are used (each digital image has 7 megapixels). The parameters used in this experiment are

- d: medium
- o: sequential
- f: normal

```
Detail: extension PhotogrammetrySession.Request.Detail{
        self =.medium
}


Sample-ordering: extension PhotogrammetrySession.Configuration.
SampleOrdering {
        self =.sequential
}

Feature-sensitivity: extension PhotogrammetrySession.Configuration.
FeatureSensitivity{
        self =.normal
}
```

Using the following code parameters in swift programming language would result in a code snippet like that shown above.

The format of the output 3D object is USDZ [25] (Universal Scene Description), which is developed by Pixar and Apple together. USDZ is a file format that displays 3D and AR content on Apple devices. USDZ focuses on the idea of unifying different AR creation tools and schemes.

4.2 3DF Zephyr

3DF Zephyr is an application that requires a lot of processing power. It utilizes all available CPU cores and Nvidia CUDA [26] technology, when it is available. Note that a video card with CUDA support can significantly increase performance (multi-GPU configurations are also supported in Lite, Aerial, Pro). Closing all open applications when completing a reconstruction using 3DF Zephyr is recommended.

There are three main phases in Zephyr:

- Beginning a new project and determining the camera attitude (also known as "Structure from Motion").
- Generating 3D model
- Extracting a Dense Point Cloud – this is also referred as 'Multi-View Stereo'
- Extracting a Surface
- Extracting a Textured Mesh [27]

Parameters

Camera orientation pre-sets

- Category – Ariel, Close Range, Human Body, Urban.
- Pre-sets – Fast – Reduced Resolution, bundle adjustment, and number of critical points: Faster than the default
- Default – ideal for most use cases
- Deep – This preset is best for more adjustment iterations, crucial point counts, and camera matches; it is slower than the default settings. When cameras are lost in default mode, this is suggested.

Depth Point Cloud Generation (This process can be done multiple times to get the most accurate results by tweaking the parameters shown below)

Cameras to use Presets

- All cameras,
- Selected cameras in the workspace
- Select cameras
- Select cameras by tag

Creation of presets category
- General – This category generally suits well for most kinds of reconstructions.
- Aerial – Natural Images: Reconstruction with a top-down view scenario, usually from a dataset obtained by a UAV or drone
- Urban – buildings, facades, or scenarios shot in an urban area.
- Human Body – Human body parts, including close-ups
- Surface scan – Scanning surfaces close-ups (e.g., terrain or ground)
- Vertical Structure – For scanning radio towers or other twin vertical structures from a drone

4.3 Alice vision's Meshroom

An NVIDIA CUDA-enabled GPU is appreciated. The binaries are built with CUDA-10 and are compatible with compute capability 3.0 to 7.5. Only "Draft Meshing" can be used for 3D reconstruction without a supported NVIDIA GPU.

For most datasets, the default parameters are ideal. Furthermore, numerous characteristics are disclosed for research and development purposes but are useless to users. A subset of these may be helpful for advanced users in improving the quality of specific datasets.

Presets
1. **Feature matching**: Enable Guided Matching This option enables a second stage in the matching procedure.
2. **Enable AKAZE** [28] to improve performance on some surfaces (like skin, for instance)
3. **Feature extraction** [29]: Change DescriberPreset from Normal to High. If the given dataset is not extensive (<300 images), one can use High preset.
4. To improve the robustness of the initial image pair selection/initial reconstruction, one can use an SfM with minInputTrackLength [30] set to 3 or 4, to keep only the most robust matches (and improve the ratio of inliers/outliers).

Table 8.4: Metrics of Apple Reality Composer, Meshroom, and 3DF Zephyr while processing 3D image reconstruction on the same image dataset.

Method	No. of images	CPU	GPU	Time taken
Apple Reality Composer	100	Apple M1 Max	M1 Max	~8 min
Meshroom	100	Intel i7	Nvidia 1060	~40 min
3DF Zephyr	100	Intel i7	Nvidia 1060	~26 min

Same input images (shown in Figure 8.6) are passed to all three tools, i.e., Apple Reality Composer, 3DF Zephyr and Meshroom.

Figure 8.8: Input images of a statue taken from a mobile camera [31].

As it is visible from the results (shown in Figure 8.7), Apple Reality Composer outperforms both Meshroom and 3DF Zephyr in terms of both time of execution (shown in Table 8.3) and the details produced by these tools.

Figure 8.9: Output produced by Apple Reality Composer, 3DF Zephyr and Meshroom (from left to right).

5 Conclusion

The biggest hurdle in developing a user-driven AR-powered platform is creating 3D objects that are augmentable in natural spaces. Existing e-commerce apps like Flipkart and Apple store provide AR views for their products, but those are not user-driven, because those objects are created and uploaded by the company itself.

Allowing users to capture images of a physical object and turn it into a virtual 3D object (that too in the cloud) opens up doors to many opportunities. Talking about social media platforms, several potential candidates like Decentraland, Snapchat, and Instagram are already utilizing augmented reality and virtual reality. Still, they do not provide a feature where users can create their own custom AR filters or their own custom AR objects.

On that note, it can be concluded that having a system that allows AR-enabled e-commerce and social-media experiences is very much needed. Hence, the proposed framework aims to develop and deploy this platform at scale with minimal latency and high availability, even in high traffic. The proposed framework is a cross-platform, fully-featured platform where users can buy, sell, rent, and socialize in a so-called "reality-virtuality continuum of the internet.

References

[1] Weller, K. (2016). Trying to Understand Social Media Users and Usage: The Forgotten Features of Social Media Platforms. *Online Information Review.*

[2] Wolf, P. R., and Dewitt, B. A. (2000). *Elements of photogrammetry: With applications in GIS*, vol. 3. New York: McGraw-Hill.

[3] Williams, T., Szafir, D., and Chakraborti, T. (2019). The Reality-Virtuality Interaction Cube. In Proceedings of the 2nd International Workshop on Virtual, Augmented, and Mixed Reality for HRI.

[4] Stuart, R. (1996). *The design of virtual environments*. McGraw-Hill, Inc.

[5] Hughes, C. E., Stapleton, C. B., Hughes, D. E., and Smith, E. M. (2005). Mixed Reality in Education, Entertainment, and Training. *IEEE Computer Graphics and Applications*, 25(6): 24–30.

[6] Hughes, C. E., Stapleton, C. B., Hughes, D. E., and Smith, E. M. (2005). Mixed Reality in Education, Entertainment, and Training. *IEEE Computer Graphics and Applications*, 25(6): 24–30.

[7] Dowling, M. (2022). Fertile LAND: Pricing Non-fungible Tokens. *Finance Research Letters*, 44: 102096.

[8] Matuszka, T., Gombos, G., and Kiss, A. (2013). A New Approach for Indoor Navigation using Semantic Web Technologies and Augmented Reality. In International Conference on Virtual, Augmented and Mixed Reality, pp. 202–210. Springer, Berlin, Heidelberg.

[9] Sato, F. (2017). Indoor Navigation System based on Augmented Reality Markers. In International Conference on Innovative Mobile and Internet Services in Ubiquitous Computing, pp. 266–274. Springer, Cham.

[10] Molla, A., and Licker, P. S. (2001). E-commerce Systems Success: An Attempt to Extend and Respecify the Delone and MacLean Model of IS Success. *Journal of Electronic Commerce Research*, 2(4): 131–141.

[11] Carpenter, J. P., Morrison, S. A., Craft, M., and Lee, M. (2020). How and Why are Educators using Instagram?. *Teaching and Teacher Education*, 96: 103149.

[12] Thomas, N. J., Thomas, N. T., Niyas, N., Kuruvilla, N. M., and Manoj, P. (2021). A Study on Customer Satisfaction of Electronic Products From Flipkart With Special Reference to Kottayam District..

[13] Habib, A. F., Ghanma, M. S., and Tait, M. (2004). Integration of LiDAR and Photogrammetry for Close Range Applications. Proceedings of the ISPRS Geo-Imagery Bridging Continents.

[14] Song, G., Kim, S., Hwang, H., and Lee, K. (2019). Blockchain-based Notarization for Social Media. In 2019 IEEE international conference on consumer electronics (icce), pp. 1–2. IEEE.

[15] Suratkar, S., Shirole, M., and Bhirud, S. (2020). Cryptocurrency wallet: A review. In 4th International Conference on Computer, Communication and Signal Processing (ICCCSP), pp. 1–7. IEEE.

[16] Mouragnon, E., Lhuillier, M., Dhome, M., Dekeyser, F., and Sayd, P. (2006). Real Time Localization and 3d Reconstruction. In 2006 IEEE Computer Society Conference on Computer Vision and Pattern Recognition (CVPR'06), vol. 1, pp. 363–370. IEEE.

[17] Asaad, R. R. (2021). Virtual Reality and Augmented Reality Technologies: A Closer Look. *Virtual Reality*, 1(2).

[18] Luetzenburg, G., Kroon, A., and Bjørk, A. A. (2021). Evaluation of the Apple iPhone 12 Pro LiDAR for an Application in Geosciences. *Scientific Reports*, 11(1): 1–9.

[19] Griwodz, C., Gasparini, S., Calvet, L., Gurdjos, P., Castan, F., Maujean, B., De Lillo, G., and Lanthony, Y. (2021). AliceVision Meshroom: An Open-source 3D Reconstruction Pipeline. In Proceedings of the 12th ACM Multimedia Systems Conference, pp. 241–247.

[20] 3DF Zephyr – Photogrammetry Software – 3d Models from Photos, 3Dflow, 2022. [Online]. Available: https://www.3dflow.net/3df-zephyr-photogrammetry-software/. [Accessed: 26 Apr 2022].

[21] Nhan, J. (2022). Reality composer: Creating AR content. In *Mastering ARKit*. Apress, Berkeley, CA, 309–335.

[22] Engberg, M., Bolter, J. D., and MacIntyre, B. (2018). RealityMedia: An Experimental Digital Book in WebXR. In 2018 IEEE international symposium on mixed and augmented reality adjunct (ISMAR-Adjunct), pp. 324–327. IEEE.

[23] Morrissey, S. (2010). History of apple mobile devices. In *iOS forensic analysis for iPhone, iPad, and iPod touch*. Berkeley, CA: Apress, 1–23.
[24] Kidd, R., How 'Virtual' Advertising Is Helping Brands Reach International Soccer Fans, Forbes, 2022. [Online]. Available: https://www.forbes.com/sites/robertkidd/2018/08/24/how-virtual-advertising-is-helping-brands-reach-international-soccer-fans/?sh=e91e9ea6b7f6. [Accessed: 30 Apr 2022].
[25] Kadyrov, R. I. (2021). Developing of USDZ models for 3D Digital Analysis Results Visualization in Augmented Reality. *Scientific Visualization*, 13: 1.
[26] Buck, I. (2007). Gpu Computing with Nvidia Cuda. In ACM SIGGRAPH 2007 courses, 6–es.
[27] Henderson, P., Tsiminaki, V., and Lampert, C. H. (2020). Leveraging 2d Data to Learn Textured 3d Mesh Generation. In Proceedings of the IEEE/CVF Conference on Computer Vision and Pattern Recognition, pp. 7498–7507.
[28] Sharma, S. K., and Jain, K. (2020). Image Stitching Using AKAZE Features. *Journal of the Indian Society of Remote Sensing*, 48(10): 1389–1401.
[29] FeatureExtraction – Meshroom v2021.0.1 documentation, Meshroom-manual.readthedocs.io, 2022. [Online]. Available: https://meshroom-manual.readthedocs.io/en/latest/feature-documentation/nodes/FeatureExtraction.html?highlight=Feature%20extraction#featureextraction.
[30] Reconstruction Parameters – Meshroom v2021.0.1 Documentation, Meshroom-manual.readthedocs.io, 2022. [Online]. Available: https://meshroom-manual.readthedocs.io/en/latest/faq/reconstruction-parameters/reconstruction-parameters.html?highlight=minInputTrackLength#sparse-reconstruction. [Accessed: 26 Apr 2022].
[31] Piazza dei Signori, D. V. (2022). Dante – Piazza dei Signori, Verona – 3D model by 3dflow (@3dflow) [53f0cbb], Sketchfab.

Shreya Soman, Neeru Sidana*, Richa Goel

9 An empirical analysis of conspicuous consumption of luxury cosmetic products via augmented reality technology concerning India's workforce

Abstract: Status and conspicuous purchasing pattern helps in developing links of customers with certain products and brands towards establishing status. Augmented Reality (AR) allows consumers to virtually try goods, and provide a "try before you buy" experience while shopping online. Primary data was collected throughout the Indian subcontinent region for this study. Around 300 people participated in the study and thereafter, the data was analyzed by forming clusters using IBM SPSS Statistics. Based on the premise theory, there were common structures in consumer perception and behavior across India, and cluster analysis was used to identify several kinds of luxury consumers in India. ANOVA Technique is applied to observe disparities in luxury value perception that occur across India. Within the national boundaries of a culturally diverse India, this study sought to examine the homogeneity of the luxury industry. The study aimed to decipher the behavioral pattern of the workforce in India in terms of conspicuous purchase intentions for luxury cosmetic products, with the assistance of AR-driven technology, emphasizing brand antecedents and psychology.

Keywords: Conspicuous Consumption, augmented reality, behavioural pattern, luxury products, dimensions of consumption, brand antecedents, Indian workforce

1 Introduction

Thorstein Veblen, in his book, laid the groundwork for conspicuous consumption, in which he mentioned that the preference for consuming luxury products is made to exhibit one's affluence in society's eyes [1]. It is rare to find a common consumer product like luxury items that meets both the material and the social good. The expanding middle-class discretionary income, luxury buyers' increased connectedness,

***Corresponding author: Neeru Sidana,** Amity School of Economics, Amity University, Noida, India, e-mail: nsidana@amity.edu
Shreya Soman, Amity School of Economics, Amity University, Noida, India, e-mail: shreya.soman07@gmail.com
Richa Goel, Amity International Business School, Amity University, Noida, India, e-mail: rgoel@amity.edu

https://doi.org/10.1515/9783110790146-009

and the internet's seemingly infinite online shopping opportunities have all contributed to the recent rise in luxury product consumption [2].

In the light of globalization, greater rivalry, and increased differentiation, brands are regarded as vital in building an identity, a purpose of accomplishment, and an affiliation for consumers [3]. Technology is advancing at a rapid pace. The challenge for retailers is in determining which technologies are most likely to disrupt the retail experience and, as a result, should be adopted to remain competitive. Several new technologies are already influencing the customer experience significantly [4]. AR and virtual reality (VR) have emerged as rapidly emerging technologies, which are used to enhance the selling environment and the buying experience in both conventional and online retailing [5]. At the highest level of prestigious brands are the luxury brands, which encompass a variety of physical and psychological values; simply using and exhibiting a particular branded good grants the value of esteem to its owner [6]. This indicates that consumer status and consumption of luxury products/services serve a role in building linkages between consumers with these attributes and specific sorts of products available that provide status [7].

Consumer decision-making processes has been stated as something that can be explained by their conceptions of their own self, as well as their self-accessed hedonism in terms of conspicuousness, uniqueness, and quality. Wiedmann, Hennigs, & Siebels [8] created a conceptual model that describes luxury purchasing from the standpoint of the consumer.

One dimension takes into account immediate financial factors like cost, resale cost, rebate, and investment, as well as the price of the product and the quantity of compromise a consumer is willing to make to obtain it [9]. The second dimension is concerned with the most important features and functions of a product, such as its reliability, uniqueness, utility, dependability, and sturdiness [8]. Meanwhile, the themes such as hedonism, materialism, and self-identity are addressed in the third dimension, which focuses on a customer's attitude toward luxury purchasing [8]. Finally, the fourth dimension refers to the psychological functionality that people derive from merchandise acknowledged within their socioeconomic group(s), such as prominence and prestige value, both of which can have a profound influence on the evaluation and willingness to buy or consume luxury brands [10]. It is possible to identify and categorize distinct sorts of luxury clients based on these value dimensions that interact with each other [9].

These days, shopping over the internet is growing popular. E-commerce websites allow customers to shop from the privacy of their own homes [11]. With 37% of the world's population already using social media, marketers are counting on AR to advertise. Customers will be able to check out a variety of products before making a purchase, leading to more educated decisions, loyal customers, and more income [12]. Customers cannot have faith in a product's performance since they cannot physically feel it (e.g., touch and try it) through online and mobile platforms. Consumers can virtually try products, such as makeup on their faces, using AR, which could help them have a "try before you buy" experience when shopping [13].

The study aims to decipher the cognitive and behavioral patterns in terms of conspicuous purchasing intention of products by consumers, primarily the workforce of India, as it is the world's least researched but a rapidly emerging workforce.

Following the 1991 liberalization changes, accelerated growth of the Indian economy has resulted in profound consumer transformation. Increased consumer funding options and lower interest rates by banking institutions over the last ten years, together with the nation's capacity to keep price increases at normal levels have increased the wages of the middle and affluent classes in India [14]. Product price titillation without jeopardizing a brand's image or identity is a new tactic employed by suppliers of high-end goods and services to entice new market segments to their offerings.

The study is undertaken due to the following reasons:

- An increase in the country's population's standard of living.
- Increased inclination to exhibit position in society and reputation by purchasing and consuming premium brands.
- Despite substantial market backing for the luxury industry, there are few studies on the purchasing of luxury cosmetic products.

The key variable under study is luxury cosmetic products. According to Epstein [15], "The urge to reclaim one's lost youth is an ancient human passion, shown most notably by sixteenth-century Spanish explorers who sought the Americas for a fabled eternal youth."

Antiaging moisturizers, serums, and eye creams, which make up the "fastest-growing area of the cosmetics business," make a fortune for cosmetic and skincare corporations. Consumers spend an estimated $470 million on these items each year, and they frequently buy many products at once, because it is typical to use different treatments in the morning and before bedtime [15].

The conclusions of this study can be advantageous to luxury product and service marketers in the domestic and international markets [16]. The formulation of relevant market orientation and segmentation strategies requires a basic and comprehensive model of consumer-based brand equity consumption of well-known luxury brands trying to enter the Indian market.

2 Literature review

Many social scientists are concerned about the widespread pressure to conform in opinion and behavior in today's society. Social exchange theory has been used to synthesize disparate viewpoints on conformity. While integrative, the perspective on conformity overlooks several key motivations, values, and costs. For example, in the research study by Fromkin & Synder [17], findings show that desire for "uniqueness," or the need to regard oneself as separate from one's peers, is a common driver of behavior. As a result, the degree and kind of opposing pressures for uniqueness may influence the consequences of conformity pressures.

Individuals obtain and exhibit material goods to feel distinct from others, and as a result, they are bombarded with a range of marketing signals directed at boosting self-perceptions of individuality. Counter-conformity motivation manifests itself in consumer ownership and display. The study by Tian, Bearden, & Hunter [18] can be used to investigate ideas and models concerning its genesis. In theory, it can be used to describe how people consume things, experience, and resolve conflicts caused by an individual's counter-conformity motivation, as a response to marketing signals that use the appeal of individuality, and then decide whether or not to replace a product.

The study by O'Cass & Frost [19] investigates the relationship among brand associations and consumption pattern in order to gain a better grasp of brands and their influence on customer behavior. According to the study's findings, the higher the symbolic aspects, the stronger the associated positive feelings, and the greater the congruence between the consumer and brand identity, the more likely the brand is perceived as exhibiting high status-conferring elements.

The study by Lindstorm a Seybold [20] focuses on the interaction between brands and so- called "tweens" (8 to 14 years old). The authors believe that tweens are becoming an increasingly important demographic target for businesses for a variety of reasons, including the influence that children of that age have on their parents' shopping decisions. Even while tweens do not make the final decision when purchasing something as important as a new car or house, their involvement in the decision-making process is influential. Furthermore, today's generation of children is thought to be highly brand aware and demanding of the companies with which they seek to associate. Much of the study's examination into the tween market is based on the impact of the internet. Since birth, these youngsters have been exposed to both domestic and foreign trends. As a result, marketers must understand the attitudes and behaviors of the tween demographic.

Another study investigates the link between luxury consumption behavior and consumer status [21]. This research also looks at particular brands, their perceived status, as well as how they are used for conspicuous consumption. As it turns out, the two notions of conspicuous consumption and luxury consumption due to status are separate. Self-monitoring and peer influence has an equal effect on status consumption, whereas only peer influence impacts conspicuous consumption.

The relevance of young people in the perspective of marketers is acknowledged in another paper [22]. Its goal is to figure out how practitioners think about and build communication techniques for young people. The findings show that practitioners have differing perspectives on youth media sophistication. It is not commonly acknowledged that young people are smart and engaged consumers of advertising. Practitioners were discovered to take various approaches to formulating solutions, based on their knowledge of young people. The findings cast doubt on the consequences of particular positions in an increasingly digital media ecosystem.

The authors Amaldoss & Jain [23] highlight the circumstances in which a consumer's need for uniqueness might raise demand as a product's price rises. The authors discover that buyers purchase high-quality products despite their demand for distinctiveness. The authors uncover evidence in a laboratory test that demand for a product among consumers seeking distinctiveness grows as its price rises.

Conspicuous consumerism among middle-aged clients (40–60) is the focus of a study by Shukla [24], which uses the context of vehicle purchasing behavior to examine the issue. There is a substantial correlation between conspicuous purchasing behaviors of middle-aged customers and the automobile industry. It is now easier for researchers to investigate the middle-aged population's tendency toward ostentatious consumption, thanks to the findings of this study.

The recent economic ascent of the working class, coupled with a rise in disposable money, has resulted in widespread luxury consumption. The research paper by Eng & Bogaert [25] examines literature-based luxury notions and offers some insights into conspicuous consumption patterns in a subcontinent like India. People in India are more likely to buy things that are visible because of cultural and psychological factors. Although the results do not support the idea of a uniform conspicuous desire, Indian consumers have similar social features such as excessive luxury consumption and wealth displayed at social gatherings. When it comes to Indian society, luxury is seen as a way to show off one's wealth and prestige, as well as to convey one's sense of self and place in society.

Conspicuous expenditure in Turkey is investigated by the researcher HIZ [26]. Those in the middle, upper-middle, or the upper classes appreciate branded products and inventive and precious possessions. An individual's desire for status and distinction is presumed in this decision. For example, according to Duesenberry, a person's consuming functions are not private, but rather linked with the consuming activities of others.

Memushi [27] draws upon relevant existing literature to discuss the various aspects of the phenomenon of luxury consumption. The paper discusses the evolution of conspicuous consumerism (and its definition) across time. Some of the causes concerning conspicuous consumerism and the incentives that drive consumers to it are explained, based on relevant studies from economists, marketers, sociologists, and psychologists. Assessing which variables could influence a household's degree of consumption of luxury goods and experience is perhaps the most challenging issue addressed in the research article.

The purpose of another research study [28] is to discover the reasons behind the acquisition of Veblen goods like high-end vehicles, luxury apparel, and precious metal and gemstone jewelry, and to examine the differences between Aryan and Dravidian cultures in terms of conspicuous consumption. Both Aryan and Dravidian civilizations differ in their reasons for acquiring certain Veblen products, but there is no significant difference in their justifications for purchasing the same products. Moreover, the study found that diamond and gold jewelry is widely used in South India to enhance one's self-esteem. While in the north of India, the purchase of high-end cars is prevalent, which shows a strong link between their self-esteem and the acquisition of such vehicles.

Sarathy [14] used the analytic hierarchy process (AHP) and discovered that marketing associates the objective level with social gain and opinion leaders at the attribute level for conspicuous consumption of low-level income consumers. From the research, we can infer that opinion leaders are important in influencing decision-makers among low-income customers.

Research by T. R. & Bhardwaj [29] looks at the effects of message framing in cause marketing initiatives, particularly on the role that favorable and unfavorable structuring play in determining a consumer's response, based on how relevant the cause is to them. In a cause marketing campaign, an experiment with fictitious businesses demonstrated that framing and relevance affect customer response to such activities, with negatively worded messaging serving as a strong trigger that influences attitude and involvement intention. Consumers who considered the issue to be extremely personal had more favorable opinions and plans to participate than those who did not.

While the Veblen Effect is a well-established construct, and all luxury professionals were fully aware of it in the subjective pre-study, the researcher [16] emphasizes that there is no confirmation of its existence and the idea has been ignored for the past half-century. Furthermore, research has focused on outward consumption while dismissing its overarching construct. The study also raises the question of whether the Veblen Effect is primarily driven by conspicuous (Veblenian) spending, or if other luxury purchasing reasons such as a hedonist, a snob, or a bandwagon play a role.

This research investigates the effect of social variables on use of automobiles and appliances [30] in urban Indian households, by stressing the "status views." Researchers employ factor analysis to classify appliances based on their latent properties and then use a bivariate-ordered probit formula to capture the consumption drivers in urban households. While income and household demographics are, indeed, the primary reasons behind the appliance and car adoption, the family's notion of status, as measured by a variable quantifying the conspicuous expenditure, emerges as a crucial social component impacting the uptake. This research suggests that how consumers perceive themselves in society has an impact on their appliance and car consumption.

3 Research methodology

As part of this research, the researchers are attempting to determine if there exist varied reasons why Indian working consumers might be interested in luxury goods and experience; if they have common values; and if they have similar motivational drivers, even though the significance of various dimensions varies [31]. In the course of this investigation, the researchers will try to determine the type of motivation

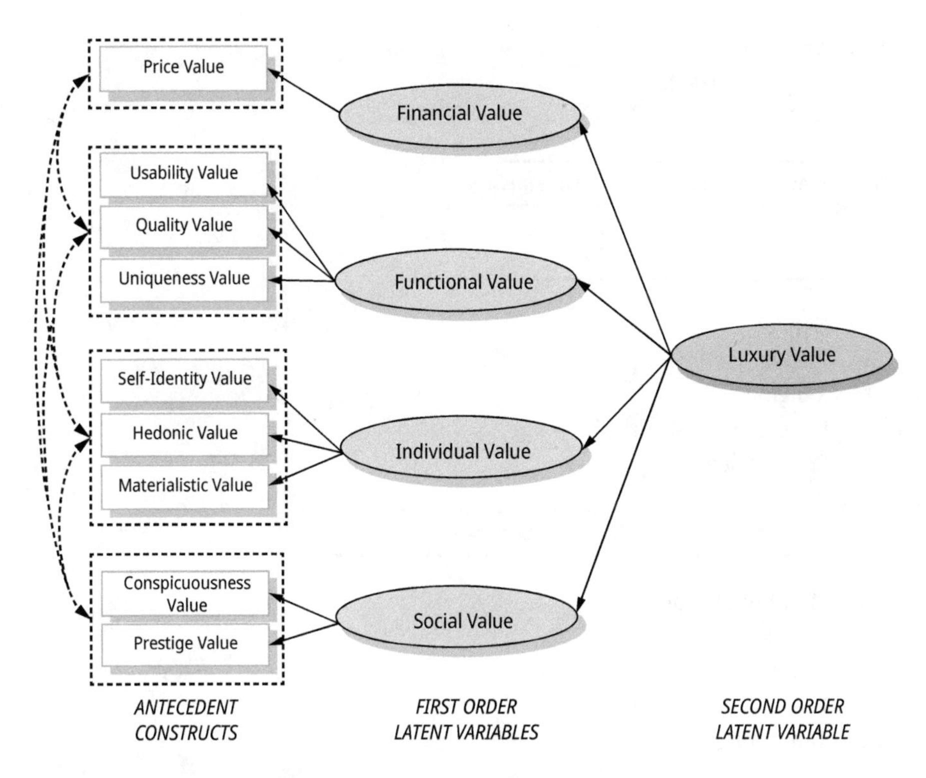

Figure 9.1: Dimensions of motivation driving conspicuous consumption.
Source: Wiedmann et al. [8]

driving conspicuous consumption, and they will also try to figure out how many dimensions there are to this motivation. As exhibited in Figure 9.1 the researchers were able to establish with the help of previous studies done in this field [8], that there are four main dimensions that drives the luxury value of conspicuous consumption.

To accomplish the objectives of the study, the following hypothesis has been proposed:

First set of hypothesis

H0: There is no significant difference regarding the motivation factors behind the purchasing decision among the working population.
H1: The motivating elements that drive purchasing decisions range significantly among the working population.

Relevant literature was painstakingly researched, and a well-structured questionnaire on numerous significant factors linked to luxury product consumption among India‘s working population was administered. The following variables are taken into consideration:

1. Luxury cosmetics
- Demographic variables
  1. Age

Young workforce	Older workforce
20–30	41–50
31–40	51–60

  2. Gender
  3. Region
  4. Religion
  5. Education level
  6. Employment

Employment status	Profession	Designation	Organization

  7. Monthly income level
  8. Marital status
  9. Family structure
  10. Luxury cosmetic products used
  11. The frequency of purchase

The researcher conducted empirical research keeping an exploratory and quantitative research design in mind. The survey questionnaire method having close-ended questions was used to gather primary data. All the questions were graded on a 5-point Likert scale, where 1 denoted strong disagreement and 5 denoted strong agreement. The questionnaire's initial version was verified twice using exploratory and expert interviews, with its pilot research being done with 55 respondents.

To increase cultural homogeneity of respondents and to minimize minor random mistakes, similar samples from distinct demographic groups were recruited. In this research, the sampling unit refers to the working individuals of India consuming luxury cosmetic products.

The questionnaire was pretested to ascertain construct validity and the scales' cross-cultural reliability and validity.

The simple random sampling technique was used to collect primary data. The researcher also constructed a verified sample size table to have data with just 5% as the margin of error.

The researcher contacted respondents directly to assess their interest in luxury brand products and their readiness to engage in the study project. In March 2022, researchers collected 300 valid and complete questionnaires. Table 9.1 outlines the demographics of the data collection participants at each site.

Table 9.1: Demographic distribution across India.

Location	%	Male (%)	Female (%)	Mean age	Income level
Bangalore	12.5	45.7	54.3	21.0	Middle income
Kolkata	8.3	27.7	72.3	23.5	Middle income
Mumbai	9.1	49.5	50.5	23.7	Middle to high income
Delhi	21.5	41.5	58.6	20.6	Middle to high income
Kochi	8.5	45.3	54.7	25.6	Middle income
Pune	6.5	46.9	53.1	23.5	Middle income
Ludhiana	7	49.5	50.5	20.3	Middle income
Ahmedabad	7.9	34.7	65.3	23.4	Middle income
Hyderabad	7.9	48.8	51.2	20.8	Middle income
Nagpur	9.8	54	46	23.3	Middle income
	100.0	44.0	53.0	22.0	Middle income

Author's own calculation.

As showcased via Table 9.1, 300 complete questionnaires were received; 53.0 per cent of individuals surveyed are those who identify as female, have a mean age of 22 years, and self-identified as belonging to a middle-income group. Considering that such a study necessitates some level of familiarity with luxury items, all final survey respondents claimed that they prefer luxury goods and services and that they regularly consider purchasing premium cosmetics brands, even if they agreed that luxury products are intrinsically pricey. According to the survey results, a majority of respondents said they might contemplate purchasing high-end items in the future. Even if the sample is not representative of the exploratory research topic, it gives a balanced range of data from different regions of India.

4 Results and discussions

IBM SPSS Statistics 26.0 was used to analyze the data to test the hypotheses and to identify potential differences and/or similarities among consumers belonging to different Indian states through data analysis. The researchers conducted ANOVA to see if there were any considerable variations across the aspects of customer perceptions of luxury value (as mentioned in the conceptual model), between states, based on averages of the variables calculated in various states. Based on the premise theory in the study that there are common structures in consumer perception and behavior across India, and to identify several kinds of luxury consumers in India, the authors used a cluster analysis. The construct measurements, ANOVA findings, and cluster segments are described below.

ANOVA was used to investigate the differences in luxury value perception that exist across India. As indicated in Table 9.2, the subsequent hypothesis is verified, as participants considerably (0.01) are in agreement with luxury value statements and their linkages with key elements of luxury denoting premium cosmetics brands.

Table 9.2: ANOVA table.

	Bangalore[a]	Kolkata[b]	Mumbai[c]	Kochi[d]	Delhi[e]	Ludhiana[f]	Pune[g]	Ahmedabad[h]
Dim1: Fn V	0.21	0.62	−0.23	0.36	−0.13	0.25	−0.04	0.21
Dim2: Fu V	−0.10	−0.09	0.43	−0.1	−0.06	0.13	−0.14	−0.07
Dim3: In V	0.26	−0.2	−0.05	−0.3	0.32	0.08	−0.22	−0.38
Dim4: So V	0.05	0.21	−0.25	0.13	0.28	−0.37	0.19	−0.23

	Hyderabad[i]	Nagpur[j]	F	Significance
Dim1: Fn V	−0.01	−0.36	11.58	0
Dim2: Fu V	−0.03	0.07	2.99	0.001
Dim3: In V	−0.63	0.4	19.13	0
Dim4: So V	−0.52	0.18	9.91	0

Author's own calculation.

When it comes to luxury value judgments in terms of money, Kolkata performs better with average scores across all statements, followed by Kochi, Ludhiana, Bangalore, and Ahmedabad. Customers in Mumbai and Nagpur offer the lowest ratings for a luxury brand and product financial evaluations of any other country's consumers in the world. Consumers in Bombay are a lot more than likely to agree with claims regarding the quality and performance of the product, when it relates to the functional dimension of luxury brand views. Consumers in Nagpur, followed by those in Delhi, Bangalore, and Ludhiana, focus largely on the hedonistic, efficacious, and materialistic features of luxury consumption. Except for Mumbai, Ludhiana, Ahmedabad, and Hyderabad, the results demonstrate that the mean scores for the social dimension are moderate. Thus, the results confirm the predicted factor structure, which supports the initial premise (H1).

4.1 Cluster analysis: segments across India

A preliminary hierarchical clustering approach was used to determine the number of feasible clusters and seed points for the k-means cluster analysis. To arrive at the correct number of clusters, a hierarchical clustering method was used to divide up the survey participants. Ward's minimum variance approach was applied to determine cluster differences and to maximize homogeneity within and between clusters.

The first prerequisite was to determine the optimal cluster size for the data collected. The results of calculations using the agglomeration method suggested the presence of four clusters, which was confirmed using cluster analysis. The second prerequisite was to check and eliminate variables based on their significance to pinpoint any data variation. Using SPSS 26.0's ANOVA table, none of the variables were removed because their significance came out to be lower than 5 per cent.

Through Table 9.3, it can be observed that the clustering results were valid for classifying consumer segments based on significantly different conceptions of luxury.

Table 9.3: k-Means cluster results.

	Cluster 1	Cluster 2	Cluster 3	Cluster 4	Significance
Dim1: Fn V	0.62	−0.32	0.38	−0.43	0
Dim2: Fu V	0.38	−2.13	0.14	0.44	0
Dim3: In V	0.55	0.15	−0.74	0.09	0
Dim4: So V	0.8	0.35	−0.61	−0.35	0

Author's own calculation.

Comparison of the four clusters on several descriptive factors, including demographic and socioeconomic features, was carried out. Consequently, the data gleaned is segregated into four groups:

Cluster 1: luxe addicts

This cluster accounts for 30% of the sample, has an average age of 23 years, and consists of 42% male and 58% female participants. This group's financial situation is average in comparison to the other groups. In terms of national origin, this group is primarily composed of consumers from Nagpur (25 per cent), Delhi (16.4 per cent), Kochi (10.5 per cent), and Bangalore (10.2 per cent). The social, personal, and financial value of guaranteed luxury products received the highest scores from respondents in this category, but the functional value was rated lower. These consumers believe that luxury items are unique and cannot be mass-produced to maintain their exclusivity. In this setting, this group's members have a great need to be distinct and distinguished from others; consumerism enables them to do so.

They are preoccupied with social approval even though they will purchase luxury brands that are tailored to their specific requirements.

Cluster 2: hedonistic purchasers

The second cluster accounts for 20% of the sample, has an average age of 24 years, and comprises 55% female respondents and 45% male respondents. In comparison to other clusters, this group comprises of those having medium level of income. This cluster is mostly made up of consumers from Delhi (22.5%), Nagpur (16.1%), Pune (13.3%), and Bangalore (10.4 per cent). Consumers in this cluster are more likely to value the societal and personal fulfilment dimensions of luxury goods than consumers in clusters 3 and 4. In this context, they concur with statements highlighting the significance of luxury brand status consumption. Additionally, they place a premium on the hedonic aspect of consumption, particularly when it pertains to luxury consumption. By contrast, these consumers place little value on the financial and functional aspects of luxury.

Cluster 3: the quality lovers
This cluster, which accounts for 26.7% of the sample, is composed of 46.6% men and 53.4% women respondents, with an average age coming out to be of 22.2 years. This cluster has a medium self-reported income level compared to other groups. In terms of origin, this group is primarily composed of consumers from Hyderabad (17.9%), Kochi (13.2%), and Ahmedabad (12.9%). Cluster 2 consumers identify luxury consumption with more emotive aspects of luxury value, such as the personal and societal dimensions, they place a lower premium on financial and functional characteristics, and consumers in cluster 3 have the opposite perceptions. When it comes to luxury, buyers in this cluster choose quality over status. Furthermore, for these customers, luxury is closely associated with exclusivity and originality. For one, these buyers shop for premium brands for the quality rather than to impress others. They routinely receive the lowest materialistic attitudes evaluations of any demographic. This group of consumers opposes consumption's hedonistic and status-related components because they believe they would be less pleased if they had nicer items.

Cluster 4: the prudent behaviorists
This cluster is the smallest and accounts for 18.0% of the sample and comprises of 54% male respondents and 46% female respondents and has an average age of 24 years. This cluster has the highest median income of any group. In terms of origin, this group is primarily composed of consumers who work in Mumbai (36.2%) and Kochi (14.8%). Although these consumers place a lower premium on the financial value of luxury, they place a much higher value on the utilitarian aspects of luxury than the average customer. Additionally, members of this cluster place a higher premium on individual considerations than on the prestige aspect of luxury consumption: they evaluate the substantive characteristics and performance of a luxury brand over the opinions of others. Additionally, members of this cluster place a higher priority on individual concerns than on the prestige aspect of luxury consumption: they evaluate a luxury brand's substantive features and performance instead of considering others' opinions. Luxury purchases are mostly motivated by the high personal quality standards of these customers. Luxury goods are more frequently purchased by people in this cluster than by those in other clusters because they provide a delightful experience, according to surveys.

5 Conclusions and implications

In today's economy, which is characterized by increasing global communication, education, fast media transmissions, and widespread mobility, luxury products are generally seen as a distinct product category. Within the national boundaries of a culturally

diverse India, this study sought to examine the homogeneity of the luxury industry as well as factors related to economic and cultural remoteness. The insights derived from this study can lead to the following key conclusions:
1. The concept of luxury value perceptions, which incorporates the various variables that influence purchase intent for such things, can be applied across a subcontinent like India. As a result, the primary drivers of luxury consumption can be generalized; only specific buyer perspectives vary.
2. When the Indian state variable is used as the primary unit of analysis, the ANOVA findings reveal significant within-country parallels and differences. These statistics reveal possible target markets as well as the relative importance of certain luxury spending parameters.
3. In the assessment of within-country disparities, the results of cluster analysis show that there are consistent trends across India. This market segmentation and orientation strategy promotes the idea of similar customer groups that cross state and cultural divides in the Indian subcontinent.

5.1 Academic contribution and future implication

- The formulation of relevant market orientation and segmentation strategies requires a basic and comprehensive model of consumer-based brand equity consumption for well-known luxury brands trying to enter the Indian market.
- This study can be of help to prospective marketers who plan to focus on this market by renovating their current tactics and designing new ones to provide high-quality products to clients rather than just burning a hole in their wallets.
- The budding designers will have a base plan to enter these markets and the areas to target.
- To increase their revenue, it is a good sign for India's national and state governments to revisit their current tax policy on luxury cosmetic products.

In light of these findings, marketing research and business operations could benefit from them. Accordingly, subsequent studies should focus on concerns of sampling and country selection, which might have an impact on the capacity to evaluate findings and generalize conclusions. Research can be conducted to further study the values that motivate this type of behavior by studying a representative sample of affluent consumers who regularly purchase high-end luxury brands [20]. Over a variety of time frames, it is possible to see patterns of long-term change and, in doing so, create a dynamic segmentation [32]. To better comprehend and interpret luxury brand views from a financial, functional, individual, and societal value standpoint, it is necessary to identify cultural values, conventions, and rituals that act as drivers and moderators [33].

Price thresholds are another important topic for future research. When it comes to price changes, all the consumers possess an upper and lower purchasing limit, at which their responses become extreme [34]. This is also true for the Veblen Effect, which is theorized to occur only in a specific section of the demand curve and ceases to exist on any other part [35]. As a result, most of the luxury shoppers keep a budget range in mind while shopping, only within which the Veblen Effect might occur [35, 36], and future study should discover these thresholds for diverse product types with differing absolute price levels.

References

[1] Veblen, T. (1899). *The theory of the leisure class: An economic study in the evolution of institutions*. New York: The MacMillian Company.

[2] Hammerl, M., and Kradisching, C. (2018). Conspicuous consumption (marketing and economics). In *Encyclopaedia of evolutionary psychological science*. Springer Nature.

[3] Watson, A., Alexander, B., and Salavati, L. (2020). The Impact of Experiential Augmented Reality Applications on Fashion Purchase Intention. *International Journal of Retail & Distribution Management*: 433–451.

[4] Lorraine, L., Tracy, M., and Smith, J. (2012). Reinventing the Customer Experience: Technology and the Service Marketing Mix. *Service Management*: 143–160.

[5] Bonetti, F., Warnaby, G., and Quinn, L. (2018). Augmented Reality and Virtual Reality in Physical and Online Retailing: A Review, Synthesis and Research Agenda. *Augmented Reality and Virtual Reality*: 119–132.

[6] Topçu, U. C. (2018). *Conspicuous consumption in relation to self-esteem, self-image and social status: An empirical study*. Boston: Springer.

[7] Whang, J. B., Song, J. H., Choi, B., and Lee, J.-H. (2021). The Effect of Augmented Reality on Purchase Intention of Beauty Products. *Journal of Business Research*, 133: 275–284.

[8] Wiedmann, K.-P., Hennigs, N., and Siebels, A. (2007). Measuring Consumers' Luxury Value Perception: A Cross-Cultural Framework. *Academy of Marketing Science*: 1–23.

[9] Hennigs, N., Wiedmann, K.-P., Klarmann, C., Strehlau, S., Godey, B., Pederzoli, D., Oh, H. (2013). Consumer Value Perception of Luxury Goods: A Cross-Cultural and Cross-Industry Comparison. *Psychology & Marketing*: 1018–1034.

[10] Liang, Y. (2018). *Exploring Chinese consumers' luxury value perceptions: Development and assessment of a conceptual model*. Bournemouth: Bournemouth University Press.

[11] Lee, K. C., and Chung, N. (2008). Empirical Analysis of Consumer Reaction to the Virtual Reality Shopping Mall. *Computers in Human Behavior*, 24: 88–104.

[12] Yadav, A. (2020). Digital Shopping Behaviour: Influence of Augmented Reality in Social. *Journal of Multidimensional Research & Review*, 1: 68–80.

[13] Javornik, A., Rogers, Y., Moutinho, A., and Freeman, R. (2016). Revealing the Shopper Experience of Using a "Magic Mirror" Augmented Reality Make-Up Application. *DIS '16: Proceedings of the 2016 ACM Conference on Designing Interactive Systems*. New York: Association for Computing Machinery (ACM), 871–882.

[14] Sarathy, P. (2017). Antecedents of Conspicuous Consumption in Decision-Making. *Contemporary Management Research*: 177–192.

[15] Epstein, S. S. (2010). *Healthy beauty: Your guide to ingredients to avoid and products you*. Dallas, Texas: BenBella Books, Inc.

[16] Dahm, J.-M. (2018). The Veblen Effect Revisited: Literature and Emperical Analysis. *Journal of Product & Brand Management*: 199–212.

[17] Fromkin, H. L., and Synder, C. (1980). The search for uniqueness and valuation of scarcity. In K. J. Gergen, M. S. Greenberg, and R. H. Willis (Eds.), *Social exchange*. Boston, MA: Springer, 57–75.

[18] Tian, K. T., Bearden, W. O., and Hunter, G. L. (2001). Consumers' Need for Uniqueness: Scale Development and Validation. *Journal of Consumer Research*: 50–66.

[19] O'Cass, A., and Frost, H. (2002). Status Brands: Examining the Effects of Non-product Related Brand Associations on Status and Conspicuous Consumption. *Journal of Product & Brand Management*, 11(2): 67–88.

[20] Lindstorm, M., and Seybold, P. B. (2003). Brandchild: Remarkable Insights into the Minds of Today's Global Kids and their Relationships with Brands. *Brand Management*, 11(1): 81–86.

[21] O'Cass, A., and McEwen, H. (2004). Exploring Consumer Status and Conspicuous Consumption. *Journal of Consumer Behaviour*: 25–39.

[22] Grant, I. C. (2004). Communicating with Young People Through the Eyes of Marketing Practitioners. *Journal of Marketing Management*, 20(5–6): 591–606.

[23] Amaldoss, W., and Jain, S. (2005). Pricing of Conspicuous Goods: A Competitive Analysis of Social Effects. *Journal of Marketing Research*, 42: 30–42.

[24] Shukla, P. (2008). Conspicuous Consumption among Middle Age Consumers: Psychological and Brand Antecedents. *Journal of Product & Brand Management*: 25–36.

[25] Eng, T.-Y., and Bogaert, J. (2010). Psychological and Cultural Insights into Consumption of Luxury Western Brands in India. *Journal of Customer Behaviour*: 55–75.

[26] Hiz, G. (2011). A Field Survey on the Conspicuous Consumption Trend in Turkey (Case Study of Mugla Province). *International Journal of Social Sciences and Humanity Studies*: 255–266.

[27] Memushi, A. (2013). Conspicuous Consumption of Luxury Goods: Literature Review of Theoretical and Empirical Evidences. *International Journal of Scientific & Engineering Research*: 250–255.

[28] Potluri, R. M., Ansari, R., Challa, S. K., and Puttam, L. (2014). A Treatise on the Cross-Cultural Analysis of Indian Consumers' Conspicuous Consumption of Veblen Products. *International Journal of Industrial Distribution & Business*: 35–43.

[29] S, T. R., and Bhardwaj, M. (2017). An Experimental Examination of Framing Effects on Consumer Response to Cause Marketing Campaigns. *IIM Kozhikode Society & Management Review*, 7(1): 1–10.

[30] Ramakrishnan, A., Kalkuhl, M., Ahmad, S., and Creutzig, F. (2020). Keeping up with the Patels: Conspicuous Consumption Drives the Adoption of cars and Appliances in India. *Energy Research & Social Science*: 1–12.

[31] Grubb, E. L. (1967). Consumer Self-Concept, Symbolism and Market Behavior: A Theoretical Approach. *Journal of Marketing*, 31(4): 22–27.

[32] Trigg, A. B. (2001). Veblen, Bourdieu, and Conspicuous Consumption. *Journal of Economic Issues*, 35 (1): 99–115.

[33] Onkvisit, S. (1987). Self-Concept and Image Congruence: Some Research and Managerial Implications. *Journal of Consumer Marketing*, 4(1): 13–23.

[34] Koen, P., Shuba, S., and Franses, P. H. (2007). When Do Price Thresholds Matter in Retail Categories? *Marketing Science*, 26(1): 83–100.

[35] Leibenstein, H. (1950). Bandwagon, Snob, and Veblen Effects in the Theory of Consumers' Demand. *The Quarterly Journal of Economics*, 64(2): 183.

[36] Monroe, K. (1971). "Psychophysics of Prices": A Reappraisal. *Journal of Marketing Research*, 8(2): 248–251.

Abhiraj Malia, Prajnya Paramita Pradhan, Biswajit Das,
Ipseeta Satpathy, Sambit Lenka

10 Application of virtual reality and augmented reality technologies to boost the technical capabilities for a constructive and effective learning in education for a performing virtual eco-system: a design thinking approach

Abstract: This research article introspects into the use of the technical capabilities of virtual reality and augmented reality to boost constructive learning in the domain of effective education. It examines a model approach for instructional and developmental design for creating a virtual educational ecosystem. It creates a team for participation, employing VR/AR technology to combat the hurdles in ease of learning, and assess its feasibility of performance. The adoption of the design thinking model handles and navigates, through its macro strategy, identifying subject topics for presentation. Similarly, the micro strategy ensures the effective and efficient presentation of the academic content, which is designed and intended for learning. The model is developed for the implementation of outcomes, along with developing the components of the virtual ecosystem of optimal education inputs, with a specialized evaluation of individual learners and small groups.

The research article has employed exploratory study with critical literature review in a mixed method approach. It engaged the method for a deeper understanding of the constructivism learning, applied to the virtual reality technologies for better knowledge.

Keywords: Virtual Reality, Augmented Reality, Educational Ecosystems, Learning, Technical Capability, Virtual Environment

Abhiraj Malia, School of Management, KIIT University, Bhubaneswar, India,
e-mail: abhirajmalia75@gmail.com
Prajnya Paramita Pradhan, School of Management, KIIT University, Bhubaneswar, India,
e-mail: prajnyapradhan11@gmail.com
Biswajit Das, School of Management, KIIT University, Bhubaneswar, India,
e-mail: biswajit@ksom.ac.in
Ipseeta Satpathy, School of Management, KIIT University, Bhubaneswar, India,
e-mail: ipseeta@ksom.ac.in
Sambit Lenka, International Business School, Jonkoping University, Sweden,
e-mail: sambit.lenka@ju.se

https://doi.org/10.1515/9783110790146-010

1 Introduction

Recent advancement in fast-moving networking and small mobile computing platforms has generated significant interest for a broader human digitalization interaction, beyond the typical flat panel displays. AR and VR headsets are positioned as the future generation's hands-on display, which is competent enough for delivering rich three-dimensional graphic experiences. Their useful applications mainly include learning, medicare, manufacturing, and gaming. Virtual Reality allows for a completely immersive experience, whereas augmented reality motivates the contact between the digital media users and the actual environment, showing computer-generated pictures while preserving capabilities.

In terms of displaying a presentation, AR and VR encounter similar issues to meet the challenging requirements of human vision, like angular resolution, eyechox, the field of view (FoV), dynamic series, correct depth cue, etc. VR and AR are two of the most viable solutions for creating learning objects and meeting their basic needs. Technological integrations in the learning area will improve deeper awareness through a multisensory atmosphere [1].

1.1 Virtual reality interactions

VR allows users to engage with computer-based surroundings, enhanced by a graphic system and a wide range of display and input devices. Virtual reality technology primarily provides prospective education, including the competence to permit beginners to visualize and interrelate with 3D virtual representations, real-time immersion in the virtual environment, visualize the effective relationship among some attributes in a virtual environment system and intellectual concepts, assimilate their understanding of phenomena by composing or controlling virtual surroundings, and permit individuals to co-operate with one another in social settings. With such capabilities, in which some are unique to this sector, VR has several educational aids that, if properly applied, will exert a strong influence on its application to education. For example, in the field of specialized vocational training, the usage of VR headsets may significantly improve vocational training.

Participants can utilize virtual reality technology to involve themselves in virtualized scenarios, based on the frequent issues faced in any particular industry/business. One of the most attractive aspects of VR as a learning aid is its capacity to present items and situations that are not ordinarily accessible to humans, but allow people to interact with them, for instance, by reaching out to 'touch' the atoms of a complicated molecule which is shown in Figure 10.1. It can also simply transform a wheelchair into a vehicle capable of transporting its owner across the world, for example, virtual safari parks or cities. Moreover, as they are based on complex simulations, VR systems allow us to explore learning domains through What-If experiments, the results of

Figure 10.1: Images on Virtual Reality.

which will encourage the development of cognitive models and learning abilities that can be utilized in a variety of circumstances. As a result, future VR systems will provide interesting new modes of learning, including experiential learning.

1.2 Augmented reality interactions

Augmented reality is a unified knowledge of a physical world situation in which physical world materials are enriched with computer-generated perceptual information, sometimes across many sensory modalities, like graphical, accrual, tactual, somaesthesia, and odorous. AR is an expanded representation of the real physical world, which is made possible by the use of digital virtual components like music or other sensory inputs framed via technology. Augmented reality mainly includes three basic features: a mixture of the physical and virtual world, actual time interface, and 3D modeling of real and virtual things which is represented in Figure 10.2. The principal benefit of augmented reality is how electronic world mechanisms blend into people's ideas of the actual world, not as a simple exhibition of data, but through the integration of immersive experiences that can be viewed as accepted features of any situation [2]. For example,

Figure 10.2: Images on Augmented Reality.

in the AR app "Dinosaur 4D +," users can view 3D dinosaurs by scanning through a set of flash cards. Students can use the application for spinning, zooming, and such other similar functions, while watching a dinosaur in action. In addition, the application gives basic information about dinosaurs.

Augmented reality utilizes technology to make a layer of information that is available to humans, allowing them to merge their vision of the real environment with digital stuff provided by computer software. Augmented reality is found to be highly active in the field of education and it may be used to stimulate the attention of students and youngsters to study the subjects, which are quite challenging. By combining these two broad perceptions of AR and mobile learning, as well as by intensely perusing the role of mobile augmented reality, theory can be transformed in an AR application, which is mobile interactive. The approach may be based on the best educational approaches, in which the communicating knowledge book will help as a marker and the mobile camera may be used as a display device to initiate a new level of learning involvement of general science concepts such as liquids, gases, solids, and liquid crystals; the different phenomena through that students experience in the nature and in the whole world; the fundamental human skeleton components, digestive, and respiratory organs, etc.

2 Review of theoretical foundation of AR and VR

(a) Virtual reality usage of education

Virtual reality displays a variety of distinct characteristics that reflect great technological advantages. This technology continues to grow at a rapid pace while providing more features that may eventually allow the development of new learning experiences. Indeed, the emergence of VR technology in education creates enthusiasm and high expectations of its abilities among instructors. Nevertheless, it is vital to highlight that technology is only a tool and it must be properly and efficiently handled to aid in the learning process. Meredith Bricken believes that VR is a strong educational instrument for productive learning, a theory given by Jean Piaget, with a special focus on the immersive uses of VR [3, 4]. The VR learning environment, according to Bricken, is immersive and instinctive; it comes up with a common knowledge background, in addition to differentiated interactivity that can be tailored to individual learning and presentation styles. Experiential learning, team working and consultations, field trips, duplications, and concept visualization are some effective teaching methods that may be supplemented by virtual reality. Bricken believes that, given the constraints of system functionality, it is feasible to build anything imagined and become a part of it. VR also allows natural engagement with information in the field of learning. Learners in virtual surroundings can walk, speak, wave, and control objects as well as the system, automatically. According to Bricken, VR is extremely robust and it has a strong effect in the area of education. Many students believe that the usage of VR has a useful instructional consequence. [5] provided a sound overview of the benefits and drawbacks of the use of VR, giving more focus on the educational context. Their findings will provide an overview to educators, with an understanding of the many viewpoints on the use of VR in higher studies. These will help to consider whether VR is an acceptable technology to incorporate into their curriculum and pedagogical approach toward their course delivery. Virtual reality offers a creative method for leadership professional development that focuses on increasing the capacity of school leaders to identify the features of academic disclosure during a classroom observation. Respondents specified that the aptitude to perform classroom observation in a safe environment is quite advantageous to them [6].

VR has captured the interest of both children and adults in its visual ability to transfer reality into the virtual world, where they may interact with any object and experience it. This capacity of transferring the real world to the virtual world may help to explain the popularity of virtual reality. Virtual appearance has been found to be a role in science learning because it allows students to engage with naturalistic explanations that are otherwise inaccessible in the actual world [7]. Virtual reality can help students to analyze facts that may be too little, too distant, or too far in the past or in the future. Cost and capital are two further characteristics of lack of availability. It could be prohibitively expensive to engage in any particular activities with limited resources, complexities, or the requirement for repetition to develop expertise. VR provides the accessories for creating its own virtual learning Environment (VLE). For students from religions and cultures, who have ethical objections to dissection, VR enables a high-quality alternative to gaining critical knowledge and skills through participation in dissection activities. According to Joseph Henderson, virtual reality may deliver an encounter that is transferrable and enhances our non-virtual world [8]. This also applies to simulations, which are computer-assisted educational formats. In many circumstances, VR technology is a model that goes beyond the typical computer-user interface to include the experience of becoming one with the simulated world. The consumer manipulates the activities that happen in ways that the standard computer-assisted instruction simulation does not allow. Simulation, by simulating the actual world, allows participants to experiment with various possibilities without the risks, expenses, or time commitment that the 'real thing' could entail. Zachert points out that the 'learner is shielded against injuring' himself or herself, as well as from 'causing damage on the actual world' [9]. Virtual reality programs, like simulations, may be extremely motivating, as passivity is nearly difficult to achieve. A student participating in a virtual reality program cannot be only a spectator. It is necessary for the user to actively participate in the action. The high degrees of engagement and individualization generate a great deal of attention. In this sense, virtual reality provides a new type of experience that may be both fun and educational.

Despite the fact that interest has risen since the team's study and throughout the previous few years, key challenges continue to stymie the adoption of VR in education. The majority of modern VR education systems are rooted in the constructivist approach, which claims that students generate their own creation of what they learn, and assume, and conclude that education is a cooperative effort. As an outcome of this approach, the structure of these systems frequently promotes rich-choice learning and exploration. However, the systems' objective is frequently to supplement or even replace classical classroom instruction [10]. Several researchers have indicated that the visual reality of visualization tools should be used with care and caution. It is unclear if more realism can enhance learning because it may distract the student from focusing on the material [11]. We also expanded the realm of investigation over instructional video with individual agents and desktop animation agents for immersive VR training, which is designed to be a more engaging learning setting [12]. Sometimes, it is found that the effect of VR on student motivation is non-immersive [13]. Meta-analysis contrasted the head-mounted displays' effectiveness in K-12 education, including additional VR machinery. These reports suggested that modest sized, head-mounted monitors remain hugely successful than the additional system [14]. According to the authors, each educational level is related to the acquisition of distinct information or abilities, which might affect the efficiency of VR. In this work, we assess the existence of potential disparities related to levels under a phase of education, K-6, that has not previously been thoroughly explored. A new study at various educational levels was necessary in order to provide a more comprehensive grasp of this technology. A meta-analysis reveals that, when focusing on K-6 levels, the advantage of VR in the case of knowledge development is considerable, irrespective of whether pupils are in preschool [15]. Furthermore, within these 2nd subcategories, research differentiates between the initial stages of early education (1st to 3rd class) and later years (4th to 6th class), demonstrating that VR is similarly beneficial across both categories. Semi-immersive VR systems offer consumers the impression that they are only partially involved in the dominion that is digital in nature. To accomplish this, some critical efforts are enhanced, or users' active participation in the virtual world is enhanced [16]

(b) Augmented reality usage of education

The application of augmented reality (AR) to achieve educational inclusion has received little attention. This systematic study summarizes the present status of employing AR as an instructional tool that considers the needs of all students, including those with disabilities. AR generally refers to the technology that frequently combines real-world experiences with smart-based digital information. Recent improvements in mobile computing technologies have made AR systems more accessible to the general public. Today, mobile AR apps utilize smartphone and tablet head-mounted displays, camcorders, Global Positioning System (GPS) sensors, and digital connections to connect real-world situations with dynamic, situation-based, and collaborative digital material [17]. The use of AR technology is no different. It is being utilized in a variety of disciplines like military, medicine, industrial design, robotics, telerobotics, industrial, conservation and repair applications, customer design, psychiatric therapies, and so on [18]. AR may be utilized for knowledge, amusement, or entertainment by increasing the user's perception and interaction with the natural surroundings. The operator may roam around the 3D virtual image and study it from any angle, much like a natural thing. The knowledge supplied by computer-generated items aids greatly in carrying out real-world tasks. AR has the ability to modify how we interact with technology. It makes the impossible feasible, and its applications in education are underway. AR interfaces allow for the seamless interaction of the real and virtual worlds. Learners engage when AR technology is combined with educational information; it generates new types of automated applications that improve the efficacy and desirability of teaching and learning in real-world contexts for learners.

AR is a revolutionary media that associates the features of pervasive figuring, physical computing, and communal computing. This platform offers unique possibilities by integrating the actual and the simulated world, along with uninterrupted and implicit user control of the topic of view and easy involvement with 3D data, items, and actions in a natural way by means of augmented reality technologies [19]. The designed and constructed AR application has proven to be beneficial in mastering computer science concepts that ordinary students struggle to comprehend or grasp. It has really pushed learning to a great height, allowing the learners to visualize smoothly about what is going on and understand the difficult topics. On the other hand, several experiments were also conducted to establish the finest set of practices for the creation and usage of such AR apps [20]. Students who adopted Mobile Learning Augmented Reality (MLAR) were able to pursue all the textbook subjects, without wasting much time browsing for information. Learners using MCAR are motivated to study plants at the education stage because they are designed to view virtual plant models using AR technology. The ability of students to access material when and where they want and use the devices they already own, has a huge potential to improve learning, and AR provides a valuable technique to easily engage learners in mobile learning. Nursing educators should continue to think imaginatively about how to best integrate these emerging tools. Students and teachers will have a more pleasant experience with AR technology if they pay close attention to the technological solutions available, learner media preferences, and accordingly choose the implementation strategy. The digital technological environment soaked in AR not only enhances the learner's and educators' capacities in terms of digesting and transmitting information but also helps to establish a synergetic association between them. The ultimate objective of combining current technology into tertiary education and research is to enable end users (both apprentices and instructors) to understand and recreate an advanced degree of engagement in a cybernetic reality as well as exploit the endless potential by developing smart ecosystems of information and instructional resources [21].

Somatic education is a curriculum that integrates strong involvement with skills. Practical coaching in AR is used in classroom physical education very occasionally. Visual coaching has also been employed in athletics, although it does not incorporate immersive practice, and does not equally represent academic learning and athletic skills [22]. The AR technique of training is effective for educating school kids in physical education improvements, notably in achieving a greater result in students' participation in sports. Nevertheless, most AR applications can work in more expensive, complex, combined units and AR simulations cannot be executed on standard computers [22]. The impact on job skills fit into routine expectations, exhaustion, public inspiration and smoothing condition on Behavioral Intention (BI) for acceptance of AR in the educational context is minimal yet progressive [23].

The current COVID-19 epidemic has created a number of difficulties and challenges for tourist education, one of which is the change from traditional to hybrid learning modes and methods. Applications of AR and VR in tertiary tourism teaching in the framework of the present pandemic are found to be immensely significant. The benefits and hindrances of using AR in the field of sports, teaching, and training were investigated. It was learned that several AR techniques should be utilized for providing learning and insights (visual, aural, and haptic information) to enhance the learner's experience and efficacy in constructing training situations [24]. It is expected that experimental research approaches will help to analyze the extent and impact of smart technology tools and apps for the adoption and usage in learning, providing a better understanding of the genuine effect of digital equipment tools in education. AR platform for digital learners to supplement their study books with images and virtual multimedia applications will give amazing results [25]. New AR technologies that take into account potential consumers' cultural variables when seeking to increase technology distribution are found to be complex [26]. For educational users, this technology provides great teaching options, such as mobility, visualization, alternative views, differentiation between many views, and a combination of numerous perspectives. AR-based education greatly enhances learners' successes and enthusiasm while minimizing cognitive load via a meaningful formative evaluation mechanism.

3 The similarity and dissimilarity between AR and VR is given below

Augmented reality	Virtual reality
1. AR is a fusion of real and virtual worlds.	1. VR generally creates the virtual world.
2. It lets people interact both with the real as well as the virtual world, and differentiate between the two.	2. It is very difficult to differentiate between what is real and what is unreal.
3. The goal of AR is to improve user understanding by incorporating virtual notions such as digital pictures and graphs or a new layer of communication with the actual environment.	3. The goal of VR is to construct its own world that is entirely computer controlled.

(continued)

Augmented reality	Virtual reality
4. It adds significant evidence to the present real world.	4. It incorporates heavy use of visuals to create a virtual environment.
5. Mostly used for demonstrating interior designing and mapping.	5. It is used in games, medicines, the military, and in allied domains.
6. Users keep their physical presence in the world.	6. The system has power over the users' senses.
7. Users can transfer, alternate, measure, and control objects in the real world.	7. In the virtual environment, users may move, rotate, and scale things.

Source: Own compilation, 2022.

4 Problem identification

AR and VR are utilized in a wide range of industries, including entertainment, manufacturing, medical, architecture, military, and education. Aside from other uses, the most common application of AR and VR in education is its direct application in classrooms for teaching and learning. It assists teachers in explaining a subject, provides a visual depiction of the content or location, and assists students in efficiently acquiring, processing, and remembering information. Nowadays no academic institution wants to be left behind, especially when it comes to adopting technologies that truly aid in its operations. All of them welcome new ideas and incorporate them into their daily operations. However, in most cases, they are insufficiently equipped or lack a sound plan to launch, develop, and support such a program. Any attempt is pointless unless it is guided by a strong vision, regardless of how encouraging the invention or technology is or how much it can affect or drive a change.

5 Research gap

This specific introspection pertains to visualizing how AR and VR are being implemented in the field of educational programs by design.

5.1 Research objectives

i) To demonstrate how AR and VR are progressing in the field of educational programs.
ii) To support education by integrating educational objectives and learning theories using AR and VR.
iii) To design a model framework for the use of AR and VR in education.

Advantages of integrated AR and VR

The benefits of the integrated application for VR and AR deployment were discovered from the plethora of surveys and analyses of relevant papers. The literature review was conducted for a theoretical framework. It cross-validated and explained the outcomes of related studies on the advantages of technology in education, and the need for a new integrated platform for new technological solutions. In this case, the practical implementation is assessed. The approach adopted was in accordance with the model Figure 10.3.

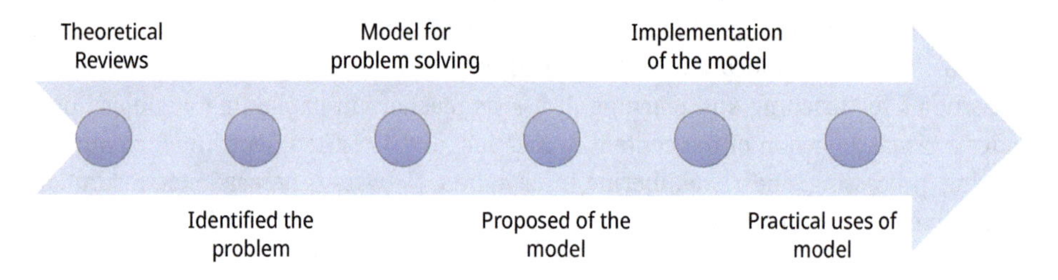

Figure 10.3: Implementation Model.

5.2 Creating an educational environment for learning with VR and AR

Several steps were planned on integrating VR and AR with the combined LO design model, during the planning stage: AR & VR prepared design, introduced to the real world, display, and comments, investigate the proposed design, uses of the proposed model, pros and cons of used model, critiques and cons of model, again development of design, and share represented below in Figure 10.4. The virtual reality-based learning object may potentially be employed in other educational content. This strategy might aid in the planning, creation, and evaluation of the learning environment.

During the growth stage, an algorithm was created that would be applied utilizing several VR technologies, for example, the researchers took two types of virtual reality: 5D technology and AR. When it comes to virtual and augmented reality, 4D technology has

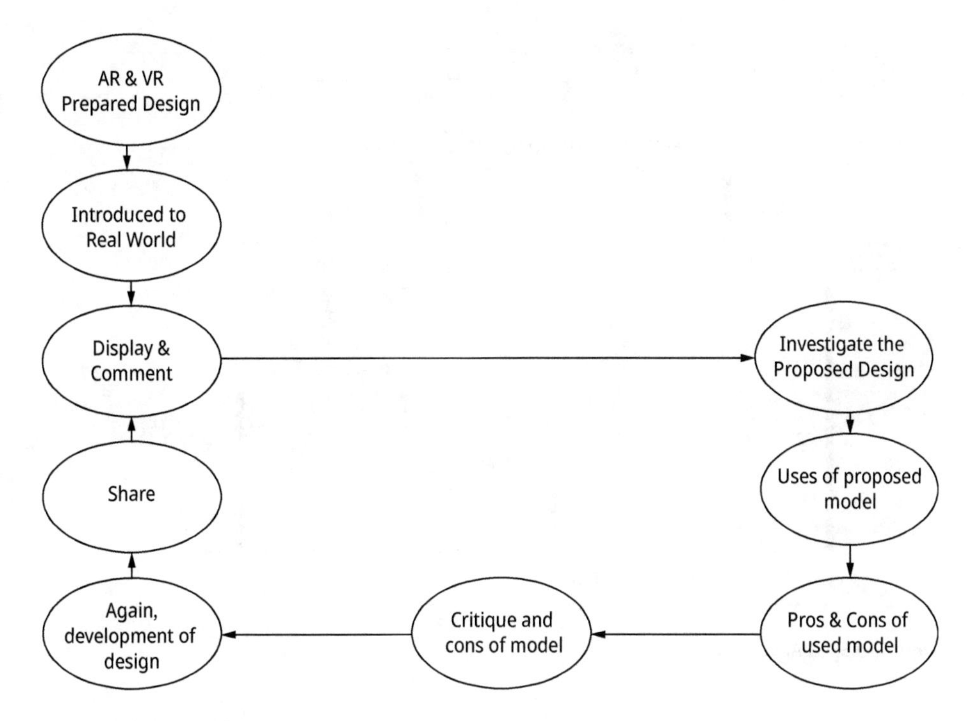

Figure 10.4: Development of AR and VR Model.

been known to suffer. With 5D technologies, VR will allow for hands-on experience, making learning more fun and inventive, and AR will allow the users to observe certain digital items in real time. It can be an effective way to get the correct information from the right people at the right moment. Augmented reality assists students in quickly acquiring, processing, and remembering knowledge. Both of these technologies require high-speed internet access and a smartphone. Learners' virtual presence in virtual and augmented reality will totally change learning, making it more attractive, inventive, and engaging. Five-dimensional (5D) technology now serves as the foundation for nearly all current computer games and virtual environments. Many educationalists and academic institutions see positive results while using five-dimensional virtual worlds for simulations and games for teaching and learning various disciplines. 5D learning objects bring students together to collaborate on project-based activities that provide real-world benefits. 5D learning objects make learning an enjoyable, dynamic group activity by fostering cooperation and strengthening critical skills such as teamwork. Augmented reality experiences may take many different shapes. A computer camera is used in webcam-based augmented reality to record a tangible real-world location and present growth on a panel, such as a projector or a computer desktop, letting handlers easily manage the augmented reality information. The flowchart (Figure 10.5) allows an insight into usage of LO with reference to AR VR in implementation of multimedia and texts.

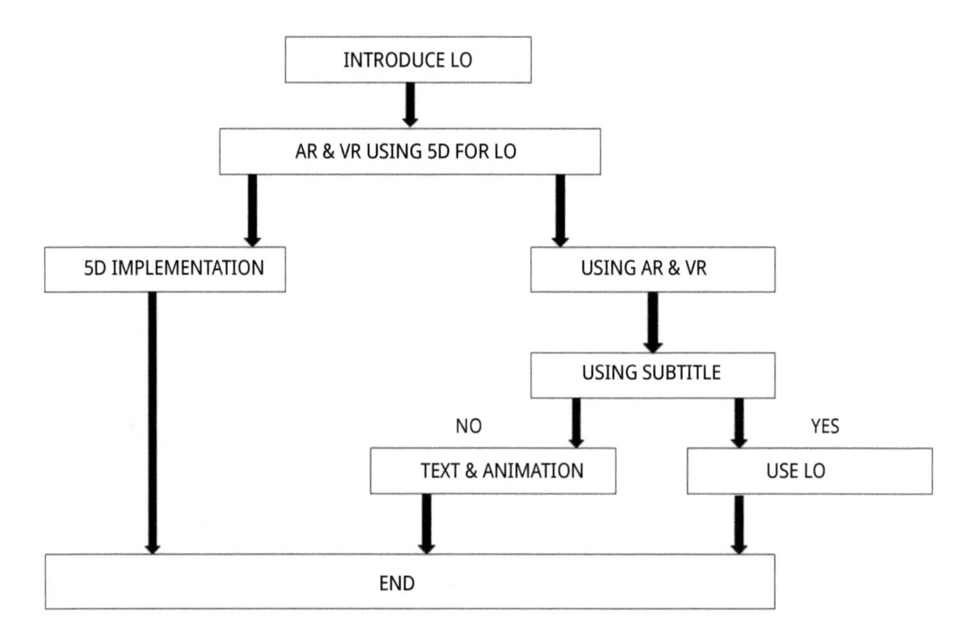

Figure 10.5: LO Algorithm.

Parents' and stakeholders' attitudes and ideas about the use of ICT have developed considerably at the same speed as research, stages of participation, and the development of 5D. The usage of ICT in an educational approach that combines games, creativity, and storytelling has helped to improve the understanding of the device's characteristics among children and parents. Children use new applications made available by the program, expanding their communication options. The emphasis on the 5D work and its tenacity as a game or a test fosters a sense of security among youngsters. Simultaneously, it aids in the learning of certain communication and computer tool usage skills [27].

Future of virtual and augmented reality

(a) Between 2019 and 2022, the use of VR and AR is going to increase 21-fold

AR and VR expenditure would range to $18.8 billion in 2020, an increase of 78.5% from $10.5 billion in 2015, with a 77.0% five-year annual growth rate (CAGR) through 2023 shown below in Figure 10.6. AR and VR will be at the forefront of technological change, with investment from businesses and consumers growing by an additional 80%.

(b) VR and AR are making their way to standard smartphones and tablets

Between 2020 and 2025, the AR and VR sector is expected to grow at a CAGR of 48.8% and its sales are expected to reach 161.1 billion by 2025 represented in Figure 10.7.

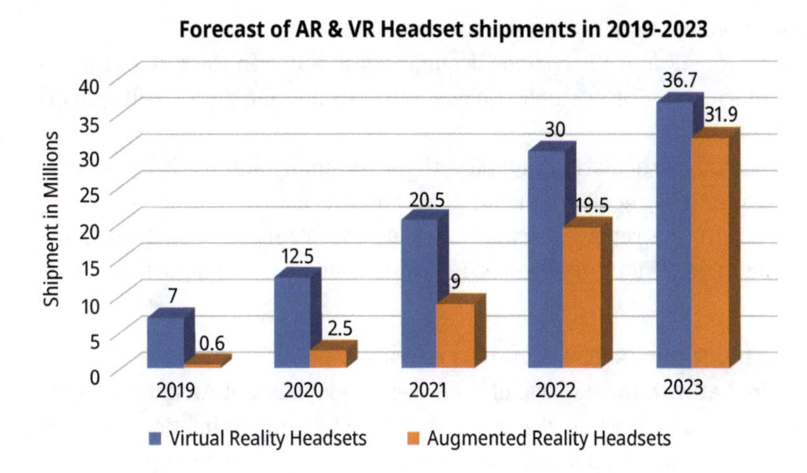

Figure 10.6: Forecast of AR & VR Headset shipment in 2019–2023.

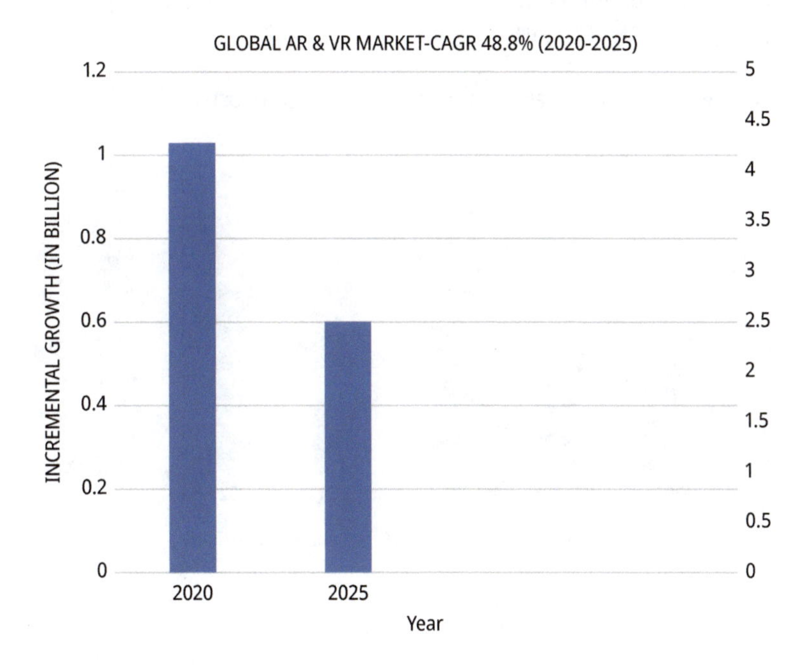

Figure 10.7: Global AR and VR Market.

Aside from technology improvement, the primary drivers of this growth will be greater usage of tablets, laptops, and smartphones, as well as the global concentration of large tech corporations in AR and VR. In terms of revenue, the hardware industry today outpaces the software industry. The software program industry, on the other hand, will grow faster because of increasing demand for mass media and performing sectors to meet demands such as AR-based game simulation [28].

(c) Increase your revenue with VR and AR content

Increased demand for AR and VR systems is subjected to a rise in the number of AR VR headset manufacturers. For example, Google, HTC, Oculus, and others will support the market content growth.

The training sector, particularly in employers for retraining and publicity reasons, is likely to be dominated by the development of VR and AR companies in the forthcoming years. According to the reports, corporations such as Walmart, Boeing, UPS, and others are utilizing AR and VR for operations, indicating that content is in high demand.

(d) Market share of AR and VR across various regions

It is clearly visible that with the passage of time, the development of AR and VR in the field of education is also increasing year after year. Graphics and animation advancements, sensory impacts, tactile feedback, and enhanced mobility are all improving the VR experience and adding to its growing popularity [29]. The VR industry has the potential to expand by $75.57 billion between 2021 and 2025, with the market's growth velocity accelerating at a CAGR of 55.34%.

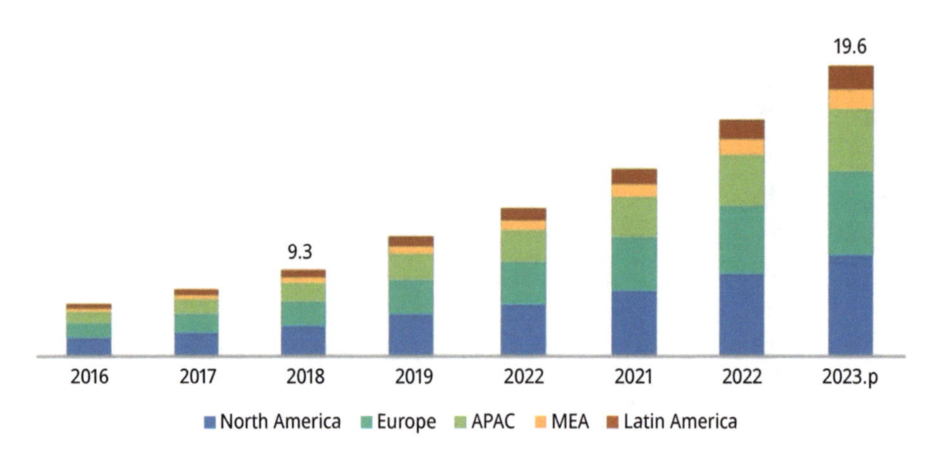

Figure 10.8: AR and VR in Education Market, by Region.

The largest market share for augmented and virtual reality in education is found in North America. The growth of the North American region as a whole is mostly due to the contributions of the United States and Canada. The education industry in North America is constantly looking for growth due to the presence of numerous significant educational institutions and the high-quality education they provide shown in Figure 10.8. In educational and business contexts, the usage of smart teaching and technologically sophisticated classrooms has increased as a result of technology advancements. These tools help create learning environments that are focused on the needs of a person or company. The adoption of electronic learning and the coming together of academic hardware and software are the main growth drivers of this market.

6 Conclusion

The primary goal of this research is to determine how far AR and VR have progressed and whether they are now mature enough to be employed in educational programs. Several advanced hardware and software developments indicate that AR and VR will be trustworthy enough as new computing platforms in the coming years. This promises significant improvements and new teaching and learning approaches that should meet the requirements of the twenty-first century student, who does not think in the same way as those of the twentieth or nineteenth centuries. It is also a fact that digital behemoths like Facebook, Google, Microsoft, and Apple consider virtual reality and augmented reality as viable investment prospects in the near future. It has been concluded that advances in e-learning are required and can provide a technology that is acceptable for customized knowledge, based on active education elements as well as cluster learning via virtual chat rooms [30].

It is apparent that the successful integration of these technologies demands various changes and modifications, not only on the part of engineers and AR and VR specialists but also on the part of teachers and all those working in this educational sector. Engineers must become more comfortable and accessible to VR and AR headsets. In fact, extended use of the recommended VR and AR equipment produces considerable discomfort. Education professionals must construct more forward-thinking educational programs that are consistent with the nature of these technologies and fulfill the needs of the student. When utilized properly, these technologies have the potential to improve the current educational environment and give students more learning opportunities. In all cases, it is certain that AR and VR will change the way we interact with the real world in the next years, and will be widely adopted across all industries.

Reference

[1] Heverton, M., Teixeira, M. M., Aquino, C. D., Miranda, L., Freita, W. C., and Coelho, A. N. D. R. E. A. (2016). Virtual Reality: Manipulating Multimedia Learning Objects. In *International Conference on Web Research. Tehran, Irã. Anais do II ICWR.*

[2] Barab, S., and Squire, K. (2004). Design-based research: Putting a stake in the ground. *The Journal of the Learning Sciences*. 13(1): 1–14.

[3] Bricken, M. (1991). Virtual reality learning environments: Potentials and challenges. *ACM Siggraph Computer Graphics*, 25(3): 178–184.

[4] Bricken, M., and Byrne, C. M. (1993). Summer students in virtual reality: A pilot study on educational applications of virtual reality technology. In *Virtual reality*. (pp.199–217). Academic Press.
[5] Baxter, G., and Hainey, T. (2019). Student perceptions of virtual reality use in higher education. *Journal of Applied Research in Higher Education*, 12(3): 413–424.
[6] Militello, M., Tredway, L., Hodgkins, L., and Simon, K. (2021). Virtual reality classroom simulations: how school leaders improve instructional leadership capacity. *Journal of Educational Administration* 59(3), 286–301.
[7] Hite, R. L. (2016). *Perceptions of Virtual Presence in 3-D, Haptic-Enabled, Virtual Reality Science Instruction*. North Carolina State University.
[8] Henderson, J. (1991). Designing realities: Interactive media, virtual realities, and cyberspace. *Multimedia Review*.
[9] Zachert, M., and Jane, K. (1975). *Simulation Teaching of Library Administration*.
[10] Taxén, G., and Naeve, A. (2002). A system for exploring open issues in VR-based education. *Computers & Graphics*, 26(4): 593–598.
[11] Wickens, C. D. (1992). Virtual reality and education. In *[Proceedings] 1992 IEEE International Conference on Systems, Man, and Cybernetics*. IEEE, 842–847.
[12] Makransky, G., and Lilleholt, L. (2018). A structural equation modeling investigation of the emotional value of immersive virtual reality in education. *Educational Technology Research and Development*.
[13] Pellas, N., Mystakidis, S., and Kazanidis I. (2021). Immersive virtual reality in K-12 and higher education: A systematic review of the last decade scientific literature. *Virtual Reality*, 25(3): 835–861.
[14] Wu, B., Yu, X., and Gu, X. (2020). Effectiveness of immersive virtual reality using head-mounted displays on learning performance: A meta-analysis. *British Journal of Educational Technology*, 1991–2005.
[15] Sarıoğlu, S., and Girgin, S. 2020. The effect of using virtual reality in 6th grade science course the cell topic on students' academic achievement and attitudes towards the course. *Journal of Turkish Science Education*.
[16] Di Natale, A. F., Repetto, C., Riva, G., and Villani, D. (2020). Immersive virtual reality in K-12 and higher education: A 10- year systematic review of empirical research. *British Journal of Educational Technology*, 51(6): 2006–2033.
[17] Azuma, R. T. (1997). A survey of augmented reality. *Presence: Teleoperators & Virtual Environments*, 6(4): 355–385.
[18] Azuma, R., Baillot, Y., Behringer, R., Feiner, S., Julier, S., and MacIntyre, B. (2001). Recent advances in augmented reality. *IEEE Computer Graphics and Applications*.
[19] Kesim, M., and Ozarslan, Y. (2012). Augmented reality in education: Current technologies and the potential for education. *Procedia-social and Behavioral Sciences*.
[20] Sungkur, R. K., Panchoo, A., and Bhoyroo, N. (2016). Augmented reality, the future of contextual mobile learning. *Interactive Technology and Smart Education*, 13(2): 123–146.
[21] Gandedkar, N. H., Wong, M. T., and Darendeliler, M. A. (2021). Role of virtual reality (VR), augmented reality (AR) and artificial intelligence (AI) in tertiary education and research of orthodontics: An insight. *Seminars in Orthodontics*, 27.
[22] Liu, Y., Sathishkumar, V. E., and Manickam, A. (2022). Augmented reality technology based on school physical education training. *Computers & Electrical Engineering*.
[23] Faqih, K. M. S., and Jaradat, M.-I. R. M. (2021). Integrating TTF and UTAUT2 theories to investigate the adoption of augmented reality technology in education: Perspective from a developing country. *Technology in Society*.
[24] Soltani, P., and Morice, A. H. P. (2020). Augmented reality tools for sports education and training. *Computers & Education*.
[25] Neffati, O. S., Setiawan, R., Jayanthi, P., Vanithamani, S., Sharma, D. K., Regin, R., Mani, D., and Sengan, S. (2021). An educational tool for enhanced mobile e-learning for technical higher

education using mobile devices for augmented reality. *Microprocessors and Microsystems*, 83: 104030.

[26] Hincapie, M., Diaz, C., Valencia, A., Contero, M., and Güemes-Castorena, D. (2021). Educational applications of augmented reality: A bibliometric study. *Computers and Electrical Engineering*.

[27] Ramos, D. S., Elena, M., and Ornellas, A. (2013). ICT collective appropriation on childhood and its impact on the community: The 5d educational model potentials and limits. *eLearn Center Research Paper Series*, 15–26.

[28] Mostafa, J., Hashemi, S. A., Sosahabi, P., and Berahman, M. (2017). The role of ICT in learning–teaching process. *World Scientific News* 72: 680–691.

[29] Floyd, B., Santander, T., and Weimer, W. (2017). Decoding the representation of code in the brain: An fMRI study of code review and expertise. In *2017 IEEE/ACM 39th International Conference on Software Engineering (ICSE)*. IEEE, 175–186.

[30] Muliyati, D., Bakri, F., and Ambarwulan, D. (2019). The design of sound wave and optic marker for physics learning based-on augmented reality technology. *Journal of Physics: Conference Series*, 1318(1): 012012. IOP Publishing.

Ambika N., Mansaf Alam

11 Role of augmented reality and virtual reality in sports

Abstract: Computer-generated reality and expanded reality are part of comparative studies of HMDs. The two advances use high-goal shows and sensors to show stereoscopic pictures in head-mounted devices. In virtual reality (VR), the client sees the virtual environment. It shuts out impacts from the real world to amplify the experience. The system uses cloud functionalities. It overlays virtual images on real-world environment impeccably. The user cannot differentiate between genuine and virtual images while using an expanded reality headset.

In augmented reality (AR) applications, virtual substances can significantly alter the present reality seen by the users by adding, eliminating, or changing a few parts of this reality. These applications need item location and acknowledgment in the preprocessing stage. The principal component of vision innovation has made headway in the last twenty years. Applications use a mix of AR and Principal Component (PC) vision. It includes game diversion applications. Enlisting AR and PC vision innovation into sports diversion applications opens new doors and comes with new challenges. This work is a survey article discussing the various contributions.

Keywords: Augment reality, Virtual reality, sports, artificial intelligence, sports, exercise

1 Introduction

Industry 4.0 [1–3] converts the existing production atmosphere into a store network. It uses computers in manufacturing, mechanization and hybrid zones. It has grid assembly and real-time knowledge business. The most advanced technologies, like IoT [4, 5] stockpiling, etc., are employed in a mixture of notions and practices of conventional enterprise manufacture. These methods allow clever manufacturers to optimize creation strategies by undervaluing production costs, increasing outcome quality, and delivering quick modifications to meet international demand. Current production processes must inevitably be reformed, changed, and enhanced by presenting new methodologies, machines, and procedures. The new manufacturing surroundings act as the spine for the evolution of intelligent, flexible, and configurable approaches.

***Corresponding author: Ambika N.**, Department of Computer Science and Applications, St. Francis College, Bangalore, India, e-mail: Ambika.nagaraj76@gmail.com
Mansaf Alam, Department of Computer Science, Jamia Millia Islamia, New Delhi, India, e-mail: malam2@jmi.ac.in

https://doi.org/10.1515/9783110790146-011

Sports [6, 7] is a demonstrating platform that encourages innovative methods. The manner in which that game is engaged by competitors, seen by customers, and adapted and directed by management is important. Mechanical advancements alter them. New advances further develop how individuals view sports and improve the game for fans and onlookers. Specialized arrangements make unique sorts of content accessible. Examples include video or article content, sports news, and game outcomes through customary media conveyance and sports streaming platforms. Technology helps fans interact with leading stars, groups, and associations. They increase faith in fans and upgrade their game experience. Innovative games allow playing cash on games and web games. The stages aim to develop participant understanding. It puts down sports wagers, supports dream sports, involves in game distribution. Advanced applications used for expert or novice sports groups, clubs, and scenes can expand the productivity of activities. It gives better experiences to sports purchasers. The help frameworks help referees to navigate. They assist with promoting forestall human assessment mistakes and decrease error choices on the pitch. They consider how advanced mechanics are open to further developed competitors preparing and detailing how mechanical technology and computerization have expanded sports.

The current review [8] intends to acknowledge instructive exercises using the web exercises encompassing new advances via expanded, digital, and mechanical technology facts, fully intent on including innovative speculations consistent with the law and deal an ideal instructive proposal for understudies. The learning techniques for the open universities foster exploration methodologies for the most exceptional creative arrangements and incorporate specialized apparatuses, for example, e-learning and videoconferences. These internet-based exercises take care of every one of the disciplines vital for accomplishing the scholarly capability. It is feasible to follow this examination back to the hypothetical references connected with virtual and automated information and the subject of expanded reality. Innovations open doors to instruction growth.

1.1 Importance of artificial intelligence

Two principal classes [9] were used as info highlights to foresee wounds – training responsibility elements and highlights of player's mental condition appraisal. Other jobs include work performed during preparation or match meetings. Inner preparation responsibility is surveyed by assessing the players' Rate of Perceived Exertion. Psycho-physiological information incorporates some details. They include body arrangement data, cardiopulmonary evaluations, injury history, etc. Information preprocessing considers the time-series information to give more insights about history of players' preparation.

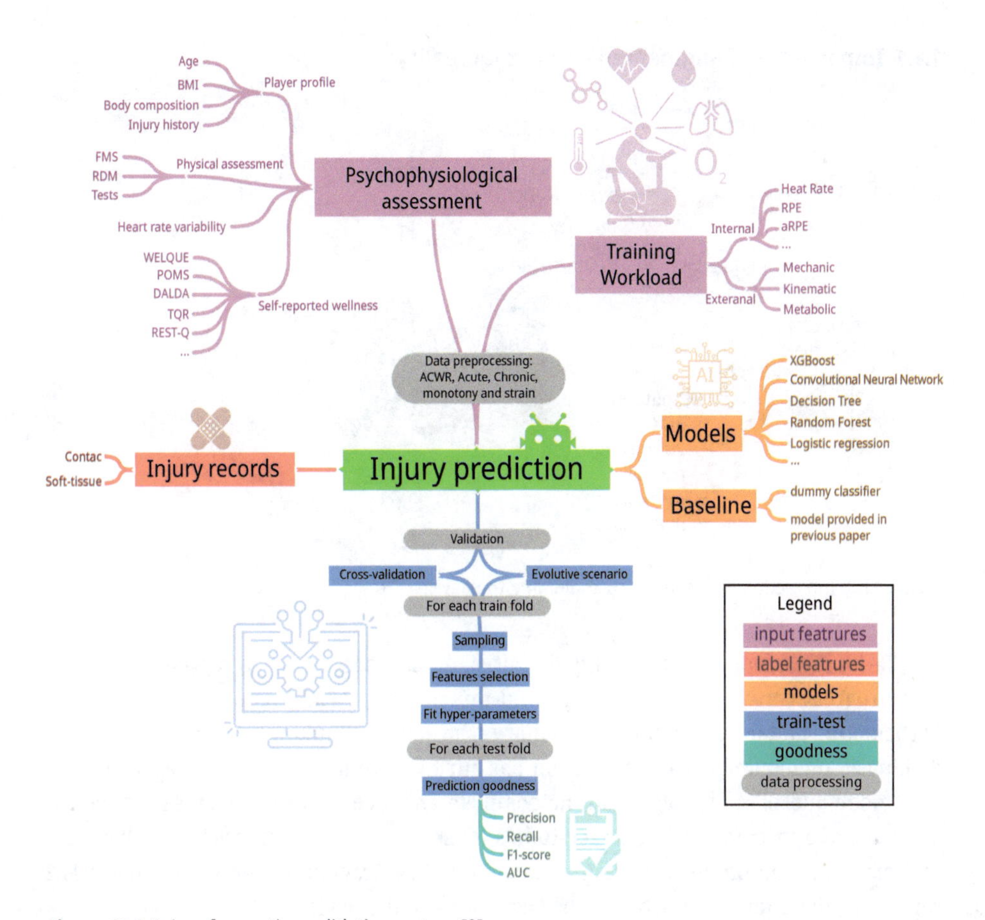

Figure 11.1: Injury forecasting validating system [9].

The primary undertaking [10] is to gather and investigate the craftsman's information as precisely as possible. For the craftsman's movement catch, the sensor [11] is introduced on the craftsman's wrist, and the finger is joined with the sensor on the brush to work. The place of a pen and the change cycle of track and variety are gathered as video recordings. The signal element extraction calculation gave a powerful brush. The work split the video into a gathering of casings and investigates the stroke age process during the brush development, afterward. It joins the model-based target following innovation. Figure 11.1 depicts the injury forecasting validation system.

PCA calculation is utilized to ascertain the pivot data of the movement of the brush. The brush movement act constrained by the painter is found and determined by the data, including the movement speed and the bearing of the brush head. The methodology slope fluctuates and it is set to regularize the system.

1.1.1 Importance of augmented and virtual reality

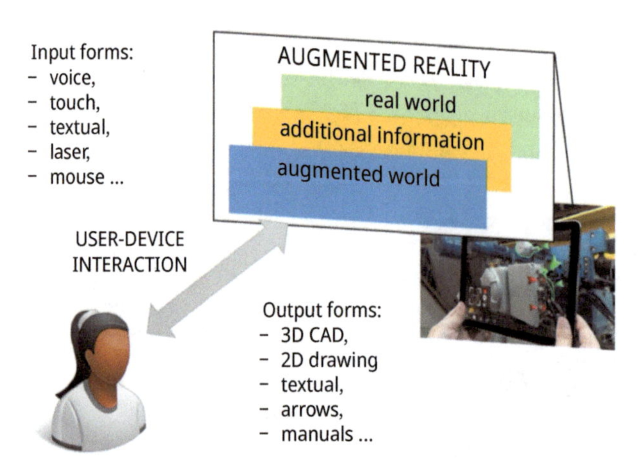

Figure 11.2: Interaction between the customer and the device [12].

AR [13–16] is a dynamic functionality that uses multisensory stages to reproduce and impart progressively. The motivation behind this design research is that client focus on participating and to lay out interdisciplinary instructing techniques helping countless individuals. Augmented reality instruction has attributes of distinctiveness. Regarding display mode and plan, it stresses on the complete enhancement of display experience. It aims to drive instructors and understudies to isolate from the customary display idea and mode and accomplish twice the outcome with a fraction of the work. Figure 11.2 represents the interaction between the customer and the device.

VR [17] exhibits three fundamental characteristics. Immersiveness implies that clients can communicate with a PC in the virtual world like in reality, unlike the conventional PC interface. Intelligence is the component that differentiates between the VRT (virtual reality therapy) framework and three-layered animation. Using intelligence, customers cannot get PC information. They can work with virtual items to change the world. The representation causes the clients to obtain better interpretation of the recognized object. The interpretation is obtained from consolidated situation of object under interest. This procedure extends origination. VRT framework is proficient image removal system having utility programming framework, input unit and exhibit hardware. Input hardware includes a protective cap marker, headphone, tail after head framework, and information glove. The virtual situation portrays dynamic characteristics, association, and intuitiveness rules. PC framework and picture-sound gear are external hardware.

1.2 Literature survey

The segment briefs the works of numerous writers towards the domain. The created EAR framework [18] is an improved variant of our previously completed application. It identifies, perceives, and shows the player significant insights that add a degree of data over the captured input picture. It is a completely computerized and comprehensive framework for game diversion applications from performer discovery to showing the valuable insights to the players. It tends to have a few downsides. It consolidates a robust picture upgrade method, especially when captured pictures are corrupted by nonuniform, solid daylight, or awful lighting. Improving player image gives a decent visual appearance and recognizable face-proof precision. The player picture improvement plot has never been used in any AR athletic framework.

The used face acknowledgment module is hearty and effectively perceives low-goal faces. Exploratory outcomes show. The face size is 5 × 5 pixels. The MSR processes the luminance channel, which is separated using the PCA. The Discrete Fourier Convert speeds up the upgrade interaction. It undergoes Contrast Stretching. Applying the MSR calculation brings about higher discovery and acknowledgment accuracy. Player and face identification is accomplished using AdaBoost calculation. A face acknowledgment calculation coordinates an informal look with the put-away faces in the data set. Face acknowledgment uses boosted LDA (latent dirichlet allocation) to include determination and Classic Nearest Neighbor Classifier to group players' faces. Measurements of each perceived player are recovered in light of the result of the player face acknowledgment module.

Hawk-eye [19] is the camera-based globe-global positioning structure that authorizes a group of actors to contest line-calling selections of the judges in noteworthy tennis events like Wimbledon. It uncovers the machine vision innovation behind the decision on the big screen. The work portrays how the International Tennis Confederation tried and certified the framework as an administering help and the specialized difficulties overcome to meet their rigid standards for exactness and dependability.

Following the investigations and exploration, the creators [20] of this exploration have given the assurance of six primary classes. It can incorporate the mixed reality and RA (réalité augmentée) in the games field. The virtual preparation is that the vast majority of the exercises are sans risk and, the athletic signal is spotless, straight, simple to get a handle on and examine. Expanded reality, likewise, assumes the primary part in advertising and methodologies pointed toward further enhancing the fan experience. It incorporates training for experts and amateurs, marketing methods, revenue expansion, follower practise, supporters assignation, legal and practicable alternatives during epidemic outbursts, size of sport consequences, and sports appraisals.

It is an expanded reality framework [21] intended to create visual improvements for TV disseminated court-net games. It targets integrating virtual game scenes from TV communicated video. It plays out a camera alignment calculation to layout planning between the soccer court arena in the picture and that of the computer-generated act. The player's stance is chosen from three essential decisions — stop, walk, or run, using the player's head movement and speed. Computer design procedures are used to create an energized scene. A probabilistic strategy in light of the Expectation-Maximization system is used to find the ideal component focus. It empowers the programmed securing of the camera boundaries from the TV picture with high exactness. An obtained simulated camera from the first camera assists with integrating an assortment of computer-generated acts, like the scene from a player's perspective. It depends on the expectation of the client. The performer's form and surface are separated from the videotape. The framework has court-net game recordings encompassing tennis, badminton, and volleyball.

Drivers [22] race for the love of the game, with rivalry taking care of habit-forming energy that permits the person in question to become one with a machine for the particular reason for triumph. Supporting every driver is an inspired group: engineers attempt to out-think top-notch educated people. Mechanism coordinates vehicles to match developing climate and track conditions. Team bosses turn language, information, and instinct into execution wizardry. Spotters keep drivers on their strict and figurative paths. Proprietors endeavor to demonstrate that they have the versatility and mental fitness and take steps to oversee better, compared any other individual. The makers of the battle portray that they use the best available. The structure close to home bonds with drivers, automobiles, tracks, groups, and varieties by viewing people model an automobile looking like their individual to triumph. It fills in as a testbed for advancements that might influence customer vehicles and the world writ large. The game develops with motor and suspension changes, streamlined advances, and security upgrades. Formula 1 is an advancement-driven sequence that will profit from mechanical technology, robotization, and connected innovations. Detecting will produce the more extravagant vehicle and natural information highlighting ongoing activity and plan enhancements, and availability will work with high-loyalty telemetry for distant vehicle diagnostics and system calls. The brainpower will robotize execution improvement, regulator element adjustment, material determination, and motorized plan. Computer-generated and increased authenticity furnishes drivers with relevant data without interruption. Mechanization will bring motorsports close by advances like a jolt, brilliant materials, and dynamic streamlined features to make a convincing bundle. Vehicle configuration might change to take on driver help frameworks. The miniature actuators have full-body streamed attributes. These materials adjust skeleton firmness progressively. Cerebrum HMI permits vehicles to eliminate perilous controlling haggles and speed response. The network opens opportunities for drivers, architects, and fans. Drivers might use the network to remotely direct vehicles securely. Designers will capture information progressively for prescient diagnostics. Assuming impacts are impending between human drivers, AI could forestall these collisions and sideline the motors involved, bringing fervor, rush, chance, and results without the danger of actual injury or insolvency. Assuming crashes are unavoidable between humanoid motorists, AI could forestall these impacts and sideline the vehicles bringing energy, rush, chance, and outcomes without the danger of actual injury.

It includes [23] a three-venture process. The initial step has to capture competitors' activities in a given game. It provides a methodology to look at the reference developments in all circumstances. The subsequent advancement takes care of activity of virtual humanoids and their transformation. it explicit limitations to alter the piece of reproduction. The third step includes the virtual climate show. Eight rugby players were playing at the public level. The recommendation recorded the developments of two distinct players: an aggressor and a protector. The assailant conveyed a ball and attempted to beat the protector by playing out a tricky move. The protector prevents the assailant from moving beyond him. Recording the two players' activities empowers a review top to the bottom biomechanical investigation of what the aggressor does to beat the safeguard effectively. The work catches the movement of 12 handball goalkeepers playing at the public level. All subjects had a typical vision.

The work is a vision of joining augment to reality [24]. The sports components convey the game encounter. The group comprises seven players. It has three follow-ups, single guardian, deuce mixers, and the searcher. The point of the game is to score several priorities over the rival group. The focus is scored when the searcher gets the brilliant nark. The Snitch is a ball that has an unmistakable overflow of energy and performs thrill seeker flight moves to shake off its followers. In the non-enchanted rendition, the searchers follow a fair-minded player to whose apparel the Snitch is connected. The game idea accommodates a battleground with three zones. Three field players per group battle get and control a drone, which has its own life.

The proposal [25] is engineering an application that will address the requirements of watchers and telecasters for improved content creation, viewing, and collaboration through increased reality for sports broadcasting. The introduced framework focuses on the quick production of visual upgrades and their installation in the recording of games. The work distinguished between the two frameworks engaged with the upgraded telecom of games handling suite on the telecaster side and the gathering unit on the watcher side. The making of improved content requires a progression of handling steps. It closes after completion. The framework supports using sensors joined to the cameras covering the occasion. It measures the container, slant, and zoom conduct of the cameras. These sensors are discretionary since the framework will likewise offer a picture-based estimation of the camera boundaries. After the production of the improved substance has occurred effectively, the material bundles into an MPEG-4 stream. It transmits over the current DVR framework. It aims to set top boxes.

The review [26] incorporates the planning and execution of a framework to deal with a games club and expanded reality advancements of the sports preparing framework from a distance and progressively. The utilization of ICT (information and communication technology) and AR empowers the execution of instruments of hierarchical education of athletic undertakings. The test system is an individual from one perspective fostering their laborers through the legitimate choice of volume and power of work and recreation. The effectiveness is conceivable in the participation of the administrative and preparing staff with IT frameworks. It anticipates a framework that virtualizes specific areas of hierarchical games clubs, and relates to improvement of employees and players. The contemporary space associations make occasions, their changes, or cycles. Occasions that comprise the space can be produced in virtual space, theoretical resources, and media. The procedure makes the media. Task 1 is arranging instructional courses and rivalries. The module permits the development of a groupings plan for preparing and contending in contests for discrete performers and groups. Task 2 conducts players and group competitors. Task 3 sorts the games of the club. Task 4 is the HR of the board.

VRabl [27] depends on AR to flawlessly link actual cooperation with virtual substances. VRabl is a serious increased reality sporting event. The players contend with each other. A green virtual battleground is put into this present reality. The game has a level and open region available. The purpose is to knockout the adversary's pursuits by tossing balls. The objectives are 3D models of rabbits. A player can impede a dose of the rival by getting the ball before it hits his dream. The balls skip on items permitting players to utilize the actual climate while focusing on the targets. A game has numerous rounds. A round closes when a characterized period or score boundary is touched.

Eight amplifiers [28] get familiarized using the counter. The sound rollers reaches all the receivers. Such information assembles and moved to the PC. The ball assaults on the table using a projector positioned on the table. Data focuses on the sent ball. The strategies are envisioned during the game circumstances by extending roundabout shapes on the tables. The range illuminates how regularly the balls fall into regions called attention to by the round shapes. This type of expansion is a valuable help for working on competitors' preparation (execution, method, and strategies). The model has three levels. Insight includes the ability to see the status, characteristics, and elements of the components of the climate. Perception alludes to the comprehension of what information and prompts apparently mean according to objectives and goals. Projection connects with the ability of projecting, sooner rather than later, the components perceived.

The framework [29] has a second-era Kinect sensor and in-house developed programming in C#. It uses a portion of the Kinect C# collections by Microsoft. The Kinect comprises two cameras, one for noticeable and one for infrared light, an infrared producer, and amplifiers. The form in infrared producer upheld the considered example. It acknowledges by evaluating the considered object. Kinect can perceive humanoid in the scene. It defines them as 3D areas of 25 joints of a virtual skeleton relegated to a body. The expert-level judo specialist records during his exhibition of a judo toss. His body is defined as indicated by the areas of the figure linkages. The areas of the multitude of joints in all-time moments are put away in the memory of the obtaining PC. It is kept into a transcript record toward the finish of securing. They will later act as the position for the exhibition of either a less knowledgeable judo professional or the expert is attempting to develop their performance. The multi-outline 3D estimation is implanted in a Cartesian direction framework whose beginning is situated on the imagination level characterized by the imaging sensor behind the focal point of Kinect.

The proposed technique [30] expands the virtual substance in the sports scenes persistently in the progressive edges. It has a proper degree of solidness for the addition of a virtual substance in the right district. The homography between the picture level and the ecosphere direction is determined by the comparing points. When the homography between the current casing and ground level is figured in the principal outline, it is expected to gauge the homography to keep away from the estimation of homography in each edge. The graphical representation of cross-section is obtained considering the two edges. When it is resolved that the ongoing casing is suitable for addition, it is important to isolate the foundation from the frontal area, to expand the virtual substance on the inclusion locale. Variety data of the field is utilized to eliminate the players and onlookers.

The framework [31] uses 22 movement catch cameras and four projectors inside the HugePlanetary. The Bulkyarea is the ecosphere's vivid virtual climate encompassing floor estimates. It covers 7.7 m in level, 25 m in width, and is 15 m long. The work uses a global positioning framework to recognize the position and bearing of the ball. It uses players with infrared intelligent materials. The recommendation uses a 3D-printed mounter having six mounting openings for infrared resources to be set in different position mixes. The framework estimate the position. It helps the players to forecast the game. This framework renders on the floor an anticipated way determined by the past ball position; it likewise delivers player bearings similarly to the ball. Six players can all the while play a scaled-down soccer match here.

The review [32] carried out an expanded reality table tennis match-up called AR Table Tennis. The racket activity of a humanoid is caught by a camera. The picture outlines to a picture analyzer. The picture analyzer gauges the racket position and course by distinguishing and following a few racket credits. A communication controller acknowledges the activity of the hominoid performer and conjures a game rationale motor to process another playing position. The sports rationale motor assesses every game physically and decides new conditions for every virtual article. The characteristics of virtual items are shipped off a picture synthesizer to form virtual articles with the genuine picture. The picture synthesizer delivers every virtual item and shows a blended scene on screen. The Player module controls the situation with every player, as indicated by the game guidelines. The Interaction Handler deals with all communications with the humanoid. It incorporates the Racket Follower to recognize the racket activity of the humanoid. The sports Judgement Machine gives actual reenactment. We do not attempt to fabricate an ideal model yet to make a genuinely practical gaming climate. The TranslationAppliance forms the refreshed digital items on the video outline.

AR Baseball Presentation System [33] is a perception arrangement of a digital ball sports. The clients using the system place the baseball on the tabletop. The clients can watch the game from their number one perspective around the field. This framework centers around envisioning the game by using scorebook information. The client can grasp the vital game point. There is a breaking point regardless of whether a humanoid is made exhaustively by CG. This enrolment strategy splits into two steps. The mathematical relationship of the markers is assessed in primary phase. For the assessment, 3D digital area is characterized by the projective remaking of two location films. The reference pictures are chosen from some up-and-comer images. The projection framework from every marker to the information picture is computed in subsequent stage. The sports performed on the area prototype is repeated and blended by input information record.

The primary framework [34] is to make simple to manager and to handle difficult issues. It empowers hikers to share their accomplishments. It is conceivable to overlay indicators around unambiguous grasps over the camera show. The viewport is constrained by moving the portable before the divider without the need for extra connection point components. In contrast to most other PC--based AR approaches, the arrangement utilizes parts of the environment as identifiable markers. The client needs to point the camera and make contact with the showcase with the tip of his finger. The raster-based hold arrangement is used. In request to monitor the preparation development, a journal component is given. One can essentially log all unsuccessful attempts and effectively resolved issues in an instructional meeting. The client can impart the finished preparation units to companions to move, analyze, and inspire one another. Clients can get together in gatherings and share difficulties, accomplishments, and ascending logs. An accomplishment can be a predefined issue within a specific period. The projected impression of Processor-Maintained Cooperative Exercise utilizes a multi-faceted methodology of PC upheld joint efforts. The journal's usefulness essentially centers around individual preparation of the issue and the objective definition highlighted is ideally appropriate for sharing.

The work [35] concentrates on the football location and the following-calculation in football match-up video, breaking down the constant picture of continuous cell phones in games video expanded realism. The picture is pre-processed by image turning gray, picture de-noising, picture binarization, etc. The portrait is pre-processed by picture grayscale, picture de-noising, and picture binarization, and a better middle sifting technique is anticipated. Hough change is utilized to find and recognize football, and as per the attributes of football, Hough change is brought to the next level. A football-following calculation in light of SIFT highlight matching is suggested. Its coordinates distinguish the football with the example football.

The proposal [36] is the first examination. The information obtained from employing various sensors can assist movement acknowledgment and provide sufficient input to the client. It has sensors to help a satisfactory movement acknowledgment in an outside environment. The accelerometer includes estimations given by three tomahawks, giving three isolated information time series for speed increase on every pivot. This detector has been utilized vigorously in examinations containing cell phone sensing elements for non-meddling action acknowledgement. If a client changes from strolling to race strolling or running, it will mirror the adjustment of the sign state of the speed increase, following the upward pivot. GPS is a satellite-based situating component that gives the position, working with the assumption that the beneficiary has an aggregate or incomplete view of GPS satellites. Brilliant information handling necessitates catches information through the required sensors to empower learning. The arrangement of new information permits framework changes for the perceived movement to give satisfactory and significant input to the client.

The idea [37] includes perceiving both the number characters and varieties from a picture taken by the observer on a cell phone, to deliver a similar image. The spectators are probably going to experience issues holding cell phones consistently without unintentionally shaking the gadget while at the same time interfacing with the increased items got from video frames. The pictures have a higher goal than picture outlines obtained from video successions. The application prompts the observer to snap a photo utilizing the gadget's camera. At that point, it continuously looks for characters and finds the most noticeable variations in the space of interest, and chooses the subsequent number-variety mix from the accessible groups given in an SQLite database. The executive would initially choose the shirt tones by catching a picture of each pullover. The data is then put away in a data set that is either cloud-based or nearby. The insights and data in each group and its players can be made promptly available. This information can be refreshed progressively by at least one individual appointed to this task. This strategy considers increased data to be collected for novice sports, where freely accessible measurements are not available.

The proposal [38] embraced a semi-trial plan with an exploratory gathering, a benchmark group. The trial bunch consolidated the course book materials with the AR-PE class programming to learn, while the benchmark group utilized a coordinated movement video to learn. Understudies in the two gatherings utilized the course book materials under the direction of educators and learned with AR-PE class. The understudies in the trial gathering could use the camera focal point of the transporter to adjust the activity pictures on the book while perusing the book materials to get down on the 3D person model in AR-PEclass for intelligent learning. Understudies in the benchmark group were driven by the educator in the course book materials and the video guidance with no communication between the two distinct materials. The understudies in the two gatherings led the trial movement two times for 90 min each time.

The suggestion [39] is a projection-based expanded reality affordance framework that projects a perception learning put together by person movement, concerning genuinely fake stone and helps a novice climbs the stone alongside the person. An educator can give climbing stances and movements immediately to the pupil. The fledglings can pick the way they need to prepare more than once without the assistance of a teacher. The projection-based increased reality affordance space is a counterfeit stone, a pillar projector component, and a person movement component. A pillar projector is on the roof of the indoor stone that doesn't obstruct the development range of the client. Amateurs can figure out how to climb the fake gravel. It is synchronized with the counterfeit stone. The activity holds in synchronization with the development of the virtual person. Table 11.1 lists the system complexity of the contributions. Table 11.2 portrays the implementation details of the contributions.

Table 11.1: System complexity of contributions.

Contribution	System complexity
[18]	O(n)*log(n/2)
[19]	$O(n^3)$
[20]	O(n logn)
[21]	Ω(n)
[22]	Ω(n logn)
[23]	$O(n^5)$
[24]	$O(n^2/2)$
[25]	$\Omega(n^2 \log(n/2))$
[26]	$\Omega(O(n^3))$
[27]	$O(n/2 \log n^2)$
[28]	$O(n^2)$
[29]	$O(n \log n^2)$
[30]	Ω(log n)
[31]	$O(n^3)$
[32]	$O(n^4 \log n)$
[33]	$O(n^5)$
[34]	O(n log n)
[36]	$O(n^2)$
[37]	$O(n^3 \log n)$
[38]	O(n log n)
[39]	O(n log n)

Table 11.2: Implementation details of the contributions.

Contribution	Implementation details
[18]	– The work uses 2,096 images. – It takes 1.0872 s. – It works on Super micro–Super-Server (SYS-7047GR-TRF) with 128 GB of RAM containing 96 cores. – Object detection requires 4.48 GB and recognition algorithms require 1.96 GB
[19]	– It uses geometric algorithm.
[21]	– The work extracts six video clips from consistent TV transmissions. – It executes on P-IV 3 GHz PC. – The average computation of frame is 30–59 ms.

Table 11.2 (continued)

Contribution	Implementation details
[23]	– 12 infrared cameras were used. – The work utilizes Oxford Metrics Group's Vicon signal seizure strategy to keep competitors' developments up-to-date in genuine circumstances.
[24]	– The visual location will work through a Simultaneous Localization and Mapping calculation or Machine Learning. It uses Raspberry Pi 4 and Coral USB Accelerator on the robot with the possibility to support the ML execution.
[25]	– The work segments the athlete by positioning 18 predefined body-joints in the initial image pair of the two cameras.
[26]	– It uses AZS AWF Wroclaw scheme.
[27]	– It uses Unity game engine and C# programming language. – HoloToolkit is combination of two Microsoft HoloLenses.
[29]	– Kabsch algorithm is used. – Dynamic Time Warping technique is used.
[30]	– SIFT algorithm is used. – RANdom sample Consensus robust estimator is utilized to compute the inter-frame homography. – The video frames are all 25 frames/second, with a resolution of 720 × 480.
[31]	– The work uses OpenGL and four PCs (CPU: Intel Core-i7 4790 K, GPU: NVIDIA Quadro K5000) in a PC cluster runs the OpenGL program.
[32]	– The system uses 2.6 GHz Pentium 4-based systems – The frame size is fixed to 320 × 240.
[33]	– The system uses web camera (ELECOM UCAM-E1D30MSV). – It uses PC (OS: Windows XP, CPU: Intel Pentium IV 3.2 GHz. – The resolution of the captured image is 640 × 480 pixels.
[34]	– The system is implemented for the Android platform using the Qualcomm Vuforia SDK for Augmented Reality. – Samsung Galaxy S was used as development device.
[36]	– The system uses a Kirin 620 CPU and 2.0GB of RAM using an Android version 6.0. – It uses accelerometer sensor vendor ROHMKX023, sensor Resolution of 0.009576807 m/sec^2, Max Range: 39,02266 m/sec^2, Min Delay: 10.000 microseconds.
[37]	– The work uses Java and XML.
[39]	– The artificial rock wall is of width of 4 m and a length of 3 m. – The system uses Windows 7 Pro 64bit (Microsoft, Redmond, WA, USA) operating system and Unity Engine 4.

1.3 Future directions

- AR stages could likewise give sharing, cooperative preparation, and social highlights to make, share, impart, and characterize objectives and difficulties for the participants. Users can likewise profit from AR to expand their consciousness of perils.
- Future frameworks ought to be lighter, less nosy, and incorporated with steady and standard connection points.
- Equipment upgrades could permit more data to be spoken with voice and hand signals.
- The timing of the virtual symbol with the visit booked by clients or the time or their appearance ought to be matched to offer an all-the-more simple and intuitive insight into a virtual historical center visit.
- The nature of the current design calculation in Augmented Reality, which manages object impediments can be essentially improved by including not many additional imperatives; one exceptional requirement can be the situation of uncertain outer marks.
- Mindfulness of the surroundings helps. Other UI methods should be examined while performing tasks inside and outside with our expanded reality frameworks, which convert the real world to a small-scale world.

1.4 Conclusion

Increased reality is only the approach to consolidating the reality furthermore; the virtual world is to help the client in playing out an undertaking in an actual setting. It comes from intuitiveness of people with virtual articles. It makes safe connection points that lay out the deception of the virtual, combining the real world and the virtual so that clients can easily cross from one to the next. The article is a survey of role of augment reality and virtual reality in sports.

References

[1] Ambika, N. (2021). A reliable blockchain-based image encryption scheme for IIoT networks. In *Blockchain and ai technology in the industrial internet of things*. US: IGI Global, 81–97.

[2] Chen, Y., Han, Z., Cao, K., X., Z., and Xu, X. (2020). Manufacturing Upgrading in Industry 4.0 Era. *Systems Research and Behavioral Science*, 37(4): 766–771.

[3] Javaid, M., Haleem, A., Vaishya, R., Bahl, S., Suman, R., and Vaish, A. (2020). Industry 4.0 Technologies and Their Applications in Fighting Covid-19 Pandemic. *Diabetes & Metabolic Syndrome: Clinical Research & Reviews*, 14(4): 419–422.

[4] Nagaraj, A. (2021). *Introduction to sensors in IoT and cloud computing applications*. UAE: Bentham Science Publishers.

[5] Ambika, N. (2022). Enhancing security in IoT instruments using artificial intelligence. In *IoT and cloud computing for societal good*. Cham: Springer, 259–276.

[6] Mitchell, J. H., Haskell, W. L., and Raven, P. B. (1994). Classification of Sports. *Journal of the American College of Cardiology*, 24(4): 864–866.

[7] Baker, W. J. (1988). *Sports in the Western World*. United States: University of Illinois Press, vol. 157.

[8] I., and Viscione, D. F. (2019). Augmented Reality for Learning in Distance Education: The Case of E-Sports. *Journal of Physical Education and Sport*, 19: 2047–2050.

[9] Rossi, A., Pappalardo, L., and Cintia, P. A. (2022). Narrative Review for a Machine Learning Application in Sports: An Example Based on Injury Forecasting in Soccer. *Sports*, 10: 5.

[10] Gong, Y. (2021). Application of Virtual Reality Teaching Method and Artificial Intelligence Technology in Digital Media Art Creation. *Ecological Informatics*, 63: 101304.

[11] Geethanjali, B., and Muralidhara, B. (2020). A wireless sensor system to monitor banana growth based on the temperature. In *Information and communication technology for sustainable development*. Singapore: Springer, vol. 933, 271–278.

[12] Reljić, V., Milenković, I., Dudić, S., Šulc, J., and Bajči, B. (2021). Augmented Reality Applications in Industry 4.0 Environment. *Applied Sciences*, 11: 5592.

[13] Carmigniani, J., and Furht, B. (2011). Augmented reality: An overview. In *Handbook of augmented reality*. cham: Springer, 3–46.

[14] Furht, B. (2011). *Handbook of augmented reality*. Switzerland: Springer Science & Business Media.

[15] Jerald, J. (2015). *The VR book: Human-centered design for virtual reality*. United States: Morgan & Claypool.

[16] Soltani, P., and Morice, A. H. (2020). Augmented Reality Tools for Sports Education and Training. *Computers & Education*, 155: 3.

[17] Zheng, J. M., Chan, K. W., and Gibson, I. (1998). Virtual Reality. *Ieee Potentials*, 17(2): 20–23.

[18] Mahmood, Z., Ali, T., Muhammad, N., Bibi, N., Shahzad, I., and Azmat, S. (2017). EAR: Enhanced Augmented Reality System for Sports Entertainment Applications. *KSII Transactions on Internet and Information Systems (TIIS)*, 11(12): 6069–6091.

[19] McIlroy, P. (2008). Hawk-Eye: Augmented reality in sports broadcasting and officiating. In *7th IEEE/ACM International Symposium on Mixed and Augmented Reality*, Cambridge, UK.

[20] Sawan, N., Eltweri, A., De Lucia, C., Pio Leonardo Cavaliere, L., Faccia, A., and Roxana Moşteanu, N. (2020) Mixed and augmented reality applications in the sport industry. In *2nd International Conference on E-Business and E-commerce Engineering*, Bangkok Thailand.

[21] Han, J., Farin, D., and de With, P. H.. (2007). A real-time augmented-reality system for sports broadcast video enhancement. In *15th ACM international conference on Multimedia*, Augsburg Germany.

[22] Miah, A., Fenton, A., and Chadwick, S. (2020). Virtual reality and sports: the rise of mixed, augmented, immersive, and esports experiences. In *Twenty-first century sports. future of business and finance*. Cham: Springer, 249–262.

[23] Bideau, B., Kulpa, R., Vignais, N., Brault, S., Multon, F., and Craig, C. (2009). Using Virtual Reality to Analyze Sports Performance. *IEEE Computer Graphics and Applications*, 30(2): 14–21.

[24] Eichhorn, C., Jadid, A., Plecher, D. A., Weber, S., Klinker, G., and Itoh, Y. (2020) Catching the drone-A tangible augmented reality game in superhuman sports. In *International Symposium on Mixed and Augmented Reality Adjunct (ISMAR-Adjunct)*, Recife, Brazil.

[25] Demiris, A., Traka, M., Reusens, E., Walczak, K., Garcia, C., Klein, K., Malerczyk, C., Kerbiriou, P., Bouville, C., Boyle, E., and Ioannidis, N. (2001). Enhanced sports broadcasting by means of augmented reality in MPEG-4. In *International Conference on Augmented, Virtual Environments and Three-Dimensional Imaging*, Mykonos, Greece.

[26] Cieśliński, W., Witkowski, K., Piepiora, Z., and Piepiora, P. (2017). The model of advanced ICT and augmented reality in sports enterprises. In *The organizational cyberspace: E-trainerism*. Cham: Springer, 167–177.
[27] Buckers, T., Gong, B., Eisemann, E., and Lukosch, S. (2018). VRabl: Stimulating physical activities through a multiplayer augmented reality sports game. In *First Superhuman Sports Design Challenge: First International Symposium on Amplifying Capabilities and Competing in Mixed Realities*, Delft Netherlands.
[28] Loia, V., and Orciuoli, F. (2019). ICTs for Exercise and Sport Science: Focus on Augmented Reality. *Journal of Physical Education and Sport*, 19: 1740–1747.
[29] Sieluzycki, C., Kaczmarczyk, P., Sobecki, J., Witkowski, K., Maśliński, J., and Cieśliński, W. (2016). Microsoft Kinect as a tool to support training in professional sports: Augmented reality application to Tachi-Waza techniques in judo. In *Third European Network Intelligence Conference (ENIC)*, Wroclaw, Poland.
[30] Monji-Azad, S., Kasaei, S., and Eftekhari-Moghadam, A. M. (2014). An efficient augmented reality method for sports scene visualization from single moving camera. In *22nd Iranian Conference on Electrical Engineering (ICEE)*, Tehran, Iran.
[31] Sano, Y., Sato, K., Shiraishi, R., and Otsuki, M. (2016). Sports support system: Augmented ball game for filling gap between player skill levels. In *ACM International Conference on Interactive Surfaces and Space*, Niagara Falls Ontario Canada.
[32] Park, J. S., Kim, T., and Yoon, J. H. (2006). AR table tennis: A video-based augmented reality sports game In *International Conference on Artificial Reality and Telexistence*, Hangzhou, China.
[33] Cheok, A., Ishii, H., Osada, J., Fernando, O. N. N., and Merritt, T. (2008). Interactive Play and Learning for Children. *Advances in Human-computer Interaction*.
[34] Daiber, F., Kosmalla, F., and Krüger, A. (2013). BouldAR: Using augmented reality to support collaborative boulder training. In *CHI'13 extended abstracts on human factors in computing systems*. Paris, France.
[35] Wang, H., Wang, M., and Zhao, P. (2021). Sports Video Augmented Reality Real-Time Image Analysis of Mobile Devices. *Mathematical Problems in Engineering*.
[36] Pascoal, R. M., de Almeida, A., and Sofia, R. C. (2019). Activity recognition in outdoor sports environments: Smart data for end-users involving mobile pervasive augmented reality systems. In *Adjunct Proceedings of the 2019 ACM International Joint Conference on Pervasive and Ubiquitous Computing and Proceedings of the 2019 ACM International Symposium on Wearable Computers*, London, United Kingdom.
[37] Bielli, S., and Harris, C. G. (2015). A mobile augmented reality system to enhance live sporting events. In *6th Augmented Human International Conference*. Singapore.
[38] Chang, K. E., Zhang, J., Huang, Y. S., Liu, T. C., and Sung, Y. T. (2020). Applying Augmented Reality in Physical Education on Motor Skills Learning. *Interactive Learning Environments*, 28(6): 685–697.
[39] Heo, M. H., and Kim, D. (2021). Effect of Augmented Reality Affordance on Motor Performance: In the Sport Climbing. *Human-Centric Computing and Information Sciences*, 11.

Bibhorr

12 Space traffic management: a simulation-based educational app for learners

Abstract: Due to the swift creation of human-centric applications and technologies, educational and training programs are progressively transforming toward newer ecosystems of modernization through the adoption of augmented and virtual reality technologies. Simulations and gamifications in engineering education and training could play a huge role in imparting knowledge with entertainment and amusement thereby bringing in positive and effortless understanding of concepts among the students. Foreseeing this, a game based on virtual reality named "Space Traffic Management" has been incepted with the purpose to disseminate space flight and astrodynamics learning environment in the current education scenario. The game, chosen in the national level toy hackathon event (Toycathon 2021), is a typical simulation of realistic space traffic schema with infotainment factor for helping to simplify space education among aspiring students and bring value in their practical understanding. The game helps in stimulating strategic understanding among the players by providing them an invigorating environment consisting of animations, graphics, sound effects, and music. This chapter reports the inception of this game prototype with the mathematical theory related to its creation.

Keywords: Game Simulation, Space Traffic Management, Traffic Triangulation, Virtual Reality, Aerospace Defense

1 Introduction

Space traffic management (STM) has become a subject of serious debate and research for the past few years and has been evolving with each passing day. STM is described as a set of methods and procedures for prudently, sustainably, and undamagingly approaching, managing activities in and returning from outer space. All manned and unmanned spacecrafts, functional as well as nonfunctional satellites, experimental machines, or any other man-made hindrances in space could be effectively termed as "Space Traffic." STM has got the following key elements: (1) activities associated with space situational awareness, including space surveillance and tracking; (2) procedural implementation of debris mitigation and amelioration; (3) the entire life cycle of space operations, including launch phase, in-orbit operations of spacecraft, and end-of-life deorbit operations; and (4) spacecraft reentry phase (both controlled and uncontrolled).

Bibhorr, IUBH University, Mülheimer Str. 38, Bad Honnef, 53604, Germany e-mail: bibhorr@zoho.com, https://orcid.org/0000-0003-0404-4601

https://doi.org/10.1515/9783110790146-012

The complexity of the airspace depends on structural and flow characteristics of airspace [1]. The next-generation air traffic control (ATC) designs must realize a capacity increment [2]. For air traffic management, airspace divisions enable distributed control. An aircraft within a specific airspace zone is managed by air traffic controller in that zone only. As the aircraft enters into the other airspace division, it is then commanded from the ground air traffic controllers of that airspace division where the control is passed on from the leaving zone's controller to the entering airspace's controller. Air traffic management involves installation of ATC towers at every airport that hosts scheduled intake and outgoing aircrafts responsible for handling aircraft take-off, landing, and ground traffic movement. In managing and controlling traffic movement in air, Air Traffic Control System Command Centre plays a central role which is responsible for monitoring and managing the ATC within the centers. Departing and approaching aircrafts within a specific airspace are handled by terminal radar approach control of that division. Flight service station is tasked for providing information on weather, route, terrain, flight plan for private pilots flying into and out of small airports and other remote areas.

Unlike the ATC paradigm, the framework for STM is wholly unique and different since the space situation is considerably more complex, advanced, and thus time consuming; as a result, it must be controlled efficiently and safely. Furthermore, because there are no precise constraints or borders that define the space, the distribution of traffic elements is unrestricted. The space traffic is different from road as well as air traffic systems as in this case the vehicular motion is completely unpredictable. The vehicular disposition involves horizontal, vertical, and diagonal motion in such a way that all the traffic components are observed to be transposing in nature, simultaneously. The space traffic control model has been historically put in an unclear, hazy, and ambiguous manner due to technical and political reasons [3]. STM definition entails complexities [4]. To build a comprehensive STM system it is essential to first develop a model that is capable of effectively and efficiently managing flightpath for collision mitigation [5].

As a result of increasing space mission activities by many nations now, it is imperative that work on STM is done and necessary measures taken beforehand. Increased use of space resources lacking discernment by several regions (primarily the USA and Europe) is resulting in space traffic which in turn is also adding to space debris. Also, there are many challenges involved like potential collisions which in turn increase space debris. It therefore becomes all-the-more important to rightly act now and create awareness of this concept across different levels, especially within academics where students can learn about the futuristic scope of improvement.

Many new actors have emerged, including new spacefaring nations and organizations, as well as novel uses such as cube satellites, on-orbit satellite servicing, and mega-constellations. While the benefits of space services are becoming more generally appreciated, the growing number of parties enhances the possibility of national activities and interests colliding in the future, potentially destabilizing the near-Earth orbital environment. The establishment of an STM framework is one of the policy choices for dealing with this problem.

Constellations and launch systems are fast growing in number in space. The viability and safety of space infrastructure and operations are being threatened by rising space traffic. More than 1 million trash objects ranging in size from 1 to 10 cm orbit the Earth, and the number is growing. All of this put together is actually causing a huge concern with respect to space, space operations, and sustainability – so STM is becoming a far crucial subject to take action at the right time before there could be disastrous times and it becomes too late for us to take any action then.

The global space economy is predicted to be worth more than $1 trillion by 2040. Commercialization of space activities is currently expanding globally, largely due to actors across different nations. This expansion has never been seen before; as a result, it introduces new actors, application areas, and business models resulting in more exploitation of space resources.

A major increase in the volume and diversity of commercial activities in space would likewise impact the future space operating environment. Emerging commercial ventures like satellite servicing, debris removal, in-space manufacturing, and tourism, as well as new technologies that enable small satellites and very large constellations of satellites are outpacing government initiatives to advance and implement policies and processes to address these new activities.

Critics argue that as the number of satellites orbiting the Earth grows, there is a greater risk of possibly catastrophic accidents and the resulting space debris, which could endanger other spacecrafts. Attempts are currently being made to predict potential space collisions in advance. However, due to the considerable uncertainty in estimating whether or not a collision will occur, the uncertainty can linger until a few minutes before the anticipated impact. As a result, there is a lot of buzz around the topic of STM at the national, regional, and global levels right now.

STM has grown in importance in the space sector, drawing interest from both the technical and legal and policy-making communities. There appears to be widespread interest in avoiding detrimental interference with other space operations through managing space operations. This is understandable, given the growing realization that as more nations and nongovernmental organizations engage in space activities, the risk presented by the rising population of space objects will increase. This issue is exacerbated by the fact that operators are now developing satellite solutions based on large constellation architecture concepts. Naturally, these new activities raise the potential of collisions between or among space objects. Simultaneously, there has been no multilateral engagement to bring about fundamental governance reforms in order to implement STM on a global basis. This section argues that the term "space traffic management," which is derived from the concept of "air traffic management," may set the bar too high in the current geopolitical climate, and that, in the short term, more emphasis should be placed on establishing "space traffic coordination" practices to improve the safety and security of the space environment.

It is becoming increasingly important that our new generations and students learn about STM from as early as possible so that we can have innovative minds to develop constant new solutions with respect to this problem that the future holds. For our global development, space sustainability is a must. Hence, the requirement to have the awareness created for STM wherever possible. STM provides a standard data collection, alerts, and avoidance maneuver recommendations to assist owner-operators who may not have in-house tracking or analysis capability. Monitoring space traffic is indeed a challenging task – the proactive approach is required to manage space traffic.

When students and children learn about such a futuristic problem, they can contribute better as they grow to become professionals to deal with this massive issue. The best way for children and students to learn about this is through gaming – that is, where STM game comes into picture. For young children and students, it is not easy to comprehend the problem until they actually get a chance to see the problem or if they get a chance to relate to it – hence, to address this very concern, STM game has been developed, for everyone who is new to space (especially children) who can then understand the concept and what it takes to have the desired solutions. The game is purely to introduce the space resources, concept of STM – and the approach, the actual way space traffic is required to be managed while maintaining and sustaining the space with the minimum debris and avoid collisions.

2 Related work

The combination of immersive virtual reality (VR) technology and game design principles results in the creation of immersive simulation that feature game elements and make it possible for students to participate in learning activities that would otherwise be prohibitively expensive, difficult, or impractical to carry out in a traditional classroom setting. The usage of these tools helps to modify the way students relate to the knowledge they are learning by fostering visualization, experimentation, and creative thinking. In addition, they extend the learners' exposure to other persons and ideas, foster cooperation, and support meaningful postgame conversation. Finally, they have the flexibility and complexity necessary to accommodate to varied learning styles. Immersive VR increases simulation outcomes, such as the amount of new information learned, the amount of time that it is retained, and clinical outcomes for rehabilitation. However, it does have certain drawbacks, such as the potential for getting motion sickness and having limited access to VR devices [6].

Chang and Yang [7] tested a mathematical educational app which focused on the promotion of learning skills regarding geometry, logic, graphics, and space among students. Although the app was not based on augmented reality (AR)/VR support, the students agreed that the app could help enhance their mathematical aptitude. Antunes et al. [8] implemented an educational game covering polarity, molecular geometry, and intermolecular forces for engineering students in a standard chemistry course.

Chittaro and Buttussi [9] proposed a head-mounted display-based aviation safety game and compared the VR simulation approach with the traditional aviation safety manual which concluded that conventional learning techniques and VR-based simulation games can both deliver a similar amount of information acquisition; however, VR simulations have the benefit of being better at helping students remember and retain what they have learnt. Kumar et al. [10] analyzed the effectiveness of game-based learning over conventional education for improving oral health awareness among children and it was concluded that the game application provided fruitful intervention aid for teaching good oral health hygiene among the sample. Dinis et al. [11] analyzed the VR applications developed by the first-year master's students of civil engineering for which trials were conducted among K-12 students in which it was found that applications help motivate the learning scenario and also helped transferability of knowledge among the participants. Munz et al. [12] showed that educational games may be used to encourage and educate undergraduate students in an introductory automatic control course. Braghirolli et al. [13] showed that through game, the students get motivated to participate and to better understand the course content. Furió et al. [14] concluded that games are useful tool in the learning process and could help teachers fulfil students' training requirements. Rondon et al. [15] concluded that the game-based learning approach is equivalent to the traditional learning method in general and in terms of short-term benefits, but the traditional lecture appears to be more effective in enhancing both short- and long-term knowledge retention. Erhel and Jamet [16] proved that a serious gaming environment could boost learning and motivation as long as it incorporates elements that encourage learners to actively absorb the instructional information.

Students often struggle with comprehension during the learning process due to the complexities, importance of abstract thinking, and concepts. Increasingly educational institutions throughout the world are introducing strong new technology-based solutions to assist in meeting the needs of the varied student population. VR has progressed beyond the realm of gaming to the realm of professional growth in recent years. It is vital in the teaching process since it provides an exciting and engaging approach of learning. The laboratory-based method, a less effective way of learning, results in students with inadequacies in foundational knowledge and practice, which may lead to a failure to adapt to challenges in future workplaces. The education system has been ever evolving. It has always adapted to the current technologies and the students' needs. The generation that is attending school now has spent their entire lives online. The digital world is just as essential and immersive as the physical one. They are digital natives, having been born into a world of mobile phones, widespread Internet, and instant access to the majority of requested information or data, whether it is music, video, or material. Educating Generation Z is difficult, and it necessitates an entirely different strategy in order to maximize efficiency and engagement. Education and gaming combination is well suited to today's student's needs. There are various documented benefits to employing VR technology in teaching. VR gives exceptional visualization that cannot be gained in a typical classroom setting. It reflects the world in which younger generations feel at ease. It provides nearly limitless access to information, books, and articles. It is utilized for very effective blended learning, enabling self-study and individual knowledge quest. Although the use of current technology in the classroom is undoubtedly advantageous, it is not without its own limitations and disadvantages [17].

3 The game concept

The concept of game revolves around the *STM* theme. The management of space traffic has become an unignorable core issue in today's time and that too when nations across the globe are turning more and more misaligned regarding their national preferences entailing defense strategies and national security issues. Each nation with a vision to secure affluent future is anticipating to growing space prowess by sending more than ever spacecrafts into space. The game registers this issue of space traffic as it is promptly reaching a stage where the complications of collision disasters and rapid actionable solutions are signaling to stimulating clueless systematic procedural implementation.

The digitally incorporated game application packed with VR technology named *STM* predominantly furnishes educational cognizance regarding space traffic concern among the kids, students, and aerospace engineering enthusiasts in a light and relaxed course of action. The game is an interactive simulation predominantly created with the motivation to instill scientific vigor regarding spaceflight schema among children, students, and adults alike. The game establishes its enthralling flow of spaceflight journey that gets concluded in nine theme-based levels, each corresponding to a specific environment associated with a planet.

3.1 About the game

The game opens with a horizontally oriented menu display visible on the VR headset screen, which reads "Space Traffic Management," the name of the game with three VR-compatible touch buttons namely (1) play, (2) quit, and (3) info, as shown in Figure 12.1.

The play button leads to a screen for the selection of planets. Out of nine planets, the player can choose any planet at a time for the game to be functional. Then a screen appears for the player to select the desired spacecraft followed by an avatar choosing screen. Originally 25 avatars were conceptualized but only 6 were incorporated in the game.

With all the selections done, the player is finally able to control the spacecraft in a virtual world with the help of simulation mimicking controls. The spacecraft experience

Figure 12.1: A VR headset screen visual of the STM game developed by the author.

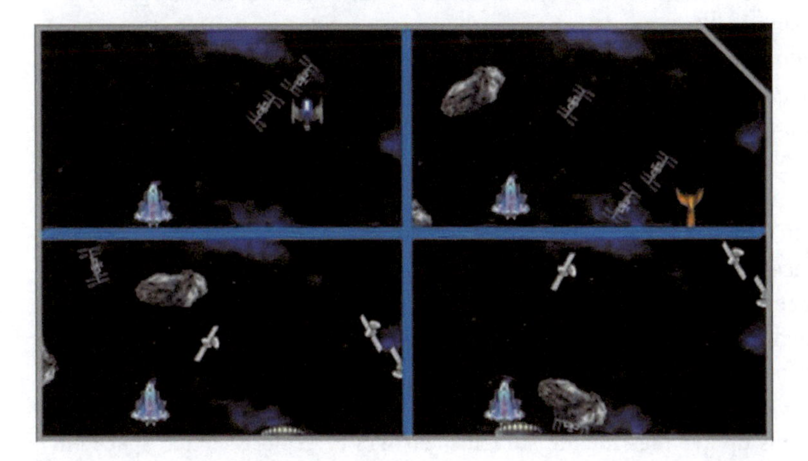

Figure 12.2: A visual from the VR headset screen in which spacecraft is completing its journey in the STM game developed by the author.

in the game appears as shown in Figure 12.2; where a spacecraft named *Rudra* is shown completing its journey in space.

After the completion of each level, the player is automatically upgraded to the next level. The players earn points by managing spacecraft safely in the space and by not harming existing satellites. If the player intentionally or unintentionally hits the satellites or uses ammunition against them, then the player is immediately deprived of *space power* with the heavy points reduced. The space power helps the player to recover spacecraft when met with an unforeseen accident. The player is able to retrieve his/her highest score and even the last session score. The player can order his/her crew to fix any error or malfunction identified onboard. The player can use predefined mathematical models to avoid traffic collisions. These include triangulation-based traffic identification, polygon shift analysis to avoid collision disaster, probability model for early-stage traffic detection, risk evaluation model, safe-trajectory planning model, and density evaluation model. The player gets additional points for proving to be a true space environmentalist by accepting "debris identification and mitigation" missions. Further to this, the player can save other cosmonauts from malfunctioned and smashed spacecrafts to earn even more.

3.2 Educational content

The purpose of the game is to impart space traffic awareness as a supplemented education learning among kids, engineering students, and aerospace enthusiasts and bring in positive scientific outlook toward the management of space as a shared resource. The key content goals of the game simulation include awareness regarding:
(1) space sustainability
(2) spaceflight safety
(3) spaceflight dynamics
(4) aerospace triangulated networks
(5) space traffic
(6) space traffic collisions
(7) space traffic disasters
(8) early space traffic detection
(9) orbital debris
(10) orbital debris mitigation
(11) rapid action tasks

The game opens up avenues for the students/learners to explore virtual space environment, take interest in space, and gain knowledge about the space-related issues leading to enhanced learning about the current and future space challenges thereby stimulating innovative ideation to help contribute relevant solutions in managing space traffic.

4 Mathematical theory: triangulated networks

Triangulation is a heterogeneous technique that prompts up an assemblage of computational operations apotheosizing geometrically aligned insubstantially figurines for obtaining accurate measurements in engineering applications. It is a fundamental procedure employed in analyzing computational geometric problems, where the complicated geometric figurines are first broken down into simplest forms like tetrahedrons and triangles depending upon the type of the problem and then analyzed further for gathering desired solutions. The game employs triangulation modeling based on a triangulation-determining formula, proposed by Bibhorr [18], while identifying traffic elements as an early-stage traffic detection scenario. The triangulated networks rendered in the game are not visible to the end user – the players but remain the mathematical basis for traffic control and management within the game simulation.

Geoinformation systems and the process of digital landscape modeling require polygon triangulation [19]. Triangulation is the process of estimating the three-dimensional position of a visible point using two pictures. This method necessitates the intersection of two known lines in the space. Nonetheless, if this intersection does not occur owing to the presence of noise, the best approximation must be determined [20]. Triangulation is the subdivision of tessellation into a number of interconnected but nonoverlapping triangles. Through points in the domain Ω, triangles inside a triangulation are formed. The collecting of points during the creation of triangulations is depicted as [21]

$$P = \{p_i\}, \quad i = 1, \ldots, N$$

The triangulation rule governs a vast array of optical approaches for the monitoring purposes. Its usual style implementation utilizes a shifted laser to examine the objective surface, with a picture sensor recognizing the position of the test spot [22]. The triangulation surveying method is based on the trigonometric assumption that if one side and the angles of a triangle are known, then it is straightforward to assess the unknown sides. In traditional surveying, triangulation approach entails measuring the base, which is a line creating one of the triangles' sides, the angles of each triangle, and the distances between two locations through consecutive triangles in regular sequence. Thus, produced triangulations are classified according to their precision of measurement as first order, second order, and third order [23].

For STM, triangulation involves space tessellation with a networked frame of numerous polygonal profile configurations such that at each adjoining node, a space vehicle is located. This triangulated tessellation of networks is the first desideratum for any kind of further analysis and outcome in the STM.

Triangulation techniques are typically utilized in astronomical calculations, but they may also be used in geodetic surveying to determine relative distances and angular radial measurements for a realistically placed polygon coherent pattern sorting. Astronomers often have to rely upon triangulation techniques for determining the distances to the nearby stars. Triangulation techniques also find their applications in medical engineering apart from aerospace and astronomical disciplines. Researchers also use the term "triangulation" in social sciences; however, there the term is used in different context with altogether different perspectives and meanings. The modern procedures involve the employment of lasers, lens, and sensors for accomplishing triangulation measurements in metrology and optics. Active optical triangulation is the most popular method for obtaining range data [24]. Optical triangulation enables distant location measurement [25].

Triangulation, in general, employs trigonometrical data collation and interpretation for basing firm and sturdy calculations. The cartographic analysis sometimes has to be relied upon the triangulation methods for gathering firm results. The position resection, GNSS, and RTL systems all rely heavily on triangulated network calculations. In GIS, planar partitions are commonly used [26]. Triangulations might potentially be employed in aerodynamic simulations to offer an error-free lattice structure during the mesh creation process.

The technique of constructing triangular tessellations within the spatial reference frame under consideration is known as triangulation in space. The time is a critical aspect in space triangulation as the data is acquired not just for a single time point but for numerous time periods since such data sets obtained at different time intervals help monitor and assess planform and structure of the tessellated patterns. Furthermore, the data in this scenario is quite sensitive because it is dependent on information from several nodal sites that are accommodated within the planform tessellation. The information about any movement can be acquired by identifying the freely dispersed collection of points and referencing to the archived time intervals. The presence of multiple potential triangulations complicates 3D spatial point triangulation [27]. The geometric satellite triangulation procedure is based on a three-dimensional triangulation of a large number of various satellite directions [28]. Previous investigations have revealed a relationship between the sort of metadata provided with a raw, basic satellite image and the algorithms used to triangulate that image [29]. The time-matricized modeling is identified as a conglomerated model with space triangulation, in which the polygon structures are detected, and the space vehicle positioned in the circumjacent pattern is monitored for safety. The model could display several phases of safety checks in which data flows from the nodes of the circumjacent polygon to the inside of the spacecraft under study. For the discovery of future trajectory approximations, time-matricized triangulation modeling augments polygon shift analysis. The shift detection time-matricization requirements for the circumjacent polygon fantastically outline the spacecraft's safety status; the data from which might be used in course modification to avoid future crashes.

5 Design and development

The development and designing process of the game followed the user experience design principles. While user experience design is crucial and of utmost relevance in the design and development of any app or digital interaction experience, it is even more significant in the designing of AR/VR-based games and digital interactions. The immersive experience in a virtual environment requires greater attention to the user experience in addition to high screen resolution, high computing capacity, and high network speed.

The user experience can normally be evaluated using one of two methods in immersive virtual environments: objective method or a subjective evaluation. However, combining the two approaches might produce more precise results. Objective methods (instrumental measures) give results based on experimental data, whereas subjective methods (noninstrumental measurements) provide outcomes from the perspective of the user. When attempting to comprehend the user's subjective opinions, attitudes, or preferences, it is common practice to adopt methodologies that are based on subjectivity. Subjective methods include things like focus groups, questionnaires, and interviews. Focus groups are techniques that help problems and concern surface through group interaction. Questionnaires are a set of items designed to obtain the user's opinion, beliefs, or preferences. Interviews are open-ended questions that help the user talk about the specific detail of his activity [30].

The app for the game went through iterative evolution to enable efficient and easy user experience design. The navigation and front-end programming was ensured to be in line with design activities with respect to user modeling, wireframing, developing low-fidelity, and consequently high-fidelity prototype, followed by optimizing the user experience to ensure an effective and satisfactory gaming experience for the players while at the same time ensuring that the core objective of learning experience while gaming is also achieved without compromising the user experience.

Preliminary survey and interviews: A preliminary survey was conducted to gather understanding of the gaming behavior and habits of students. This included students from 9 to 19 years of age across different geographies. Information on the current gaming behavior and ideal games of the surveyed students was collected to understand if students were engaged in gaming only for recreation and/or for leveraging gaming to achieve their educational or knowledge enhancing goals as well.

User personas: User personas were created to serve as a direction for design and navigation mapping, exemplify usability, create models for character–object interaction, and enhance audiovisual aesthetics during the iterative prototype evolution. A total of five learner/student personas were created based on multiple criteria including devices used/accessible, smartphone usage, games/apps frequently used, edutainment app usage, personality, and motivation among other factors such as age, grade, location, and goals were taken into consideration.

Interface wireframing: The final game design was a result of typical iterations and passed through phases from sketching, wireframing, and prototyping. In order to facilitate early visualization of various design ideas, low-fidelity prototype was initially built keeping into consideration the preliminary survey results and the user personas. This, in turn, drove innovation and improvement over the previous designs. Crude sketches allowed users to feel more at ease providing feedback on potential improvements. However, to ensure usability testing, a high-fidelity prototype was built using Adobe XD in conjunction with appropriate plugins that provided a depiction of the user interface as accurate as possible given the limitations of the medium. The high-fidelity prototype was proved to be more effective in gathering accurate performance data and user–interface interaction.

Heuristic evaluation: A list of seven heuristics was established on which evaluation from two usability experts in gaming industry was sought. The evaluators were briefed about the learning and recreation aspects and objectives of the game. The evaluation process entailed recording of problems and suggestions to mitigate problems on the basis of heuristics post two runs of the game by the evaluators. In the first run, the evaluators identified specific elements in the game that they wanted to evaluate while getting familiar with the game controls and navigation, and in the subsequent run, applying the chosen heuristics to the identified elements.

6 Research methodology

6.1 Research outline

The research employed a quantitative design analysis to determine how well the gaming app can practically help project interactive learning environment regarding STM to students. The research was predominately established to empirically evaluate knowledge enhancement quotient among the STM game players and the rote learners. The objective was to highlight the effectiveness of the game without evaluating gender-based data. Hence, the data was gathered irrespective of gender identity as the research did not require it to fulfill its goal.

6.2 Participants involved

A total of 87 students participated in the research from around 8 countries, namely, India, Russia, Germany, Spain, Thailand, Japan, Britain, and the USA. Only the students from fifth to eighth grade were allowed to participate in the research. The students were primarily divided into two groups: group A consisting of 43 participants and group B consisting of 44 participants. For each group, the research was carried out for 4 weeks and the participants were made to devote their time for 120 min/week. The students were made to learn on their own without getting support from peers or parents. Group B participants were allowed to play STM game while the group A was provided with digital learning material related to space traffic.

6.3 Data collection

After 4 weeks, a quiz regarding STM awareness was circulated among all participants, and the scores were registered for each participant. The average scores of both the groups were evaluated, and the percentage difference between the groups was analyzed for assessing the practical knowledge enhancement comparison between the groups.

7 Results

The results for the knowledge quotient (KQ) scores attained by each participant for groups A and B are shown in Table 12.1, where the scores are registered against the participant ID number of each participant. The KQ scores are provided on a 10-unit scale as follows.

Table 12.1: The data for the knowledge quotient (KQ) scores against the participant's ID number.

Group A		Group B	
Participant Id no.	KQ score	Participant Id no.	KQ score
1	1	1	7
2	2	2	6
3	2	3	8
4	0	4	5
5	1	5	6
6	1	6	7
7	1	7	8
8	0	8	9
9	0	9	8
10	4	10	10
11	3	11	7
12	0	12	8
13	1	13	5
14	0	14	7
15	2	15	8
16	1	16	7
17	0	17	7
18	2	18	7
19	1	19	9
20	1	20	10
21	1	21	5
22	1	22	8

Table 12.1 (continued)

Group A		Group B	
Participant Id no.	**KQ score**	**Participant Id no.**	**KQ score**
23	1	23	9
24	0	24	10
25	0	25	7
26	2	26	8
27	1	27	6
28	0	28	9
29	0	29	5
30	1	30	5
31	2	31	7
32	1	32	9
33	2	33	10
34	3	34	6
35	3	35	7
36	4	36	8
37	5	37	6
38	4	38	6
39	2	39	9
40	1	40	9
41	0	41	5
42	0	42	10
43	3	43	7
		44	10

From the above data, the average KQ score for group A is computed as 1.39 and the average KQ score for group B is evaluated as 7.5. The difference between the average scores of two groups is 6.11, signaling that group B scored 6.11 points higher than group A.

This means that group B scored 4.395 (≈4.4) times higher than group A. This corresponds to approximately 440% increment in scores as recorded for group B compared to group A.

8 Conclusion

The newly developed *STM* game is incepted, and a research study regarding the effectiveness of the simulation experience is performed. The research describes the inception of game and the mathematical basis of its development for the enhancement of KQ among kids and students. The game is envisioned discerning the rapid transformation of rote learning system to gamification-based learning environment packed with AR/VR assistance. The research concludes that the students who played *STM* game saw a substantial growth in their KQ regarding STM in comparison to the ones who had access only to the written digital learning materials. The knowledge graph sharply increased 4.4 times or 440% for the *STM* game players in comparison to the nonplayers. The game application proved worthy enough to be employed as a full-fledged learning resource among students to cut down on the rote learning system with which students usually struggle throughout their studies.

References

[1] Sridhar, B., Seth, K. S., and Grabbe, S. (1998). Airspace Complexity and its Application in Air Traffic Management. In *The 2nd USA/Europe ATM R&D Seminar*, Orlando, FL.

[2] Erzberger, H., and Paielli, R. A. (2002). Concept for Next Generation Air Traffic Control System. *Air Traffic Control Quarterly*, 10(4): 355–378. doi:10.2514/atcq.10.4.355.

[3] Airspace Complexity and its Application in Air Traffic Management, Johnson, N. L. (2004). Space Traffic Management Concepts and Practices. *Acta Astronautica*, 55(3–9): 803–809. doi:10.1016/j.actaastro.2004.05.055.

[4] Bonnal, C., Francillout, L., Moury, M., Aniakou, U., Dolado Perez, J.-C., Mariez, J., and Michel, S. (2020). CNES Technical Considerations on Space Traffic Management. *Acta Astronautica*, 167: 296–301. doi:10.1016/j.actaastro.2019.11.023.

[5] Ailor, W. H. (2006). Space Traffic Management: Implementations and Implications. *Acta Astronautica*, 58(5): 279–286. doi:10.1016/j.actaastro.2005.12.002.

[6] Menin, A., Torchelsen, R., and Nedel, L. (2018). An Analysis of VR Technology Used in Immersive Simulations with a Serious Game Perspective. *IEEE Computer Graphics and Applications*, (March/April), 38(2): 57–73.

[7] Chang, R.-C., and Yang, C.-Y. (2016). Developing a Mobile App for Game-Based Learning in Middle School Mathematics Course. *International Conference on Applied System Innovation (ICASI)*. doi:10.1109/icasi.2016.7539807

[8] Antunes, M., Pacheco, M. A. R., and Giovanela, M. (2012). Design and Implementation of an Educational Game for Teaching Chemistry in Higher Education. *Journal of Chemical Education*, 89(4): 517–521. doi: 10.1021/ed2003077.

[9] Chittaro, L., and Buttussi, F. (2015). Assessing Knowledge Retention of an Immersive Serious Game vs. a Traditional Education Method in Aviation Safety. *IEEE Transanctions on Visualization and Computer Graphics*, April 21(4): 529–538.

[10] Kumar, Y., Asokan, S., John, B., and Gopalan, T. (2015). Effect of Conventional and Game-based Teaching on Oral Health Status of Children: A Randomized Controlled Trial. *International Journal of Clinical Pediatric Dentistry*, 8: 123–126. doi: 10.5005/jp-journals-10005-1297.

[11] Dinis, F. M., Guimaraes, A. S., Carvalho, B. R., and Martins, J. P. P. (2017). Virtual and Augmented Reality Game-Based Applications to Civil Engineering Education. *IEEE Global Engineering Education Conference (EDUCON)*. doi:10.1109/educon.2017.7943075

[12] Munz, U., Schumm, P., Wiesebrock, A., and Allgower, F. (2007). Motivation and Learning Progress Through Educational Games. *IEEE Transactions on Industrial Electronics*, 54(6): 3141–3144. doi:10.1109/tie.2007.907030.

[13] Braghirolli, L. F., Ribeiro, J. L. D., Weise, A. D., and Pizzolato, M. (2016). Benefits of Educational Games as An Introductory Activity in Industrial Engineering Education. *Computers in Human Behavior*, 58: 315–324. doi: 10.1016/j.chb.2015.12.063.

[14] Furió, D., González-Gancedo, S., Juan, M.-C., Seguí, I., and Rando, N. (2013). Evaluation of Learning Outcomes Using an Educational iPhone Game vs. Traditional Game. *Computers & Education*, 64: 1–23. doi: 10.1016/j.compedu.2012.12.001.

[15] Rondon, S., Sassi, F. C., and Furquim de Andrade, C. R. (2013). Computer Game-Based and Traditional Learning Method: A Comparison Regarding Students' Knowledge Retention. *BMC Med Educ*, 13: 30. Doi org: 10.1186/1472-6920-13-30.

[16] Erhel, S., and Jamet, E. (2013). Digital Game-Based Learning: Impact of Instructions and Feedback on Motivation and Learning Effectiveness. *Computers & Education*, 67: 156–167. doi: 10.1016/j.compedu.2013.02.019.

[17] Kamińska, D., Sapiński, T., Wiak, S., Toomas Tikk, R. E., Haamer, E. A., Helmi, A., Ozcinar, C., and Anbarjafari, G. (2019). Virtual Reality and Its Applications in Education: Survey. *Information*, 10(10): 318.

[18] Bibhorr. (2019). Analytical Study of Measurement Methods in Engineering Applications Using Triangulation Techniques, *National Seminar on Use of Scientific and Technical Terminology in Science and Technology, Commission for Scientific & Technical Terminology, MHRD.*

[19] Saračević, M., Plojović, Š., and Bušatlić, S. (2020). Iot application for smart cities data storage and processing based on triangulation method. In M. Alam, K. Shakil, and S. Khan (Eds.), *Internet of things*, 317–334. Cham: Springer.

[20] Vite-Silva, I., Cruz-Cortés, N., Toscano-Pulido, G., and de la Fraga, L. G. (2007). Optimal triangulation in 3D computer vision using a multi-objective evolutionary algorithm. In M. Giacobini (Eds.), *Applications of evolutionary computing. EvoWorkshops 2007. Lecture Notes in Computer Science*, vol. 4448. Springer Berlin Heidelberg.

[21] Hjelle, Ø., and Dæhlen, M. (2006). *Triangulations and applications*. Berlin, Heidelberg: Springer Science & Business Media.

[22] Donadello, S., Motta, M., Demir, A. G., and Previtali, B. (2019). Monitoring of Laser Metal Deposition Height by Means of Coaxial Laser Triangulation. *Optics and Lasers in Engineering*, 112: 136–144.

[23] U.S. Coast and Geodetic Survey, *Triangulation*. U.S.: Government Printing Office.

[24] Curless, B., and Levoy, M. (1995). Better Optical Triangulation Through Spacetime Analysis. *IEEE International Conference on Computer Vision – Cambridge, MA, USA*, 987–994. doi:10.1109/iccv.1995.466772

[25] Ji, Z., and Leu, M. C. (1989). Design of Optical Triangulation Devices. 21(5): 339–341. doi:10.1016/0030-3992(89)90068-6.

[26] Ohori, K. A., Ledoux, H., and Meijers, M. (Oct. 2012). Validation and Automatic Repair of Planar Partitions Using a Constrained Triangulation. *Photogramm. – Fernerkund. – Geoinfo*, 2012(5): 613–630.

[27] Morris, D. D., and Kanade, T. (2000). Image-Consistent Surface Triangulation. *Proceedings IEEE Conference on Computer Vision and Pattern Recognition. CVPR 2000* 1, 332–338. doi:10.1109/CVPR.2000.855837

[28] Schmid, H. H. (1974). Worldwide Geometric Satellite Triangulation. *Journal of Geophysical Research*, 79 (35): 5349–5376. doi:10.1029/jb079i035p05349.

[29] Jeong, I.-S., and Bethel, J. (2010). A Study of Trajectory Models for Satellite Image Triangulation. *Photogrammetric Engineering & Remote Sensing*, 76(3): 265–276. doi:10.14358/pers.76.3.265.

[30] Tcha-Tokey, K., Loup-Escande, E., Christmann, O., and Richir, S. (2016). A Questionnaire to Measure the User Experience in Immersive Virtual Environments. *Proceedings of the 2016 Virtual Reality International Conference (VRIC '16)*. Article 19, 1–5.

Raj Gaurang Tiwari, Ambuj Kumar Agarwal, Mohammad Husain

13 Integration of virtual reality in the e-learning environment

Abstract: In our ever-changing world, learning is an essential part of everyone's daily routine. The traditional approach to education relies on students applying what they have learned from textbooks and professors in the classroom to real-world circumstances. The use of cutting-edge technology in teaching and learning techniques is critical in today's digital world. Information and communication technology (ICT) is a major focus for universities. These are important scientific instruments with the potential to have an impact on how people learn and teach science. Three-dimensional (3D) virtual reality (VR) interfaces give an experience via e-learning activities, software games, and simulated labs. Using VR, users may see, modify, and interact with computer systems and massive amounts of data. The term "visualization" refers to a computer's ability to provide the user with sensory inputs such as visual, aural, or any other combination of these. E-learning may benefit from the realistic virtual environments that VR and web technologies can create. 3D graphics have been made possible by the development of web technologies such as HTML, JAVA, and the Virtual Reality Modeling Language. Issues linked to employing a VR application as a medium for students to study and gain practical knowledge about a place shown in a virtual environment are discussed in this chapter.

Keywords: Virtual Reality Modelling Language, Virtual Reality, E-Learning, Computer-based training

1 Introduction

As technology advances, we tend to acquire the necessary abilities to keep up with the changing environment. The educational landscape has been profoundly altered by the rapid expansion of the information age. Computer-based training (CBT) supplements instructor-led instruction. In academic and industrial settings, technology is revolutionizing how people learn. In today's technological era, a learning organization has emerged. As digital technology advances, the web's media and interaction will continue to be enriched. According to Urdan and Weggen [1], e-learning arose from the

Raj Gaurang Tiwari, Chitkara University Institute of Engineering and Technology, Chitkara University, Punjab, India, e-mail: rajgaurang@chitkara.edu.in
Ambuj Kumar Agarwal, Department of Computer Science and Engineering, Sharda University Greater Noida, India, e-mail: ambuj4u@gmail.com
Mohammad Husain, Department of Computer Science, Islamic University of Madinah, Saudi Arabia, e-mail: dr.husain@iu.edu.sa

https://doi.org/10.1515/9783110790146-013

information-based economy, a paradigm change in education, and knowledge gaps. According to them, individualized learning may help students retain more information since technological solutions cater to a wider range of learning preferences. Student productivity and teamwork were cited as benefits of the online environment, which included case studies, demonstrations, role-playing, and simulations.

With e-learning, students may study at their own pace, from any location, at anytime. Many people have been able to learn new things as a result of this. The majority of the information in e-learning is two-dimensional in nature. Large and small organizations of various kinds, including corporations, universities, the government and the military, and nonprofits, are steadily increasing their use of e-learning projects. Improvements and advances in the design and delivery of e-learning programs have led to increasingly complex, sophisticated programs with a wider range of applications. In VR, the idea of three-dimensionality is introduced. When it comes to supporting education, virtual reality (VR) technology is often seen as a game changer. VR and computer simulations have been educational tools for years. Since their introduction into the workplace, these technologies have found their way into classrooms and other educational settings. There have been several studies into the development of virtual environments (VEs) for educational and training uses. This chapter focuses on prior research on the application of VR in educational settings.

1.1 E-learning

There are essentially no limits to what you can learn and develop from using the internet. Students from all around the world may now take advantage of educational opportunities available 24 h a day, 7 days a week, because of the Internet's capacity to cut over traditional boundaries of distance and time. Now, students may learn in a variety of mediums, including audio, video, and text, at their own speed. Connecting people and resources for educational objectives through electronic means is what we mean by "e-Learning" [2]. When it comes to performance enhancement, incidental learning is just as important as planned learning when it comes to e-learning, which values both the planned and the unplanned aspects of the student's experience. The remarkable expansion of the Internet has played a significant role in the rise of e-learning activities. When it comes to information, the World Wide Web has grown into a global platform that serves a variety of customers across a wide range of geographic and cultural settings.

Internet and World Wide Web-based information distribution have become normal practice for universities, institutions, and businesses in recent years [3, 4]. E-learning is something that many of them provide. Hypertext and multimedia have the potential to revolutionize how information is shared and how it is consumed. E-learning environments rely increasingly on communication, visualization, and organizational technology, leading to VR in teaching and learning. People of days are expected to be well educated throughout their whole lives. An e-learning system allows users to study on their own time and at their own pace. Engaged learning is taking the place of traditional passive learning.

As web design progresses and e-learning trends emerge, e-learning platforms must be more than "functional" and "easy to access." On the e-learning platforms, techniques, and technologies are examined as vehicles of communication and prospective suppliers of sensory experiences [5]. Examples of computer simulations include a skyscraper or chemical compounds, as well as processes like population expansion or biological degradation. 3D multimedia formats are becoming more popular for presenting these types of information to the public. Computer simulations may range from 3D geometric representations to interactive computerized laboratory experiments. To do this, e-learning systems emphasize visual communication, sensory components, and cultural and aesthetic artifacts. When digitally provided information and learning support and services are combined, e-learning may be described as an effective learning process.

1.1.1 E-learning environments

In e-learning, communication and support technologies are often interwoven. They are places where students may engage with each other, based on the premise that information is not merely passively absorbed from a teacher but is actively generated by the learner himself or herself. Learning experiences may be tailored to suit the needs of the whole class as well as an individual. As a result, students will be able to spend more time on the subject matter at hand than they would otherwise have. Students may benefit from well-designed e-learning environments in grasping ideas that may be difficult for teachers to explain or depict. There should be an emphasis on teamwork in e-learning dispersed systems.

E-learning environments may give learning help as follows:

- Observing an expert and building a mental model.
- Coaching that offers tips and new assignments to improve performance.
- Scaffolding to assist the student complete the assignment.

The development of cutting-edge and relevant e-learning environments is easier in fields where technology is quickly transforming professional practice. Using tiny camera technology, doctors can do laparoscopic surgery remotely while their students watch from anywhere around the globe. E-learning content, tools, and settings may be successfully developed by a quickly expanding, technology-rich result-based profession. Students might benefit from visual presentations of topics or phenomena in subjects such as mathematics, biology, genetics, and physics. Fractal geometry and Fermat's theorem [6] may be used to describe and convey mathematical and scientific concepts and connections.

Simulations and models may help students make connections between theory and practice, whether they are shown in class or made accessible to them on their own time. When it comes to complicated systems, Conway's Game of Life is an excellent illustration. A Java applet allows students to experiment with their own patterns and behaviors to see how complex patterns may develop from the simplest of principles. There is a lot of discussion in the e-learning literature on how many factors, such as learner involvement and engagement, feedback quantity and quality, the learning environment, and technology, all affect the overall pleasure and experience of students. The design of an e-learning course is the most important component, but many other factors contribute to its success. User interface and learning environment design are regularly highlighted in the literature as essential elements of successful e-learning systems [7].

1.1.2 E-Learning in the present day

E-learning is presently transitioning from a pioneering stage to a more serious, long-term approach. Many sectors, including education, are seeing an increase in demand for e-learning. People are looking for alternatives to Moodle and other standard learning management systems (LMSs), and this has led to the evolution of LMSs. Current e-learning systems may be classified based on the information kind. In these days, the majority of e-learning systems are purely theoretical. They typically transfer text to the internet. "Experimental E-learning systems" must have aspects like practice and interaction if students are to gain proficiency and expertise. To achieve these outcomes, several E-learning systems used a substantial amount of multimedia. There are many different types of multimedia technologies that may be used in a variety of ways. Authors consider interaction essential to learning via the Internet, yet these may explain the material better than only writing.

VR has been used to address this issue and enhance the interactivity of online courses. It is the purpose of VR to immerse the user in a 3D world that can be directly controlled, allowing them to have a sense of interactivity with the environment rather than the computer. Immersive VR, semi-immersive VR, and desktop VR are the three primary study areas in VR. Desktop VR may be used for e-learning purposes. Virtual communities are used in e-learning to encourage more engagement and immersion. This format has the potential to draw in a larger audience for e-learning while also enhancing the learning experience. In a classroom, they help to increase the degree of student engagement, encourage pupils, and foster a sense of excitement about learning.

Using plug-ins that can comprehend and generate interactive 3D scenes in standard languages like VRML [8, 9], web browsers that are commonly used may combine diverse types of media in an e-learning application, which can enhance both the program's content and the learning experience.

1.1.3 Advantages of e-learning

E-learning provides distinct benefits as follows:

Anytime: The learning tools may be accessed at anytime suitable to the participants, rather than the predetermined hours of a typical course.

Any place: Participants need not even be in the same country as the instructor to participate in this course. No matter where they live, instructors and students may collaborate virtually. It is possible to share knowledge with people all over the world, and doing so frequently enhances the learning experience for everyone involved. Logging on from a hotel or a community learning center is as easy as going to a library or working from your desk at home.

Asynchronous interaction: E-mail correspondence does not need a prompt response like face-to-face or telephone talks. Consequently, communication may be more short, more thoughtful, and more direct. In this course, students have plenty of time to think about what they want to say and how to say it. These discussions may then be more serious and innovative.

Group collaboration: The use of electronic messaging opens up new ways for groups to collaborate, allowing them to have more intelligent and long-lasting talks through the internet. These kinds of encounters give a wealth of opportunities for both learning and problem solutions.

New educational approaches: By taking online courses, students get access to a wide range of new educational opportunities. New technologies allow us to use academics from across the world and establish teams of great teachers, researchers, and scientists. Additionally, instructors may benefit from the support of online forums and professional academics when they share new ideas in their own work.

Enriched learning through simulations, gaming, and interactivity: Immersive learning activities boost students' understanding and retention in the classroom and the workplace. A growing number of e-learning programs make extensive use of these types of activities.

Integration of computers: It is possible to employ computer programs with online learners since they all have access to a computer. So that all attendees may run, explore, and edit a spreadsheet model generated in the lecture before sharing their findings and improvements.

Performance support: Workers may get on-the-job coaching to help them perform better on tasks that are directly tied to corporate objectives and the learning experience from certain online education providers.

2 Virtual reality

Jaron Lanier pioneered VR at VPL Research and constructed the first commercially available head-mounted display (HMD). Alternative immersive systems, such as Fakespace's BOOM and the University of Illinois at Chicago's CAVE, began to appear on the market in the following years. Users of these immersive viewing devices see a computer-generated 3D world in full size and in stereo. With head-referenced viewing, you can glance about, move around, and fly around a virtual world (VW) with a natural interface for navigation in 3D space.

VWs may be controlled, operated, and manipulated with data gloves and other similar technologies. Nonvisual technologies such as sound and haptics are often used to improve the virtual experience. Real-time responsiveness is a fundamental necessity for viewing and other operations, which requires tremendous computer capacity. The outcome is a compelling illusion of being completely immersed in a computer-generated environment.

Computer visuals and human–computer interaction are also part of VR. To construct a 3D VE, it makes use of computers. Users may then interact with the VW as if it were real. VEs may now be created on a standard home computer thanks to advances in computer technology. VR is said to have its power because it entices the user's attention, resulting in a sensation of immersion. Passively monitoring the movement of the user's head is frequently used with a display to enable them to gaze in any direction. Interactive 3D graphics may be used to generate VWs on a desktop monitor without the need for head tracking. Human–computer interfaces in VR are extremely interactive because they allow for interaction between real and VWs [10, 11]. It provides a new level of involvement for e-learning consumers that have never been possible before.

Humans may use VR to engage with computers and incredibly complicated data in a manner that is completely new to them. The term "visualization" refers to the process in which a computer creates sensory outputs for the user to experience a VE. VR has a broad range of uses, from gaming to architectural design and teaching. VR has recently been used to enhance e-learning applications [12, 13]. It is feasible to gain a feeling of a 3D world with this technology. A small number of websites provide the ability to see 3D objects. When using a 3D-capable browser, the whole VW may be seen in 360°, with the option to zoom in and out. Users can also swiftly move between different locations in the VW and see the world from various points of view. User attention and interaction with 3D-modeled things are increased, much as in the real world.

At least in certain cases, Ainge [14] and Song and Lee [15] show that VR encounters may give an advantage over more conventional teaching experiences. Ainge found that students who used desktop VR software to build and explore 3D solids improved their capacity to identify 3D forms in daily situations, but classmates who made 3D solids from paper did not.

The research also found that students who used the VR software were more engaged throughout the length of the project. According to Song et al. [15], students who spent some of their geometry class time investigating 3D solids performed better on tests that asked them to visualize the solution to the issue. These two studies suggest that VR benefits are generally limited to very specialized talents. VR-taught geometry students did not do better on nonvisual activities.

2.1 Virtual model

There are various CAD and CAM technologies out there that can help you create a 3D computer model for your VR experience. A polygonal representation of every geometry is required for the model. Only this kind of representation is capable of real-time rendering. The boundary representation of an object generated using constructive solid geometry must first be constructed (B-rep). Using a tessellation technique, a polygonal approximation may be generated if the borders are curved surfaces (usually triangles). Since the rendering performance is dependent on the number of polygons in the model, the "polygon count" is a crucial consideration in any VR application. A minimum of 20–30 frames (perspective views) per second is required for real-time rendering. In real time, a laptop computer can render up to 10,000 polygons. Complicated models need powerful computers with specialized graphics technology (up to 1 million polygons). As the number of polygons in a model increases, so does the amount of computing power required to run it.

An accurate computer model must include information on how things look (color, reflective qualities, and textures), how lighting affects them, how they could interact, and how they might sound before they can be seen or interacted with. "Virtual model" refers to this augmented polygonal computer model.

Theoretically, a virtual model may be used with any VR system. Data formats and lack of standards make switching between VR systems difficult, however, this is improving as new tools are introduced. Nonimmersive VR may be seen on a desktop PC, but it can also be used with an HMD or CAVE to get full-scale representation, stereoscopic vision, and head-referenced navigation.

Still, the process of creating a virtual model is a time-consuming and labor-involved one. Existing 3D models (such as CAD models) need extra work to convert them into polygonal representations with an appropriate polygon count and to specify appearance, lighting, functionality, and other features. It is still a hot issue in the field of rapid virtual prototyping, or the creation of a virtual model quickly.

In addition to being an exciting new addition to the World Wide Web, VRML (Virtual Reality Modeling Language) is also a viable answer for a standardized virtual model. In contrast to HTML, the Hypertext Markup Language, VRML facilitates the dissemination of 3D models via the Internet. As previously said, they are virtual models with all of the aforementioned properties. All of this is possible because of the usage of a polygonal representation that may be animated as well as dynamically changing.

VRML files describe a virtual reality model (VRML). A VRML file is a text file that provides virtual model data. All web browsers need a VRML plug-in to view interactive VRML files (like Netscape or Internet Explorer). Many of these plug-ins are free and may be downloaded from many places.

Activating the plug-in and seeing a first glimpse of the virtual model occurs when a user clicks on a link to a VRML file on the Web and downloads it to their computer. Navigational tools such as walk-through and flyover are included in the plug features. These features allow the user to freely navigate the model or follow a predetermined route. The mouse is often used to control both navigation and interactions.

Using VRML on the web is a great way to share virtual models with distant users and facilitate teamwork and concurrent engineering. As long as you have access to a networked computer and the VRML plug-in, this is a very low-cost method of delivering VR content. VRML files representing sophisticated virtual models take time to download, and the user's local computer must be fast enough to render and interact with the VW in real time. Those constraints will be progressively removed in the near future, thanks to the present development trend toward high-capacity networks like the Internet and increasingly powerful desktop and laptop computers with 3D graphics acceleration.

VRML models may be seen on a display through the Internet, but the experience is limited to a nonimmersive one. While the Virtual Reality Modeling Language's syntax, data structures, and other capabilities allow for the definition of a complex VR application, they are also strong and comprehensive modeling tools. As long as translators are accessible, this description should suffice for most immersive systems.

2.2 Virtual reality simulations

VR-based training systems have been employed in a variety of contexts [16]. Using a VE, the user may experience real-life instruction and interactions. There are three or more degrees of freedom in VR [17], which is a term used to describe real-time computer graphics systems that allow for human interaction and movement. VR has evolved into a new branch of research that brings together several formerly separate disciplines, such as computer science, robotics, graphics, engineering, and cognitive science. 3D computer-generated settings, known as VR worlds (VR worlds), allow one or more users to interact with virtual objects in a fully or partly immersive manner. Since presentation and haptic devices are so expensive, most applications just employ a stereovision solution or do not use any equipment at all, instead projecting the image onto a standard, flat screen.

As VR technology developed, less immersive applications appeared. Desktop VR applications are cheaper and easier to build than immersive ones. A mouse, joystick, or space/sensor ball is used to traverse 3D computer images. Video arcade games used desktop VR first. Realistic, adaptive, interactive, and easily changed screen-based settings opened up new possibilities for unwired VR.

Even in the early stages of computer graphics research, it was expected that the cheap cost and relatively simple technological requirements of VR would make it a reality for the general public. High-end computer graphics software and low-cost consumer PCs with high-end graphics hardware make VR accessible to everyone. Installing and using desktop VR systems is simple and may be done through the Internet or on a CD. This sort of VR may usually be operated on a regular PC with only a viewer piece of software.

2.3 Learning outcomes of VR simulations

Computer simulations have been shown in several studies to be an efficient method of increasing student learning. The following are the three key takeaways:

Conceptual change: The creation of conceptual transformation is one of the most exciting educational uses of computer simulations. Misconceptions might be historical, mathematical, grammatical, or scientific, and students commonly have them. To assist students face and fix these errors, which frequently include fundamental learning ideas, computer simulation has been studied. Zietsman and Hewson studied how a microcomputer simulation affected pupils' comprehension of velocity and distance (*A Computer Microworld to Introduce Students to Probability | Journal of Computers in Mathematics and Science Teaching*, n.d.). Most of these studies are scientific, but one is mathematical.

Students' mental models in science and mathematics may be greatly enhanced through computer simulations.

Skill development: The development of skills is a more commonly studied outcome metric in the computer simulation literature. More than half of the research surveyed found that computer simulations improved students' ability to do a certain task. These simulations were primarily focused on math and science, but a few included other subject areas, such as history and creativity, which allowed students to make decisions that influenced the outcome of the simulations (a digital text that simulated historical events and allowed students to make decisions that influenced outcomes) (a simulation of Lego block building). Several abilities have been claimed to have been enhanced, such as reading comprehension, problem-solving, scientific method, 3D visualization, mineral identification, abstract thinking, and algebraic skills. Students' motivation and involvement are emphasized in this descriptive research, according to what is read [18]. Computer-simulated experiments (CSE) and problem-solving techniques impact high school students' chemistry performance, scientific process abilities, and attitudes about chemistry. The investigation made use of the following instruments: ChemAchievement, the Science Process Skill Test, the Chemistry Attitude Scale, and the Logical Thinking Ability Test are all chemistry-related assessments. CSE and problem-solving were shown to be more effective in teaching chemistry and scientific process skills than the traditional technique. With the CSE technique, students were more optimistic about chemistry than with the other two ways. The traditional approach was the least successful. The effectiveness of computer simulations to develop diverse abilities, notably those in science and mathematics, is so well supported on the whole.

Content area knowledge: Computer simulations might also be used to teach students about a certain subject matter. Using computer programs that simulate frog dissections, a lake's food chain, microbe development, and chemical compounds might help students learn in relevant areas of the curriculum, as shown by the study literature. Eleven studies examined the impact of computer simulations on subject knowledge. Students' scores on content-area assessments increased dramatically when they used computer simulations [19–21]. When it came to generating topic knowledge, using computer simulations was as successful as or even more effective than using conventional, hands-on materials in practically every scenario. The use of computer simulations as a complement or replacement for conventional methods of teaching topic knowledge has a fair deal of support [22, 23].

3 Virtual environments

Three-dimensional VR models may be found in nonreal VEs known as VWs. Standard PC input and output devices, such as a mouse, keyboard, and monitor, may be used to represent and interact with complicated engineering models. With VEs, there is a great deal of promise for educational and training reasons since they may be used to avoid physical, safety, and financial limitations. For those who want to learn how to execute certain activities or grasp ideas, VEs may provide a safe and repeatable setting where the job can be performed as many times as necessary and in a controlled atmosphere.

VEs replace the one-dimensional views of the Earth and space provided by books and maps. VEs are valued for their interactive capabilities. First-person experiences are quick, nonreflective, and sometimes unconscious. Third-person experiences are more difficult to understand and offer less depth of knowledge than first-person experiences because they involve an intermediate interface. As a replacement for genuine experiences, a VE may provide first-hand information and enable students to learn at their own pace, with less cognitive strain than conventional educational methods.

A growing body of evidence in the field of human learning processes suggests that when more senses are included in the acquisition process, people are more responsive to the information being presented to them. Multisensory stimuli, such as 3D spatialized sound or haptic stimuli, may be used in VEs to make use of this human potential (e.g., vibration and force). More realistic and comprehensive representations of subjects may be achieved by using 3D visuals instead of 2D ones. A variety of experiences may be provided by an educational VE, some of which are not feasible in the real world due to factors such as distance, expense, risk, or practicality.

In a VE, users may alter their size to have a better perspective on the material they are studying. So, for example, they may expand their vision into extraterrestrial realms or contract their vision down to the level of individual atoms and molecules. Transcription, on the other hand, is a more sophisticated idea. Basically, a transducer is an electronic device that transforms data into a form that can be interpreted by human senses. Data may be transformed into forms, colors, motions, noises, or vibrations that can be perceived by the human senses. This means that VEs may be seen as transducers that expand the spectrum of information that can be gleaned from a first-person experience. Even if there is no physical shape in the actual world, users may sense it via transduction and scaling. Finally, the term "reification" refers to the process of making abstract notions tangible. These three knowledge-building events do not yet have educational potential.

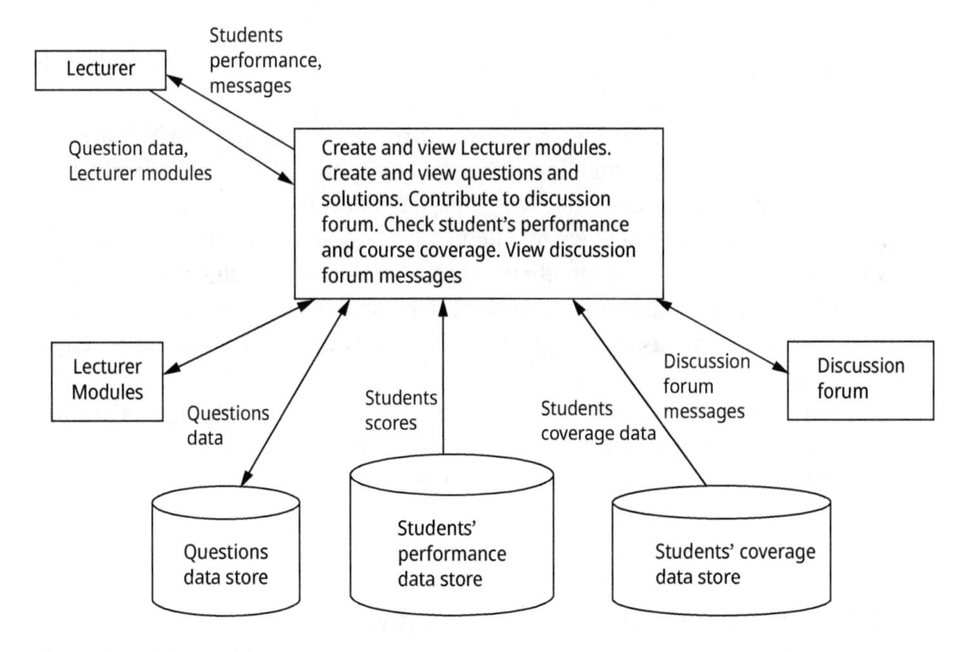

Figure 13.1: VLE essentials.

An increasing number of students are using virtual learning environments (VLEs) that may be tailored to their learning styles. A VLE's key components are shown in Figure 13.1. VLEs are becoming more adaptable to the demands of diverse learners and environments. Web-based technologies' adaptability and the necessity to reach people wherever they are to blame. As is customary, typical adaption mechanisms create student profiles based on a variety of factors including learner preferences, portfolios, prior experience and expertise, educational goals, and even distinct learning styles. The term "LMS" refers to a kind of VLE that enables instructors and students to communicate, monitor each other's progress, and distribute information easily via the internet. Apart from being a viable substitute for face-to-face instruction, they are also increasingly being used in conjunction with it in a variety of settings.

VEs may be an interesting option for decreasing training costs from an industrial standpoint. Rather than interacting with actual items, virtual representations are used, which may then be put in a shared VE. In this way, users may practice operating the actual equipment by manipulating and interacting with the simulated equipment via the use of avatars. It is also possible to simulate dangerous situations in the VW. By shutting the door during milling, the user was able to see the machining conditions and obtain a sense of the necessity of safety.

3.1 Classification of VE

The setting in which VR takes place must be meticulously built to provide a believable experience. VWs come in a variety of shapes and sizes:

- First, a VE can be static or dynamic. Those that do not have any moving parts are known as "static" (e.g., a building). Dynamic environments are those in which an object's location varies over time (e.g., a pigeon that is flying).
- Interactive or noninteractive dynamic VEs are available. The user may interact with the environment and modify the status of the things in the environment. If these items are part of the environment, they may or may not have AI capabilities. VWs that allow users to add or remove things from the scene are a unique subset of VEs.
- Single-user or multiuser interactive VEs are also possible. When there is just one user, the world may be explored, but no other users can be interacted with. In fact, the user's computer contains a VE. It is like reading a web page in HTML: each user has their own copy of the material.

3.2 Educational virtual environments (EVEs)

EVEs give visual, immersive, self-directed learning.

Students can

- immediately perceive the form, size, and duration of things and events;
- perceive the world from a different viewpoint;
- discover and explore the object's hidden parts or assess the impact of operations by interacting with it.

Moreover, EVEs for training can:
- reduce the expense of developing physical training situations on a grand scale;
- present a broad range of situations, including ones that have never before been seen in the actual world;
- repeatable simulations of training situations;
- assess the abilities of trainees by keeping track of their progress during the course.

Some researchers provide an in-depth analysis of the educational underpinning and reasons for EVEs [24, 25]. Several new EVEs have been developed in the recent decade as a result of improvements in computer graphics and hardware performance.

For example, wave propagation and relative velocity may be learned via VRPS (Virtual Reality Physics Simulation). Application interfaces combine 3D representations of actual equipment with interactive visualizations of physical conditions. An AC/DC generator may be investigated by students, who can use a voltmeter to monitor the resulting voltage as the magnetic field's direction, strength, and frequency are changed and adjusted in real time.

Using VRML and Java, Chittaro and Ranon [26] have created an intriguing Web3D EVE; the suggested web-based tool enables students to experience CNC control machine tools, which principally deal with the numeric control of cutting tool motion in manufacturing. In order to try out new code, students no longer have to create a new workpiece. Instead, they may use this application.

Zayas [27] developed a desktop VR system called VEST-Lab in Germany for the purpose of instructing chemical lab workers on the importance of safety. Accident investigation and prevention were all part of the exercises, as we are seeing potential dangers and responding to emergencies. The representation of objects and surroundings in VEST-Lab was excellent.

A study by Roussou [28] has contrasted the educational potential of immersive and nonimmersive VR settings. The scope of the project has grown as a result of creating and implementing 3D learning activities for youngsters. Vygotsky's Zone of Proximal Development (ZPD), which deals with children's ability to internalize social norms, is another method for analyzing children's activities. A 3D VE, for example, allows users to work together and learn from one another, yet they "cannot yet finish the job unassisted."

VR was used by Bricken and Byrne in an experimental research with K-12 education students [29]. Students have been engaged in hands-on technological exploration. It is possible to enter into VWs and interact with them by hearing, seeing, smelling, and touching them. Student behavior and views have been documented while they utilized VR to create and explore VEs, and the results have been analyzed. Individuals have been equipped with tools such as a head-mounted audiovisual display, position and orientation sensors, and tactile interface devices to participate fully in a computer-generated world that includes them.

A significant issue is a high cost of producing and providing VR programs, which prevents many students from using them. Virtual environments (VEs) may now be delivered through the Internet to huge numbers of students anywhere in the globe at anytime, thanks to Web3D open standards (such as VRML and X3D). Web3D VEs may also be presented without the need for a platform-specific plug-in. Desktop VR environments are often accessed with a keyboard, mouse, wand, joystick, or touch screen on an ordinary computer screen.

4 Development platform

Depending on how interaction and simulated behavior are designed, there are a variety of platforms for creating a 3D VE.

Using an API to write program code directly is time-consuming and complicated, but it delivers the best performance. In this method, the user interface is designed from the ground up, giving the user more control. Using a game engine instead of writing the environment's code is usually simpler. Along with a built-in ability to launch bullets, explosions, and so on, the engine provides a user interface for motion control. Game engines are known for their high level of rendering performance. It becomes a lot more difficult to develop when specialized object behaviors or customizing the motion control or object manipulation interface is needed.

To express the geometry, lighting, and material attributes of a 3D environment as well as the behavior of objects, VRML uses a platform-independent file format. VRML dispersed 3D worlds are likewise supported by the VRML standard. Web browser plug-ins or ActiveX controls are often used to run the VRML rendering software, known as a VRML browser. Views may be rendered using graphics hardware thanks to the VRML browser, which also offers a motion control interface. Java and JavaScript are the programming languages of choice for scripting object behavior and enhancing the user experience.

5 VRML

In short, VRML is the Web's 3D language. To give online sites more visual appeal, it incorporates a 3D element. A 3D modeling language known as the HTML of VR is VRML, an ASCII-based language. They may be seen from any angle, including close-ups since the setting is 3D. An object may be represented by gathering cones, cubes, cylinders, and spheres, as well as indexed face sets, custom shapes. You may make a small, compact model by arranging shapes with varied textures, reflectivity, and colors. After being produced, the object may be used as a macro.

When it comes to studying how to display static 3D scenes more effectively and efficiently in the past, much of the research on 3D worlds concentrated on how to make them more accessible and usable in the present. Installing a plug-in on the learner's computer is a current obstacle to the broad distributed use of VRML in e-learning. VRML standards are now being created by the Web3D Consortium [30] to address this issue.

VRML is an excellent tool for e-learning since students are engaged in VWs with multimedia material. Students may think that downloading multimedia information online is required before they may see it. This difficulty is not limited to 3D graphics programs, but to any nonstreamable material that may be combined with virtual models to enhance the student's virtual scene experience.

There are currently no restrictions on how you may interact with the environment or other users in VRML applications. Aside from the fact that a large number of websites seek to exhibit this potential, the animations and exploration are choppy even with high bandwidth and high-speed PCs. New dynamic 3D objects may be specified in VRML thanks to an extensibility approach that enables compatible expansions of the basic standard by application communities. There are mappings between VRML objects and API features that are often utilized in 3D programming.

5.1 The VRML technology

As a VR platform, VRML aims to bring the benefits of 3D settings, known as worlds, to the Internet (and using the file suffix.wrl). Because of this, they are designed to be shared by a large number of people. Simple 3D graphics may be modeled using VRML's collection of objects and methods. Situated hierarchically, they are referred to as "scene graphs." The usage of separator nodes may restrict the effects of a top-down structure in which nodes defined earlier in a scene impact those described later. When coding from an external file, you may utilize inline nodes to reuse objects or use to reuse objects inside the code.

VR in VRML (virtual reality modeling language) is based on the human space concept that it pursues. The user's movements, perceptions, and interactions with the space are all influenced by the 3D nature of the environment. A means to describe the geometry that forms the objects and environments the user moves about in, as well as light, texture, and sound, are all included in VRML. Objects may be seen and heard from a variety of perspectives, and the sound can be heard from various angles.

An ASCII file is read by the browser and turned into a 3D representation of the environment represented in the VRML file. Virtual Reality Modeling Language (VRML) is meant to integrate into the current Internet and WWW infrastructure. To the greatest extent feasible, it makes use of already-existing standards, even if they have significant drawbacks when applied to VRML. It is considerably simpler for the Web developer to utilize current tools to build VRML content than to design new, incompatible standards. Libraries of code for popular standards already exist, making it considerably simpler for someone to implement the VRML standard. Many common formats may be referenced in VRML files. Texture maps may be applied to objects using JPEG, PNG, GIF, and MPEG files. It is possible to define the sound to be broadcast using WAV or MIDI files. It is possible to reference and utilize Java or Javascript code files to create programmed behavior for the world's objects. VRML was selected because of its broad usage on the Internet, and each of them is a separate standard.

Text files describing the scene's layout and the properties of objects are used (position, location, color, and orientation). Even more, flexibility is possible by including Java script into the VRML source code. Events allow nodes to interact with each other. Triggers for events include proximity to the user (like an automated door), other things (like a moving node indicating a vehicle), and time (like a chiming clock). Triggers may also be triggered by the user, other objects, time, or a purposeful action (the user clicks on the object with the mouse). An event is a kind of interaction between the user and the VE. Create your own behaviors for forms to rotate, scale, blink, and more with the help of VRML. Animation circuits may be built with almost any node. Events may be sent and received by nodes, which are like virtual electrical components. Nodes are connected using wired pathways.

Avatar is the name given to the avatar used in the VRML environment to represent the user. A variety of modes are available, including WALK, FLY, and EXAMINE. In WALK mode, the Avatar can only move by foot. Clearing gravity allows the Avatar to soar through the air, but only if it doesn't come into contact with any objects. Allows for exploration within and outside of objects by disabling collision detection in EXAMINE mode.

5.2 The functionalities of VRML

Every object in the game is capable of acting independently or in response to the actions of the player. In many circumstances, this will be necessary for the context of schooling.

Since an interactive VE can only be created using VRML, it is evident that this technology is ideal for the task. Already, several VRML-based VWs have been released, which are experimenting with a wide range of new features. Scripting and building capabilities, as well as further textual, audio, and even video communication, are all included. Most of these accept VRML models and enable them to be utilized in shared settings with many users. In the future, VRML-based shared worlds will be very significant. Exciting for students is the prospect of not just visiting another VRML to learn about it but also seeing its creators in person. Communication among students and between students and faculty members will continue to play an important role in education, and social worlds may help make this possible.

5.3 The role of VRML in education

There are several ways that ICT may be used to enhance learning in the classroom, including teaching students how to move more effectively, incorporating graphics into movement lessons, and creating engaging visual experiments. Rather than merely learning from a book, ICT has the biggest impact on the educational process because of its ability to model and simulate phenomena and processes that encourage scientific investigation and exploration of the actual world.

Simulated environments may aid instructors in enhancing their students' understanding by helping pupils picture material in fresh and realistic ways, giving abstract ideas a realistic taste, and encouraging cross-cultural, global communities.

6 Virtual reality e-learning in education

The utilization of cutting-edge VR simulations and animations has improved students' understanding of tough topics. As strong and sophisticated tools and technology have advanced, e-learning is always evolving. With these enhancements, e-learning now has the potential to soar to new heights. These technology advancements might have a profound impact on education for a variety of reasons, including:

- Change the customary terrain profoundly using them.
- They come up with something innovative and more effective.

In 1993, basic Common Gateway Interface (CGI) applications were developed that could produce HTML code dynamically. CGI scripts were made more difficult to build with the advent of the PHP programming language in 1995. Active Server Pages (ASP) was developed by Microsoft in 1996. ASP could be implemented on the most prevalent operating system, Windows, using scripting languages that were reasonably simple to learn (Visual Basic Script or JavaScript). This release boosted dynamic website development. Sun Microsystems presented Servlets and Java Server Pages during the 1999 JavaOne conference (JSP). JSP's Java basis gives it object orientation, scalability, and cross-platform compatibility.

Immersive VR is engaging because it allows for a high level of participation, which keeps students engaged. It is also engaging since it incorporates noises, graphics, and text, making it a multisensory experience. By enabling users to explore and construct their own mental models depending on the information they find, this method is highly individualizable. It is a great way to keep people engaged and aware of what is happening around them. Active learning, as opposed to the passive learning that is characteristic in conventional training environments, is a hallmark of online education. Active learning is characterized by the fact that students are actively engaged throughout instruction, completing activities, putting their newfound skills to use, and conceptualizing in real time.

The e-learning environment is meant to encourage, inspire, and pique the students' interest in completing real-life simulated activities. Basic object manipulation is supported by the VRML standard. When using VRML, you may specify a variety of mouse sensors and then have the events produced by those sensors routed to various objects in the environment. By clicking and dragging the mouse over an object, you may move it along an animated path, rotate it, or drag it inside a plane.

To connect a VRML world to an external environment, the VRML External Authoring Interface (EAI) is employed. You may use it to interact with a 3D scene in real time and dynamically change it using methods provided by an external program. Users' actions in VRML may trigger events that can be intercepted and dealt with. We can manipulate the VRML scene from outside the application by employing the EAI. In other words, the application may add and delete objects (VRML nodes) from the scene, as well as change the scene settings. We must first create a connection between the program and the VRML browser to begin any EAI action.

7 Analysis of e-learning techniques

Bridge the gap between present knowledge, skills, and attitudes (KSAs) and desired KSAs is a key purpose of training and education Measurement of learner accomplishment is necessary to determine whether or not intended KSAs have been met. It is critical to evaluate e-learning courses based on student satisfaction, which is a significant component of effective training and especially crucial to e-learning courses. At least in the first stages of training, a simulation may serve as a good alternative for the actual world.

According to the Kirkpatrick model for training assessment, there are four levels:

Participants' reactions and feelings about a training program are the focus of the Level 1 assessment, which is called Reaction. This is an indicator of how satisfied customers are with their service. An example of a reaction-level assessment is the use of "smile sheets" at the completion of a training event.

A participant's growth as a learner is assessed at the second assessment level, called learning. An example of a level two assessment design is the use of before and posttests to evaluate learning.

After completing a training program, how much behavior has changed as a result of the experience? This is the focus of the level 3 assessment, Behavior. To put it another way, this level focuses on gauging changes in performance on the job after training. When doing research at this level, using a control group is one of the most effective strategies to acquire data.

Level 4 assessments, referred to as "Results," are based on the actual outcomes that were achieved as a direct consequence of the training that personnel underwent. As a result, there may be an increase in output and a reduction in expenses as well as an increase in profits and a decrease in turnover.

There must be an evaluation of the learning process as well as the learning results. Students' progress is tracked during the course of study through virtual exercises. One method of keeping tabs on students' progress was to have them make verbal predictions about a certain task and then compare those predictions with the results of the task. User feedback is obtained using methods such as questionnaires and interviews. Focusing on students' experiences and learning helps us enhance the user interface and better understand VR's strengths and limitations when it comes to expressing difficult scientific topics to a larger audience.

8 Conclusion

E-learning is a modern teaching method that is becoming more popular in today's society. A comfortable human–machine interface (HMI) may be built with the help of VR technology. The use of VR in education has had a significant impact on the way students learn and teach. As a result of using VR and e-learning technologies, students' capacity to analyze and solve issues is enhanced. e-learning and e-training may benefit greatly from the use of VR. Study results on VLEs' impact on education are relevant to today's emerging virtual learning environments and may serve as guidance and instructions for future VLE-based teaching. VLEs have been studied extensively.

References

[1] Urdan, T. A., and Weggen, C. C. (n.d.). *Corporate E-Learning: Exploring A New Frontier.*

[2] Tiwari, R. G., Agarwal, A. K., Kaushal, R. K., and Kumar, N. (2021a). Prophetic analysis of bitcoin price using machine learning approaches. In *Proceedings of IEEE International Conference on Signal Processing,Computing and Control, 2021-October*, 428–432. https://doi.org/10.1109/ISPCC53510.2021.9609419.

[3] Tiwari, R. G., Husain, M., Gupta, B., and Agrawal, A. (2010). Amalgamating contextual information into recommender system. In *Proceedings – 3rd International Conference on Emerging Trends in Engineering and Technology, ICETET 2010*, 15–20. https://doi.org/10.1109/ICETET.2010.110.

[4] Tiwari, R. G., Khullar, V., and Garg, K. D. (2022). *Web Recommendation by Exploiting User Profile and Collaborative Filtering.* 335–345. https://doi.org/10.1007/978-981-16-8248-3_27

[5] Kumar, A., Tiwari, R. G., Anand, A., Trivedi, N. K., and Agarwal, A. K. (2021). new business paradigm using sentiment analysis algorithm. In *Proceedings of the 2021 10th International Conference on System Modeling and Advancement in Research Trends, SMART 2021*, 419–423. https://doi.org/10.1109/SMART52563.2021.9676199.

[6] Casti, J. L. (1997). *Would-be worlds: How simulation is changing the frontiers of science*. 242.

[7] Tiwari, R. G., Misra, A., Khullar, V., Agarwal, A. K., Gupta, S., and Srivastava, A. P. (2021b). Identifying microscopic augmented images using pre-trained deep convolutional neural networks. In *Proceedings of International Conference on Technological Advancements and Innovations, ICTAI 2021*, 32–37. https://doi.org/10.1109/ICTAI53825.2021.9673472.

[8] Patle, D. S., Manca, D., Nazir, S., and Sharma, S. (2018). Operator training simulators in virtual reality environment for process operators: A review. *Virtual Reality*, 23(3): 293–311. https://doi.org/10.1007/S10055-018-0354-3.

[9] Xie, B., Liu, H., Alghofaili, R., Zhang, Y., Jiang, Y., Lobo, F. D., Li, C., Li, W., Huang, H., Akdere, M., Mousas, C., and Yu, L.-F. (2021). A review on virtual reality skill training applications. *Frontiers in Virtual Reality*, 0: 49. https://doi.org/10.3389/FRVIR.2021.645153.

[10] Guo, Q., and Ma, G. (2022). Exploration of human-computer interaction system for product design in virtual reality environment based on computer-aided technology. *Computer-Aided Design & Applications*, 19(S5): 87–98. https://doi.org/10.14733/cadaps.2022.S5.87-98.

[11] Sutcliffe, A. G., Poullis, C., Gregoriades, A., Katsouri, I., Tzanavari, A., and Herakleous, K. (2018). Reflecting on the design process for virtual reality applications. *35*(2), 168–179. https://Doi.Org/10.1080/10447318.2018.1443898. https://doi.org/10.1080/10447318.2018.1443898.

[12] Baabdullah, A. M., Alsulaimani, A. A., Allamnakhrah, A., Alalwan, A. A., Dwivedi, Y. K., and Rana, N. P. (2022). Usage of augmented reality (Ar) and development of e-learning outcomes: An empirical evaluation of students' e-learning experience. *Computers & Education*, 177: 104383. https://doi.org/10.1016/J.COMPEDU.2021.104383.

[13] Raja, M., and Lakshmi Priya, G. G. (2022). *Using Virtual Reality and Augmented Reality with ICT Tools for Enhancing Quality in the Changing Academic Environment in COVID-19 Pandemic: An Empirical Study.* 467–482. https://doi.org/10.1007/978-3-030-93921-2_26

[14] Ainge, D. J. (2016). Upper Primary Students Constructing and Exploring Three Dimensional Shapes: A Comparison of Virtual Reality with Card Nets: Http://Dx.Doi.Org/10.2190/KR4E-TUNN-GYVD-JR9U, *14*(4), 345–369. https://doi.org/10.2190/KR4E-TUNN-GYVD-JR9U.

[15] Song, K. S., and Lee, W. Y. (2002). A virtual reality application for geometry classes. *Journal of Computer Assisted Learning*, 18(2): 149–156. https://doi.org/10.1046/J.0266-4909.2001.00222.X.

[16] De Andrade, V., Gomes, L. A., Gomes, C., Hoinoski, F. R., Ferreira, G., Schoeffel, P., and Vahldick, A. (2022). *Virtual Reality Applications in Software Engineering Education: A Systematic Review.* https://doi.org/10.48550/arxiv.2204.12008.

[17] Thatte, J., Lian, T., Wandell, B., and Girod, B. (2018). Stacked Omnistereo for virtual reality with six degrees of freedom. In *2017 IEEE Visual Communications and Image Processing, VCIP 2017, 2018-January*, 1–4. https://doi.org/10.1109/VCIP.2017.8305085.

[18] Anupam, A. (2022). Teaching scientific inquiry as a situated practice: A framework for analyzing and designing Science games. Https://Doi.Org/10.1080/17439884.2021.2020285, *47*(1), 125–142. https://doi.org/10.1080/17439884.2021.2020285

[19] Celik, B. (2022). The effects of computer simulations on students' science process skills: Literature review. *Canadian Journal of Educational and Social Studies*, 2(1): 16–28. https://doi.org/10.53103/CJESS.V2I1.17.

[20] Erol, O., and Çırak, N. S. (2021). The effect of a programming tool scratch on the problem-solving skills of middle school students. *Education and Information Technologies*, 27(3): 4065–4086. https://doi.org/10.1007/S10639-021-10776-W.

[21] Rutakomozibwa, A. M. (2022). *Effect of computer simulations on female students' motivation for and engagement with physics learning: a case of secondary schools in Tanzania*. https://doi.org/10.14288/1.0406285.

[22] Egara, F. O., Eseadi, C., and Nzeadibe, A. C. (2022). Effect of computer simulation on secondary school students' interest in algebra. *Education and Information Technologies*, 2021: 1–13. https://doi.org/10.1007/S10639-021-10821-8.

[23] Uzun, C., and Uygun, K. (2022). The effect of simulation-based experiential learning applications on problem solving skills in social studies education. *International Journal of Contemporary Educational Research*, 9(1): 28–38. https://doi.org/10.33200/IJCER.913068.

[24] Jin, S., Xu, M., Chen, Z., Huang, T., Tan, L., and Wang, W. (2021). Research on case virtual simulation teaching system for automobile engine based on web 3d. *Journal of Physics: Conference Series*, 1976(1): 012082. https://doi.org/10.1088/1742-6596/1976/1/012082.

[25] Son, N. D. (2021). The application of web-3d and augmented reality in e-learning to improve the effectiveness of arts teaching in Vietnam. *Journal of Physics: Conference Series*, 1835(1): 012071. https://doi.org/10.1088/1742-6596/1835/1/012071.

[26] Chittaro, L., and Ranon, R. (2007). Web3d technologies in learning, education and training: Motivations, issues, opportunities. *Computers & Education*, 49(1): 3–18. https://doi.org/10.1016/J.COMPEDU.2005.06.002.

[27] Zayas, B. (2001). *AI-ED 2001 Workshop External Representations in AIED:* Multiple Forms *and* Multiple Roles Learning from 3D VR representations: *learner-centred design, realism and interactivity*.

[28] Roussou, M. (2004). Learning by doing and learning through play: An exploration of interactivity in virtual environments for children. *ACM Computers in Entertainment*, 2(1): http://www.aec.at/.

[29] Roussou, M., Oliver, M., and Slater, M. (2007). Exploring activity theory as a tool for evaluating interactivity and learning in virtual environments for children. *Cognition, Technology & Work*, 10(2): 141–153. https://doi.org/10.1007/S10111-007-0070-3.

[30] Flotyński, J., Malamos, A. G., Brutzman, D., Hamza-Lup, F. G., Polys, N. F., Sikos, L. F., and Walczak, K. (2020). *Recent Advances in Web3D Semantic Modeling*. 23–49. https://doi.org/10.4018/978-1-5225-5294-9.CH002:.

Shantanu Trivedi, Saurabh Tiwari

14 The resurgence of augmented reality and virtual reality in construction: past, present, and future directions

Abstract: For a long time, augmented reality (AR) and virtual reality (VR) have been used in gaming and entertainment. However, construction is turning to AR and VR in an increasing number of applications, so it is not the only place where these technologies are gaining traction. The most current developments allow complete teams to plan a project meticulously, from enhancing safety to creating intricate designs and choosing the best materials for the job. In addition, project managers can accurately convey their vision to stakeholders since VR and AR makes construction projects come to life. The study examines how AR and VR can be used more frequently in the construction industry. The objectives are to determine the amount of knowledge about AR hardware and software, look into AR application areas, and spot construction companies lagging. Though numerous readings have investigated the use of AR/VR in various contexts and subject areas, more research into its benefits in construction and manufacturing scenarios is required. The study explores how AR and VR might be used in the building process. Construction-related AR and VR applications are still in their early stages. By using visualization and simulation, this technology will enhance the construction management process, lower health, and safety concerns, and boost worker productivity. Practitioners, management, and implementers is able to understand the diverse perceptions of AR and VR assumption in the construction industry thanks to the research findings. This will assist them in determining which AR and VR tool is appropriate to integrate into their construction processes.

Keywords: Construction, Industry 5.0, Augmented reality, Virtual reality, Emerging technology

1 Introduction

Studies have shown that using specialized augmented reality (AR)/virtual reality (VR) development services is one of the most efficient ways to upgrade the construction sector. However, most of the works are still dangerous, inefficient, and unattractive to

Shantanu Trivedi, Centre for Continuing Education, University of Petroleum and Energy Studies, Dehradun, India, e-mail: s.trivedi@ddn.upes.ac.in
Saurabh Tiwari, School of Business, University of Petroleum and Energy Studies, Dehradun, India, e-mail: tiwarisaurabht@gmail.com

https://doi.org/10.1515/9783110790146-014

new workers despite significant improvements in building materials, equipment, and safety procedures over the past few decades [1].

The phases of a construction project are separated, as the design stage is the first phase; during this phase, a committed team works together on the project, managing flaws to find fixes before construction begins [2]. Additional simulations and studies of the building's structural stability, lighting, and anticipated future weather effects are performed during the design stage. The second and most challenging stage is construction. Both those in engineering and design offices and numerous laborers on the job site are involved. Keeping a tight grip on things and communicating as necessary during each deployment phase is essential. Most logistics and supply chain operations also take place during this stage. Progress should be uninterruptedly supervised, and any aberrations from the novel plan should be documented [3]. Numerous risks could jeopardize the project's timeline, the structure's quality, and the personnel's safety at a busy construction site. The life span of a facility has a postconstruction period that lasts about 30 years. Each deployment step includes maintenance operations to "impart an enjoyable living and functioning surrounding as well as to maintain instruments to block practical breakdowns" [4].

BIM, drones, AR, robotics, big data, and artificial intelligence are the technological advancements linked with Industry 4.0 (I4.0) [5]. For instance, matching the industry's key performance indicators to the UN 2030 Agenda for Sustainable Development, the 5.0 dimension combines the social element of digitalization, such as dedication to the sustainable development goals (SDGs). Industry 5.0 is increasing human–robot collaboration due to increased product customization [6]. Industry 5.0 offers mass customization at scale, whereas I4.0 lays the underpinning for product actualization and tailored. Robotization, AI, big data, and AR are anticipated to enhance the ability to monitor construction projects and the construction sector's performance in producing sustainable and intelligent buildings and infrastructure [7].

A study by Georgia Pacific Building Products [8] suggests that construction technology firms invest in technology to supplement three-dimensional (3D) design and modeling tools. According to one estimate, the AR and mixed reality market distribution will increase to 14% in engineering and 7% in real estate by 2022. Contractors and building owners can use VR and AR to see what a building should look like before construction and compare that image to what the building looks like in real time and this might completely alter the landscape of the sector.

2 Augmented versus virtual

Understanding the differences between AR and VR is essential to fully grasp how these immersive technologies may revolutionize the construction sector. A full-scale, holographic image of the building is projected onto an empty construction site in AR instead of VR, which submerges you in a virtual, 3D environment.

A study conducted by Tulane University (2020) [9] explained the differences between VR and AR in terms of the equipment needed and the incident itself: While VR is completely virtual, AR uses an actual-world environment.

- Customers of AR have command over their actual appearance in the world, in contrast to VR customers.
- VR simply improves a fictional or made-up world, whereas AR, which can be retrieve through a smartphone, improves both the virtual and actual-world experience.

VR creates an immersive experience by displaying 360° images and producing stereo sound, giving the impression of being in a virtual world.

On the other hand, AR improves what you see. For example, furniture retailers such as IKEA have incorporated AR into their apps so that customers can see how a piece of furniture will look in a room by pointing the camera in the direction where the furniture is placed.

AR in construction is a similar concept. To determine where a building should be placed or oriented, project team members can use AR. A headset or smart glasses can be used to view AR, but a tablet or smartphone app is more convenient.

Existing research in this area only covers limited scopes, such as safety and training in construction [10]. The earlier studies analyzed and categorized research on AR and VR in the construction industry and support construction [11]. For organizations looking to perform I4.0, AR technologies are essential due to their capacity to produce immersive user experiences and increase process efficiency [5]. Ivan Sutherland first proposed the concept of VR in 1965, using the phrase "a (virtual) world in the window looks real, feels real, sounds real, and acknowledge realistically to the users actions" as its definition [12]. Applications for VR can be found in a number of organizations, including manufacturing, productions, logistics and supply chain, gaming, military, and the science and technology. A recent article on the utilization of AR and VR in the construction industry is in its infancy stage though AR and VR technology in the construction industry has not received much attention. There are not many studies on (i) frequent issues with AR and VR implementation in the construction industry, (ii) prospective future study directions in this area, or (iii) how AR and VR technologies can be employed in the construction sector and what benefits they can give. This chapter's research seeks to fill that gap. The study's overarching objective is to construct a way forward and research strategies for maximizing AR and VR potential in the construction sector. The main objectives are:

1. Discuss applications and potential benefits in the construction industry.
2. To give a comprehensive overview of the issues with using AR and VR in the construction industry.
3. To outline possible research trajectories for the construction industry's use of AR and VR.

3 Methods

This study aims to shed light on how popular VR–AR technology is today and how it is being used in construction management. In this study, the content analysis-based review methodology is employed. Engineering and construction management have extensively used this technique, which is well known for evaluating and synthesizing literature and justifying results [13]. Using a literature review as a starting point, the methodology began with identifying the study issue about the knowledge of and use of developing immersive machinery in construction management. Then, the analysis of the literature included the analysis and evaluation of academic papers, books, reports, blogs, and other publications in the field of AR and VR technology in construction management.

A two-stage review process is offered (from 2006 to 2021) to conduct a focused and organized review of the literature. Scopus, Google Scholar, and WoS were used to conduct a thorough search in stage 1. The terms "VR in construction," "AR in construction," "mixed reality in construction," "immersive technology in construction," "interactive construction technology," and others were used to describe a broad variety of related fields.

After a quick visual review of the article's content, publications without the keywords mentioned above in their titles or abstracts and less pertinent and irrelevant articles were discarded. The two-stage search might not find all the publications that are worthwhile reading. However, this method is adequate because it offers a significant number of notable cutting-edge works from which the study can draw its conclusions and make recommendations for further research.

A set of data collecting criteria was employed to guarantee the value of the data for this study, as suggested by Moher et al. [14] and Van Eck and Waltman [15]. The criteria are as follows:

- Contemporary: Between 2006 and 2021, the research publications chosen for selection were published within the previous 15 years.
- Relevance: The title, abstract, and keywords were manually assessed to determine their applicability to this study.
- Only studies that appeared in reputable journals were considered to assure quality.
- Inclusiveness: This review study includes conference papers to ensure that no important research works were missed.

The first section of the chapter seeks to provide an overview and background on the study. Then, the research perspective used is mentioned in the following segment. The third section examines the technology in the construction business, the use of AR and VR, and its advantages, problems, and challenges associated with their implementation in the sector. Lastly, the study's conclusion and other essential details are covered.

4 Construction sector

Construction includes the construction of roads and public utility infrastructure, as well as the erection, maintenance, and repair of buildings or other fixed structures. Additions and structural modifications, such as bearing walls, beams, and outside walls, are also included. Every construction project also needs planning, financing, designing, carrying it out, building it, repairs, upkeep, and work enhancement. The sales revenues generated by businesses like sole proprietorships, companies, and partnerships that construct structures and engineering projects like motorways and utility systems make up the construction market [16]. The construction market has a value of USD 7.28 trillion in 2021 and is anticipate to grow at a compound annual growth rate (CAGR) of 7.3% in between 2022 and 2030 to touch USD 14.41 trillion. According to the Construction Intelligence Center in India, global construction is expected to grow by 3.7% in 2022. However, labor shortages and supply chain disruptions have significantly affected the industry, causing project delays, soaring costs, and further margin erosion. Simultaneously, digital technology continues to permeate the industry, making it critical for construction firms to prioritize data and analytics as core competencies [17].

By utilizing ICT, the construction industry is able to work more efficiently and address unique issues like sustainable design and management. ICT is essential to developing the construction sector as part of digitalization.

Smart buildings are now being created in response to requests for intelligent building management, adjustable energy systems, dependable technology, remote monitoring, and other characteristics [18].

The innovations are gradually impacting the construction industry brought about by I4.0, which is quickly sweeping through all other industries. Due to I4.0's extensive prospective to enhance the achievement of building projects and organize their fundamental management processes, the utilization of I4.0 technologies in the construction industry, also known as "Construction 4.0," has enhanced in recent years [19].

Smarter resources are needed to improve processes in modern construction and boost productivity and safety. Unfortunately, there is still very little proof that information technology is used in the building business, even though it has a lot of potential advantages for construction [20]. Therefore, the construction business, which has long been regarded as the least productive, is used in this chapter to illustrate the application, advantages, and difficulties of combining AR and VR.

The project's focus has shifted to building information modeling (BIM), which was crucial when construction digitalization was introduced in IR 4.0. BIM is hailed as an extremely potent and cutting-edge application for the engineering and construction industry because it offers extra data layers that can interrelate and cooperate in real time during the design phase. To improve material handling, BIM's innovation provides a new way to forecast, manage, and monitor material quality and quantity. BIM utilization during the building phase can be enhanced by I4.0's essential elements, including cyber-physical systems, the Internet of things, VR and AR, artificial intelligence, big data, and intelligent production applications. This improves waste management operations, which boosts productivity. By using the BIM model to monitor material handling on-site, segregate waste on-site, monitor project operation, and supervise workers, expands the capabilities of BIM [21].

Luthra and Mangla [22] claim that many other businesses than construction are suffering from a significant lack of awareness of the potential advantages of implementing I4.0. However, because the construction industry is project-based and project teams and features are dynamic, rigorous analysis of the advantages and improvements realized through I4.0 implementation is even more challenging. Adopting I4.0 becomes difficult when businesses cannot anticipate or forecast its benefits and rewards [23].

5 AR/VR in application in construction industry and its benefits

From $70.91 billion in 2021 to $102.6 billion in 2022, the market for AR services is projected to grow at a CAGR of 44.8%. The recovery from the effects of COVID-19, which had previously forced them to adopt restrictive containment measures like social isolation, remote work, and the suspension of commercial operations, which complicated operational issues, is largely responsible for the growth. At a CAGR of 44.8%, the market is projected to touch $451.72 billion in 2026 [24].

The worldwide construction business has traditionally been one of the slowest to accept latest technologies, whether it be digital twins, prefabrications, 3D printing, or AR/VR – all of these buzzwords have been in other industries for quite a long time, but now they are being looked at from the viewpoint of the construction industry in the recent years, and today, this change is occurring at a faster speed [25]. When contrasting VR and physical reality, the first differs between the categories of content and learning actions. Both are then operationalized into several factors and consequences for using VR and physical reality in learning derived from each factor. The category content includes four factors such as complexity, dynamics, networked news, and transparency, whereas the category learning action consists of three factors such as reversibility, cost dependence, and time dependence.

According to a survey of 150 construction companies conducted by data analytics firm Global Data, 55% of companies are considering implementing the technology in the next two years. According to the study, attitudes toward AR are improving, with 51% of respondents saying they are more optimistic about the technology than previously. According to the report, the global AR sector will grow by more than a third by 2030. AR technology use has increased in the construction sector recently, with companies such as Multiplex, M Group, and Eurovia using it on their construction sites [26].

Between 2019 and 2023, the global AR/VR market is anticipated to expand at a CAGR of 77%. By letting buyers see how a product (such as a piece of furniture or appliance) may look in their own house, retailers are adopting AR to enhance the online shopping experience. Examples of cutting-edge AR applications in the building industry include GAMMA AR, Akular AR, ICT Tracker, Arvizio, The Wild, and VisualLive [27].

One of the fields where AR and VR technology is most applicable is construction. AR in the construction industry has many benefits, including reducing errors, marketing, project reviews, costs, and time [28].

As per the study by Chai et al. [29], combining BIM with AR is thought to influence its applicability to fieldwork significantly. Furthermore, the AR–BIM-integrated system also saves time because when a device is mounted on a component, users can access that component's information without accessing the entire BIM model.

A growing trend in the industry is the use of AR, which can offer real-time information, boost overall project confidence, and productivity, and enhance safety [30].

A global data research claims that giving construction workers access to AR technology will not only assist reduce the cost of resources but also help draw young people into careers in the industry. In the next two years, almost 55% of construction sector executives surveyed by Global Data in late 2021 plan to invest in AR, while only 33% have no plans. Building models can be placed on a construction site utilizing headsets or smart glasses and then contrasted with the actual site. As a result, a user will easily detect errors before proceeding with further work, avoiding unnecessary rework and using more expensive materials [31].

Using a computer to simulate a realistic environment, VR enables users to interact digitally and physically with their surroundings. Simulations based on VR are intended to provide users with exclusive insights into VR-based simulations are designed to give users unique insights into how the environment they have created operates [32]. Therefore, it is essential to have respondents, content producers, hardware, and game engine software to enjoy VR. Examples of VR tools used in construction research data from a single component while installing the device include Oculus, HTC, and Samsung Gear.

AR is being utilized in the industry to raise project confidence generally, increase efficiency, and improve safety because it offers real-time information [33].

3D modeling and VR technologies have the potential to enhance communication in professional practice, education, and training. Models of the construction process were created using 3D modeling and VR technology.

The 3D models developed to assist restoration design demonstrate promise as a vital tool for identifying structural anomalies and assisting choices based on visual analyses of possible courses of action. The VR model created to aid in managing building lighting allows for the visible and interactive transfer of data concerning the physical behavior of the elements, demarcated as a utility of the time variable. Interactive construction models can also be created [34].

The foundation of 4D (3D + time) models is the combination of geometric depictions of a building and arranging data about construction plans. VR technology has been used to increase the realism and interaction of 4D models with the surroundings of construction sites. This field combines 3D models and project schedules to create 4D models [35].

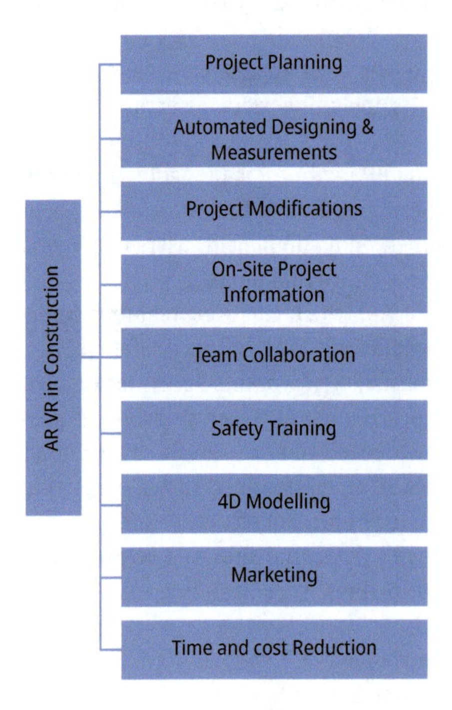

Figure 14.1: AR and VR phases in construction.
Source: Authors own creation

According to recent studies, AR and VR represent the impending construction management. Without the need for tangible media, AR and VR offer an excellent stage for project participants to interact and exchange crucial information in real time. To oversee quality and flaws in construction projects, AR and VR are also used. QA/QC management researchers above AR prefer VR, however, each technology has a significant and valuable place in the field [36]. The construction business is undergoing a revolution, and VR, AR, and mixed reality are enabling architects to create unparalleled buildings. According to economists, the most recent VR technology will cut building costs by almost 90% and prevent up to $15.8 billion in errors from incomplete data [37].

According to their findings [38], various AR features, such as comfort of use, training and learning time, field view, hardware and software outcome, obstruction, and enveloping, they aided in the implementation and completion of construction projects.

According to Albahbah et al. [39], AR and VR applications for scheduling the projects and tracking construction progress have a high potential to outperform current practices. It is proposed that they could play a promising role in advanced construction project management. It has been shown that AR and VR can enhance data collection and communication. In the early stages of projects, VR technology has been projected as a valued tool for multidisciplinary communication. In contrast, AR technology has demonstrated its ability to speed up field reporting, increase the effectiveness and efficiency of workers' tasks, and improve on-site information retrieval. Since proactive prevention of many flaws on a construction site is offered by AR, which differs from conventional approaches because it enables inspectors and site managers to check errors more proficiently and distantly, researchers highly regard AR's effectiveness in defect and quality management components.

In a study, AR and modeling of construction data were suggested by Kazemzadeh et al. [40]. These methods reduce decision-making time during the design stage, enhance document comprehension during the planning stage, and monitor the project's execution to ensure it is carried out following its stated objectives. Furthermore, the advent of AR technology has opened up new avenues for disseminating knowledge.

The construction sector can use AR to overcome several project challenges, such as eliminating the need for visualizing and enabling the designed project to be seen on the site at its true scale before execution. Time and money are significantly reduced as a result of this problem. The implementation process also facilitates a very high level of safety. Workers can, for instance, view the required training before beginning, which is also very helpful in monitoring and management. For example, you may measure the project's progress by placing the model next to the amount of digital work completed. It also has a lot of uses during the transfer phase, such as making it possible to gauge how well the project is going. In the stage of repairs and maintenance, AR allows us to readily observe elements of the project that are otherwise invisible, such as components connected to the project's facilities. AR also has various applications in the renovation process.

It might be argued that incorporating AR into these procedures increases accuracy and safety, minimizes project costs and time, and reduces mistake rates at each step of construction and in the project's final outcome [41]. It demonstrates how the implementation of AR technology in the construction industry may increase and address the goals of the sector by several factors, including lowering project costs using digital technologies;

- The ability to fulfill deadlines, save time, and reduce project delays thanks to an integrated, live schedule.
- VR has shown many advantages in design reviews, collaboration, and decision-making.

A VR scene can be created from virtual models or spherical images. BIM is often used to create VR scenes. On the other hand, unlike model-generated scenes, which present artificial content, spherical 360° images show a realistic view of the physical site, which opens more possibilities for construction inspection as shown in this study.

Design improvements can be made quickly, verified, and validated by an expert group anywhere in the world [42]. Such reviews can be carried out concurrently, separately, or sequentially. Multiple people can participate in the virtual space; each represented by their avatar, and complete a thorough design review and obtain stakeholder buy-in in much less time and at a fraction of the cost. VR technology makes design processes more efficient and saves money, which is passed on to clients. In the construction industry, getting it right the first time is critical. The developed VR models and solutions can be used to educate all personnel on the construction process and the final product. It can also explain how to meet customers' design and functional requirements, resulting in improved coordination between design and construction teams. Effective training programs for all personnel can be designed to ensure high-quality construction with optimal resources.

Using VR and AR during the construction phase can be advantageous because they can enhance quality, safety, and time-budget management. Effectiveness is measured as a component of all knowledge areas of construction management because it is the primary focus of leadership, particularly in terms of time, money, and quality. The cluster sizes for using VR and AR in construction management research are learning, cost management, and safety management, in decreasing order of significance [43].

Joshi [44] stated that building contractors may employ AR and VR technology to make 3D models of homes and an AR overlay of furniture for potential purchasers. It is possible to provide a virtual tour of a furnished apartment, providing visitors a clear idea of what the finished house would look like. Fast and effective communication is crucial for keeping construction projects on time. The design team may classify and fix any concerns by using AR headsets to convey site data that engineers or supervisors have acquired. Contracors or managers can conduct a virtual walkthrough of a site for inspection and assessment using an AR overlay of a building information model over the physical model. He said that AR helps reduce cost, errors, and time by guiding workers with the proper material alignment. Nine hundred seventy-one fatalities in the construction sector were recorded in 2017, making up 20.8% of all fatalities in the private sector. The precise situation of potentially hazardous equipment on a construction site can be determined using AR and VR technology. AR and VR can be used to train employees while reducing any physical risks. VR safety training programs has been introduced by an organization in United States in order to train, enhance and educate processes related to employee safety.

It is anticipated that VR and AR technologies would be used to offer services to purchasers in more elements of the home buying process, according to a report [45]. For example, the expansion of VR tours to include neighborhood and walkability characteristics was recommended by two realtors. In addition, another realtor proposed that VR and AR technologies could streamline the time-consuming home inspection process.

6 Application of AR and VR in the construction industry: issues and challenges

Construction professionals frequently view 3D modeling as a pointless impediment [46]. In their study, they found and emphasized a discrepancy between the level of information for building information model and AR. Since AR only partially use the sophisticated multifacet BIM data model's data for display, there may be a gap in the level of detail that calls for intelligence and interface mapping. Future BIM should be information-critical, intelligent, and context-aware to better integrate AR on the job site. Contractors still prefer to stay in their "comfort zone," adopting tried-and-true technologies without question rather than evaluating and adopting new technologies like AR. This raises a serious concern about the development and adoption of AR tools regarding the social phenomenon associated with them because technologies are inextricably linked to cultural and social contexts.

Fenais et al. [47] identified the technicalities and barriers to AR and VR implementation in construction and classified them as follows: (1) data collection issues; (2) modeling alignment roadblocks; (3) hardware limitations; and (4) data storage and management difficulties.

In their survey Noghabaei et al. [48], observed that the lack of a cost/benefit analysis because of the lower profit margins on building projects is a significant barrier hindering the industry from embracing AR/VR technologies. Furthermore, owners and enterprises are reluctant to spend (i.e., time and cost savings) without initially assessing the actual costs and advantages.

Also, Joshi [44] highlighted the two significant challenges:

1. Lack of experience and education: Few professionals are familiar with or have experience with AR and VR due to the low adoption rate. Due to workers' frequent familiarity with traditional construction practices and their lack of technical knowledge, AR and VR training can be time-consuming.
2. Lack of IT resources: Construction has seen fewer technological advancements than other industries due to its reliance on traditional methods. Since developing new technologies requires a lot of time, money, and human resources, technology adoption in the construction industry is relatively slow, accounting for less than 1% of annual construction sales.

The three biggest challenges identified while implementing innovative technologies like AR, and VR are regulatory compliance, data ownership, and information sharing in the construction industry [49]. However, the same study found that skilled workforce training, communication, government incentives, and change management are the utmost effective methods for overcoming obstacles and implementing smart technologies in the construction sector.

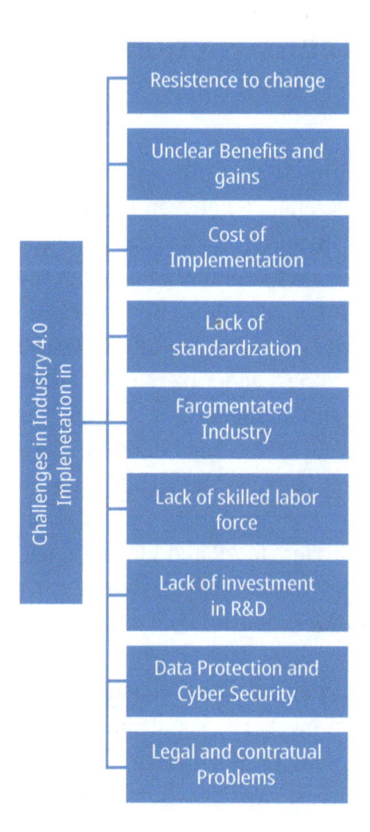

Figure 14.2: Challenges in Industry 4.0 implementation.
Source: Demirkesen & Tezel, 2021

As illustrated in Figure 14.2, there are nine significant challenges in adopting I4.0 in the construction sector. Initially it was found that older companies were lagging in comparison to younger companies. First, younger companies were found to be more successful than older companies in dealing with a lack of standardization. Companies with a higher annual revenue appear to be better able to deal with the cost of implementation challenges for I4.0 implementation. In terms of dealing with the challenges associated with implementation of I4.0 and the resistance to change, younger companies are better well off to handle pressure and challenges [23].

7 Future research direction

Even though significant progress in VR/AR-CS research has been examined in detail, it should be noted that this study is still lacking and is only concerned with the construction industry. The research in the future should focus on how other industries use VR and AR to manage safety. A potential research direction is establishing a standard or reference point for various tasks and investigational evaluation methods of VR and AR safety systems in multiple projects or job tasks [10]. The research also includes studying how AR and VR can improve performance and efficiency in the construction supply chain. An examination can consist of analyzing how other I4.0 technologies can be clubbed with AR and VR to enhance construction projects.

Furthermore, additional research can be conducted to improve and upskill construction workers [11]. More research can be done to link 3D and 4D models in construction projects [50]. Future studies would assess how well the suggestions made by this study performed in comparison to actual field data and recorded organizational data [51]. To increase the number of recommendations for the industry's use in the future and to help in the development of operative reaction plans that construction organizations can use when confronted with future pandemics, the impact of strategies adopted on key performance pointers for construction organizations, for example, must be appraised for a broader range of companies [52]. Understanding and analyzing the legal aspects, compatibility problems, and necessary skills and knowledge to help organizations move into the next phase of digital construction delivery can be another area of future research.

Additionally, scheduling, simulation modeling, and AI advances can improve the accuracy of feasibility studies during the preconstruction period. Early and more precise defect detection during construction will save time and money. Along with the industry, academic institutions must include VR–AR developments in construction management into their courses [43]. However, there is little research on intelligent decision support systems for circularly managing building and demolition waste [53]. The results of the study, "Trend Analysis on adoption of AR and VR in the Architecture, Engineering, and Construction Sector [54]," show that there is still a sizable knowledge gap between the industry and academia in terms of AR/VR awareness. As a result, the sector has to be educated and comprehend use cases to become familiar with AR/VR technologies.

Artificial intelligence is another technology making significant inroads into our daily lives, along with VR and AR. All have shown great potential in the future [55]. Companies working with these technologies are investigating the possibility of combining them. Advanced machine learning (ML) algorithms have the potential to help computers and intelligent devices see and understand data in entirely new ways. As a result, more engaging workspaces and advanced image recognition capabilities is created.

Finally, every realtor agreed that the selling process in their profession would eventually incorporate digital technologies. More investigation of the difficulties associated with incorporating such technology into a current sales process would be helpful from a managerial perspective [45].

8 Conclusion

The utility of VR/AR methods is likely to be greatest during the building's planning stages, involving solutions such as immersive VR visualization, augmented scale models, and mobile increasing on-site. The field of visualized construction management has seen a lot of development. However, there are still not enough practical and effective visualization management systems for the construction sector [56]. Through AR and VR, people can access scenarios or simulations that would otherwise be difficult to imagine or inaccessible to those who are not specialists. A virtual model enhances interaction and communication with designs and procedures [57]. Although the infrastructure and construction sectors have not yet seen as much change as other sectors, this is changing, and the shift will be profound and inevitable. As emerging technologies mature and new technologies are developed, we will be able to deliver services faster, better, and safer thanks to AR and VR, as well as more evidence-based judgments and a greater understanding of the client. Since design and other issues were resolved during the preconstruction stage, 3D modeling and simulation have increased technology's emphasis. The usage of VR and AR technology has significantly increased workplace safety. Managers have made another step toward assuring the safety of their staff as well as on-site personnel, thanks to VR immersion safety training classes. Top construction companies are actively utilizing AR and VR as part of their digital transformation strategy because it is revolutionary in the construction sector. The writing is on the wall, with the explosive growth of VR everywhere! AR/VR technologies are critical for success in the twenty-first century.

References

[1] Prograam Ace. (2021, April 15). *Blog: Program-ace*. Retrieved from Program-Ace Website: https://program-ace.com/blog/vr-in-construction/

[2] Zaker, R., and Coloma, E. (2018). Virtual reality-integrated workflow in bim-enabled projects collaboration and design review: A case study. *Visualization in Engineering*: 1–15. https://doi.org/10.1186/s40327-018-0065-6.

[3] Neenu, S. K. (2019, January 18). *Construction: The constructor*. Retrieved from The Constructor: https://theconstructor.org/construction/construction-project-life-cycle-phases/14283/#:~:text=Life%20Cycle%20Phases%20in%20Construction%20Project,-A%20standard%20construction&text=life%20cycle%20phases%3A-,Initiation,Performance%20and%20monitoring

[4] Koch, C., Neges, H., Konig, M., and Abramovici, (2012). Bim-based augmented reality for facility maintenance using natural markers. In *EG-ICE international workshop on intelligent computing in engineering*. Plymouth: European Group for Intelligent Computing in Engineering, 1–10.

[5] Tiwari, S. (2020). Supply chain integration and industry 4.0: A systematic literature review. *Benchmarking: An International Journal*, 28(3): 990–1030.

[6] Tiwari, S., Bahuguna, P. C., and Walker, J. (2022). Industry 5.0: A macro perspective approach. In *Handbook of research on innovative management using AI in industry 5.0*. IGI Global, 59–73.

[7] CICA. (2021, Jan 15). *Working groups: CICA*. Retrieved from CICA website: http://www.cica.net/activities/working-groups/construction-5-0/

[8] Georgia Pacific Building Products. (2020, March 18). *Blog: Build GP*. Retrieved from Georgia Pacific Building Products: https://buildgp.com/blog/

[9] Tulane University. (2020, Jan 20). *Blog: Tulane University*. Retrieved from Tulane University: https://sopa.tulane.edu/blog/whats-difference-between-ar-and-vr

[10] Lia, X., Yib, W., Chia, H.-L., Wang, X., and Chan, A. P. (2018). A critical review of virtual and augmented reality (VR/AR) applications in construction safety. *Automation in Construction*: 150–162.

[11] Delgadoa, J. M., Oyedelea, L., Demianc, P., and Beach, T. (2020). A research agenda for augmented and virtual reality in architecture, engineering and construction. *Advanced Engineering Informatics*: 1–21.

[12] Mandal, S. (2013). Brief introduction of virtual reality & its challenges. *International Journal of Scientific & Engineering Research*, 4(4): 304–309.

[13] Mok, K. Y., Shena, G. Q., and Yang, J. (2015). Stakeholder management studies in mega construction projects: A review and future directions. *International Journal of Project Management*: 446–457.

[14] Moher, D., Liberati, A., Tetzlaff, J., and Altman, D. G. (2009). Preferred reporting items for systematic reviews and meta-analyses: The PRISMA statement. *PLOS Medicine*: 1–12.

[15] Van Eck, N., and Waltman, L. (2010). Software survey: Vosviewer, a computer program for bibliometric mapping. *Scientometrics*: 523–538.

[16] ResearchAndMarkets. (2022, January 17). *News releases: Global news wire*. Retrieved from Global News Wire: https://www.globenewswire.com/news-release/2022/01/17/2367744/28124/en/Worldwide-Construction-Industry-to-2030-Featuring-Grupo-Bouygues-and-Skanska-Among-Others.html

[17] Hussain, A. (2022). *2022 engineering and construction industry outlook*. New York: Deloitte Research Center for Energy & Industrial.

[18] Vasista, T. G., and Abone, A. (2018). Benefits, barriers and applications of information communication technology in construction industry: A contemporary study. *International Journal of Engineering & Technology*: 492–499.

[19] Perrier, N., Bled, A., Bourgault, M., Cousin, N., Danjou, C., Pellerin, R., and Roland, T. (2019). Construction 4.0: A survey of research trends. *Journal of Information Technology in Construction*: 416–437.

[20] Kozlovska, M., Klosova, D., and Strukova, Z. (2021). Impact of industry 4.0 platform on the formation of construction 4.0 concept: A literature review. *Sustainability*: 1–12.

[21] Maskuriy, R., Selamat, A., Maresova, P., Krejcar, O., and David, O. O. (2019). Industry 4.0 for the construction industry: Review of management perspective. *Economies*: 1–14.

[22] Luthra, S., and Mangla, S. K. (2018). Evaluating challenges to industry 4.0 initiatives for supply chain sustainability in emerging economies. *Process Safety and Environmental Protection*: 168–179.

[23] Demirkesen,, and Tezel, A. (2021). Investigating major challenges for industry 4.0 adoption among construction companies. *Engineering, Construction and Architectural Management*: 1470–1503.

[24] Global News Wire. (2022, March 22). *News release: Global news wire*. Retrieved from Global News Wire: https://www.globenewswire.com/news-release/2022/03/22/2407979/0/en/Augmented-Reality-Services-Global-Market-Report-2022.html

[25] Konaraddi, S. (2022, April 26). *Pulse: Linkedin*. Retrieved from Linkedin: https://www.linkedin.com/pulse/how-ar-vr-transforming-construction-/?trk=organization-update-content_share-article

[26] Stein, J. (2022, May 10). *Tech: Construction news*. Retrieved from Construction News: https://www.constructionnews.co.uk/tech/third-of-contractors-snub-augmented-reality-tools-10-05-2022/

[27] Ellis, G. (2022, January 16). *Blog: Autodesk construction*. Retrieved from Autodesk Construction: https://constructionblog.autodesk.com/augmented-reality-ar-construction/

[28] Agarwal, S. (2016). Review on application of augmented reality in civil engineering. In *ICIDRET*. New Delhi: DSIIDC (Delhi State Industrial and Infrastructure Development Corporation), 1–8.
[29] Chai, C., Mustafa, K., Kuppusamy, S., Yusof, A., Lim, C. S., and Han Wai, S. (2019). BIM integration in augmented reality model. *International Journal of Technology*: 1266–1275.
[30] Stannard, L. (2022, March 2022). *Blog: Big rentz*. Retrieved from Big Rentz: https://www.bigrentz.com/blog/augmented-reality-construction
[31] Brown, A. (2022, May 5). *News: Internationnal construction*. Retrieved from International Constructions Website: https://www.international-construction.com/news/Augmented-reality-can-help-construction-attract-workers-/8020351.article
[32] Moore, H. F., and Gheisari, M. (2019). A review of virtual and mixed reality applications in construction safety literature. *Safety*: 1–16.
[33] Afolabi, A. O., Nnaji, C., and Okoro, C. (2022). Immersive technology implementation in the construction industry: Modeling paths of risk. *Buildings*: 1–24.
[34] Sampaio, A. Z., Ferreira, M. M., and Rosário, D. P. (2010). 3D and VR models in civil engineering education: Construction, rehabilitation and maintenance. *Automation in Construction*: 819–828.
[35] Construction Placements. (2022, April 1). *Articles: Construction placements*. Retrieved from Construction Placements: https://www.constructionplacements.com/virtual-reality-in-construction/
[36] Ahmed, S. (2019). A review on using opportunities of augmented reality and virtual reality in construction project management. *Organization, Technology and Management in Construction: An International Journal*, 1839–1852.
[37] Animation, S. B. (2021, May 10). *Blog: SB animation*. Retrieved from Sliced Bread Animation: https://sbanimation.com/how-vr-and-ar-is-transforming-the-construction-industry/
[38] Asmar, P. G., Chalhoub, J., Ayer, S. K., and Abdallah, A. S. (2020). Contextualizing benefits and limitations reported for augmented reality in construction research. *Journal of Information Technology in Construction*: 1–19.
[39] Albahbah, M., Kıvrak, S., and Arslan, G. (2021). Application areas of augmented reality and virtual reality in construction project management: A scoping review. *Journal of Construction Engineering, Management & Innovation*: 151–172.
[40] Kazemzadeh, D., Nazari, A., and Rokooei, S. (2021). Application of augmented reality in the life cycle of construction projects. In *CONVR*. North Yorkshire: Teesside University, 16–26.
[41] Hajirasouli, A., Banihashemi, S., Drogemuller, R., Fazeli, A., and Mohandes, S. R. (2022). Augmented reality in design and construction: Thematic analysis and conceptual frameworks. *Construction Innovation*: 1–32.
[42] Bharathy, C. (2022, May 27). *Blog: Fusion VR*. Retrieved from Fusion VR: https://www.fusionvr.in/blog/2022/05/27/vr-in-construction-industry-the-future-of-the-construction-business/
[43] Guray, T. S., and Kismet, B. (2022). VR and AR in construction management research: Bibliometric and descriptive analyses. *Smart and Sustainable Built Environment*: 2046–6099.
[44] Joshi, N. (2020, November 20). *Technology: BBN times*. Retrieved from BBN Times: https://www.bbntimes.com/technology/benefits-challenges-of-augmented-reality-and-virtual-reality-in-construction
[45] Sihi, D. (2018). Home sweet virtual home the use of virtual and augmented reality technologies in high involvement purchase decisions. *Journal of Research in Interactive Marketing*: 398–417.
[46] Wang, X., Truijens, M., Hou, L., Wang, Y., and Zhou, Y. (2014). Integrating augmented reality with building information modeling: Onsite construction process controlling for the liquefied natural gas industry. *Automation in Construction*: 96–105.
[47] Fenais, A., Smilovsky, N., Ariaratnam, S., and Ayer, S. K. (2018). A meta-analysis of augmented reality challenges in the underground utility construction industry. In *Construction research congress*. New Orleans: American Society of Civil Engineers, 1–10.

[48] Noghabaei, M., Heydarian, A., Balali, V., and Han, K. (2020). Trend analysis on adoption of virtual and augmented reality in the architecture, engineering, and construction industry. *Data*: 1–18.

[49] Hwang, B.-G., Ngo, J., and Teo, J. Z. (2022). Challenges and strategies for the adoption of smart technologies in the construction industry: The case of singapore. *Journal of Management in Engineering*: 1–10.

[50] Kang, L. S., Dawood, N, M. H., and Kang, M. S. (2010). Development of methodology and virtual system for optimized simulation of road design data. *Automation in Construction*: 1000–1015.

[51] Raoufi, M., and Fayek, A. R. (2022). New modes of operating for construction organizations during the COVID-19 pandemic: Challenges, actions, and future best practices. *Journal of Management in Engineering*: 1–12.

[52] Harper, C. M., Tran, D., and Jaselskis, E. (2022). Implementation of visualization and modeling technologies for transportation construction. *Journal of Civil Engineering and Construction*: 29–40.

[53] Oluleyea, I., Chana, B., D. W.,Sakaa, A. B., and Olawumib, T. O. (2022). Circular economy research on building construction and demolition waste: A review of current trends and future research directions. *Journal of Cleaner Production*, 2022.

[54] Trend Analysis on Adoption of Virtual and Augmented Reality in the Architecture, Engineering, and Construction Industry. (2020). *Data*, 1–18.

[55] Q3 Technologies. (2022, March 14). *Blog: Q3 tech*. Retrieved from Q3 Technologies: https://www.q3tech.com/blogs/the-9-biggest-virtual-and-augmented-reality-trends-in-2022/

[56] Rohani, M., Fan, M., and Yu, C. (2014). Advanced visualization and simulation techniques for modern construction management. *Indoor and Built Environment*: 665–674.

[57] Engineering Antycip, S. T. (2021, June 16). *Blog: ST engineering antycip*. Retrieved from ST Engineering Antycip: https://steantycip.com/blog/the-future-of-virtual-reality-in-aec/

Editors' biography

Dr. Richa Goel is an Associate professor in economics and international business at SCMS, Symbiosis International University, Noida. She is a gold medalist in master of economics with dual specialization at master level accompanied with an MBA in HR and also with dual specialization at graduate level with gold medalist in economics honors also with bachelor of law. She is PhD in management where she had worked for almost 6 years in the area of diversity management. She has a journey of almost 21+ years in academics. She is consistently striving to create a challenging and engaging learning environment where students become lifelong scholars and learners. Imparting lectures using different teaching strategies, she is an avid teacher, researcher, and mentor. She is working in joint collaboration with few international researchers on various projects related to E-Shiksha, women empowerment, and SDG. Her fourth project, which is on business model for inclusive banking sector, has been appreciated by the Ministry of Finance, New Zealand, and has been proposed further to Reserve Bank of New Zealand. She has to her credit more than 50 research papers in UGC, Scopus, and ABDC publications in reputed national and international journals accompanied with hundreds of research participation in international/national conferences including FDP, MDP, and symposiums. She is writing numerous book chapters for leading publishing books like Springer, Emerald, IGI, Bloomsbury, and Taylor and Francis. She is serving as a member of review committee for conferences and journals. She is handling many Scopus international peer-reviewed journals as lead editor for regular and special issue journals. She is acting as the special issue editor of *Journal of Sustainable Finance and Investment*, which is abstracted and indexed in the Chartered Association of Business Schools Academic Journal Guide (2018 edition) and Scopus (print ISSN: 2043-0795; online ISSN: 2043-0809).

Prof. (Dr.) Sukanta Kumar Baral is working as a professor at the Department of Commerce, Faculty of Commerce and Management, Indira Gandhi National Tribal University (a central university of India), Madhya Pradesh, India. As an active academician, he has been closely associated with several Indian and foreign universities to deliver different academic assignments. He has earned six Indian copyrights in his favor. He has authored 14 books and edited 11 books with most current, comprehensive, and state-of-the-art analysis to fulfill the needs of commerce, economics, and management students, professionals, and executives working in related fields. He has been conferred with 27 numbers of prestigious national and international awards. He has contributed 146 research papers and chapters in different referred national and international journals and books (including Scopus, IGI Global, ABDC, Sage, Springer, and Taylor and Francis) to his credit. He is the chief editor of *Splint International Journal of Professionals*, a quarterly published international journal (ISSN 2349-6045 (P), 2583-3661 (O) abstracted and indexed @ ProQuest, USA), J-Gate, Indian Citation Index and International Scientific Indexing since 2013. He is also been recognized as a CEGR-certified academic leader. He has served 27 years in academics by holding many key positions at various levels.

Prof. (Dr.) Tapas Mishra is currently a professor and head of banking and finance at the Southampton Business School, where he lectures in the subject areas of quantitative finance and advanced time series modeling. He is a fellow of the Higher Education Academy. Previously he held academic position (as associate professor of economics and time series econometrics) at the Economics Department in Swansea University where he taught a range of subjects, which include advanced macroeconomic analysis, econometric theory, mathematical economics, and applied economics. Prior to joining academia, he served as a scientist within the World Population Program at the International Institute for Applied Systems Analysis (IIASA) in Laxenburg, Austria, as a senior researcher at the Institute for Future Studies, Stockholm, Sweden, and as a senior research officer at the Institute of Economic Growth, New Delhi, India. He has led several research projects awarded by both national and international funding bodies, such as the Leverhulme Trust, the Ford Foundation and Winrock International, the Swedish Research

https://doi.org/10.1515/9783110790146-015

Council, and the European Commission's FP6 Framework. His papers have been published in leading scientific journals, such as *Nature*, *World Development*, *Social Indicators Research*, *Journal of Forecasting*, *Journal of Macroeconomics*, *European Journal of Finance*, *The Manchester School*, and *Economics Letters*. His recently published book on *Dynamics of Distribution and Diffusion of New Technologies* (published by the Springer) has been downloaded over 2,800 times. His research has been presented at refereed national and international conferences, such as the Royal Economic Society, the American Economic Association, Economic History Association, EURAM, Econometric Society, and Science and Public Policy Conference at the Kennedy School of Government, Harvard University.

Dr. Vishal Jain is presently working as an associate professor at the Department of Computer Science and Engineering, School of Engineering and Technology, Sharda University (Greater Noida, Uttar Pradesh, India). Before that, he has worked for several years as an associate professor at Bharati Vidyapeeth's Institute of Computer Applications and Management (BVICAM), New Delhi. He has more than 15 years of experience in academics. He obtained PhD (CSE), MTech (CSE), MBA (HR), MCA, MCP, and CCNA. He has authored more than 120 research papers in reputed conferences and journals, including Web of Science and Scopus. He has authored and edited more than 35 books with various reputed publishers, including De Gruyter, Elsevier, Springer, Apple Academic Press, CRC, Taylor and Francis Group, Scrivener, Wiley, Emerald, and IGI-Global. His research areas include information retrieval, semantic web, ontology engineering, data mining, ad hoc networks, and sensor networks. He received a Young Active Member Award for the year 2012–13 from the Computer Society of India, Best Faculty Award for the year 2017, and Best Researcher Award for the year 2019 from BVICAM, New Delhi.

Index

https://doi.org/10.1515/9783110790146-016

Also of interest

Augmented Reality.
Reflections on Its Contribution to Knowledge Formation
José María Ariso (Ed.), 2017
ISBN 978-3-11-049700-7, e-ISBN 978-3-11-049765-6

Artificial Intelligence for Virtual Reality.
Volume 14 in the series De Gruyter Frontiers in Computational Intelligence
Anett Jude Hemanth, Madhulika Bhatia and Isabel De La Torre Diez (Eds.), 2023
ISBN 978-3-11-071374-9, e-ISBN 978-3-11-071381-7

Interacting with Presence.
HCI and the Sense of Presence in Computer-mediated Environments
Giuseppe Riva, John Waterworth and Dianne Murray (Eds.), 2014
ISBN 978-3-11-040967-3, e-ISBN 978-3-11-040969-7

Advances in Industry 4.0.
Concepts and Applications
M. Niranjanamurthy, Sheng-Lung Peng, E. Naresh, S. R. Jayasimha and Valentina Emilia Balas (Eds.), 2022
ISBN 978-3-11-072536-0, e-ISBN 978-3-11-072549-0

Soft Computing in Smart Manufacturing.
Solutions toward Industry 5.0
Tatjana Sibalija and J. Paulo Davim (Eds.), 2022
ISBN 978-3-11-069317-1, e-ISBN 978-3-11-069322-5